Classical and Quantum Models and Arithmetic Problems

LECTURE NOTES

IN PURE AND APPLIED MATHEMATICS

1. *N. Jacobson,* Exceptional Lie Algebras
2. *L. -Å. Lindahl and F. Poulsen,* Thin Sets in Harmonic Analysis
3. *I. Satake,* Classification Theory of Semi-Simple Algebraic Groups
4. *F. Hirzebruch, W. D. Newmann, and S. S. Koh,* Differentiable Manifolds and Quadratic Forms (out of print)
5. *I. Chavel,* Riemannian Symmetric Spaces of Rank One (out of print)
6. *R. B. Burckel,* Characterization of C(X) Among Its Subalgebras
7. *B. R. McDonald, A. R. Magid, and K. C. Smith,* Ring Theory: Proceedings of the Oklahoma Conference
8. *Y.-T. Siu,* Techniques of Extension on Analytic Objects
9. *S. R. Caradus, W. E. Pfaffenberger, and B. Yood,* Calkin Algebras and Algebras of Operators on Banach Spaces
10. *E. O. Roxin, P.-T. Liu, and R. L. Sternberg,* Differential Games and Control Theory
11. *M. Orzech and C. Small,* The Brauer Group of Commutative Rings
12. *S. Thomeier,* Topology and Its Applications
13. *J. M. Lopez and K. A. Ross,* Sidon Sets
14. *W. W. Comfort and S. Negrepontis,* Continuous Pseudometrics
15. *K. McKennon and J. M. Robertson,* Locally Convex Spaces
16. *M. Carmeli and S. Malin,* Representations of the Rotation and Lorentz Groups: An Introduction
17. *G. B. Seligman,* Rational Methods in Lie Algebras
18. *D. G. de Figueiredo,* Functional Analysis: Proceedings of the Brazilian Mathematical Society Symposium
19. *L. Cesari, R. Kannan, and J. D. Schuur,* Nonlinear Functional Analysis and Differential Equations: Proceedings of the Michigan State University Conference
20. *J. J. Schäffer,* Geometry of Spheres in Normed Spaces
21. *K. Yano and M. Kon,* Anti-Invariant Submanifolds
22. *W. V. Vasconcelos,* The Rings of Dimension Two
23. *R. E. Chandler,* Hausdorff Compactifications
24. *S. P. Franklin and B. V. S. Thomas,* Topology: Proceedings of the Memphis State University Conference
25. *S. K. Jain,* Ring Theory: Proceedings of the Ohio University Conference
26. *B. R. McDonald and R. A. Morris,* Ring Theory II: Proceedings of the Second Oklahoma Conference
27. *R. B. Mura and A. Rhemtulla,* Orderable Groups
28. *J. R. Graef,* Stability of Dynamical Systems: Theory and Applications
29. *H.-C. Wang,* Homogeneous Branch Algebras
30. *E. O. Roxin, P.-T. Liu, and R. L. Sternberg,* Differential Games and Control Theory II
31. *R. D. Porter,* Introduction to Fibre Bundles
32. *M. Altman,* Contractors and Contractor Directions Theory and Applications
33. *J. S. Golan,* Decomposition and Dimension in Module Categories
34. *G. Fairweather,* Finite Element Galerkin Methods for Differential Equations
35. *J. D. Sally,* Numbers of Generators of Ideals in Local Rings
36. *S. S. Miller,* Complex Analysis: Proceedings of the S.U.N.Y. Brockport Conference
37. *R. Gordon,* Representation Theory of Algebras: Proceedings of the Philadelphia Conference
38. *M. Goto and F. D. Grosshans,* Semisimple Lie Algebras
39. *A. I. Arruda, N. C. A. da Costa, and R. Chuaqui,* Mathematical Logic: Proceedings of the First Brazilian Conference

40. *F. Van Oystaeyen,* Ring Theory: Proceedings of the 1977 Antwerp Conference
41. *F. Van Ostaeyen and A. Verschoren,* Reflectors and Localization: Application to Sheaf Theory
42. *M. Satyanarayana,* Positively Ordered Semigroups
43. *D. L. Russell,* Mathematics of Finite-Dimensional Control Systems
44. *P.-T. Liu and E. Roxin,* Differential Games and Control Theory III: Proceedings of the Third Kingston Conference, Part A
45. *A. Geramita and J. Seberry,* Orthogonal Designs: Quadratic Forms and Hadamard Matrices
46. *J. Cigler, V. Losert, and P. Michor,* Banach Modules and Functors on Categories of Banach Spaces
47. *P.-T. Liu and J. G. Sutinen,* Control Theory in Mathematical Economics: Proceedings of the Third Kingston Conference, Part B
48. *C. Byrnes,* Partial Differential Equations and Geometry
49. *G. Klambauer,* Problems and Propositions in Analysis
50. *J. Knopfmacher,* Analytic Arithmetic of Algebraic Function Fields
51. *F. Van Oystaeyen,* Ring Theory: Proceedings of the 1978 Antwerp Conference
52. *B. Kedem,* Binary Time Series
53. *J. Barros-Neto and R. A. Artino,* Hypoelliptic Boundary-Value Problems
54. *R. L. Sternberg, A. J. Kalinowski, and J. S. Papadakis,* Nonlinear Partial Differential Equations in Engineering and Applied Science
55. *B. R. McDonald,* Ring Theory and Algebra III: Proceedings of the Third Oklahoma Conference
56. *J. S. Golan,* Structure Sheaves over a Noncommutative Ring
57. *T. V. Narayana, J. G. Williams, and R. M. Mathsen,* Combinatorics, Representation Theory and Statistical Methods in Groups: YOUNG DAY Proceedings
58. *T. A. Burton,* Modeling and Differential Equations in Biology
59. *K. H. Kim and F. W. Roush,* Introduction to Mathematical Consensus Theory
60. *J. Banas and K. Goebel,* Measures of Noncompactness in Banach Spaces
61. *O. A. Nielson,* Direct Integral Theory
62. *J. E. Smith, G. O. Kenny, and R. N. Ball,* Ordered Groups: Proceedings of the Boise State Conference
63. *J. Cronin,* Mathematics of Cell Electrophysiology
64. *J. W. Brewer,* Power Series Over Commutative Rings
65. *P. K. Kamthan and M. Gupta,* Sequence Spaces and Series
66. *T. G. McLaughlin,* Regressive Sets and the Theory of Isols
67. *T. L. Herdman, S. M. Rankin, III, and H. W. Stech,* Integral and Functional Differential Equations
68. *R. Draper,* Commutative Algebra: Analytic Methods
69. *W. G. McKay and J. Patera,* Tables of Dimensions, Indices, and Branching Rules for Representations of Simple Lie Algebras
70. *R. L. Devaney and Z. H. Nitecki,* Classical Mechanics and Dynamical Systems
71. *J. Van Geel,* Places and Valuations in Noncommutative Ring Theory
72. *C. Faith,* Injective Modules and Injective Quotient Rings
73. *A. Fiacco,* Mathematical Programming with Data Perturbations I
74. *P. Schultz, C. Praeger, and R. Sullivan,* Algebraic Structures and Applications Proceedings of the First Western Australian Conference on Algebra
75. *L. Bican, T. Kepka, and P. Nemec,* Rings, Modules, and Preradicals
76. *D. C. Kay and M. Breen,* Convexity and Related Combinatorial Geometry: Proceedings of the Second University of Oklahoma Conference
77. *P. Fletcher and W. F. Lindgren,* Quasi-Uniform Spaces
78. *C.-C. Yang,* Factorization Theory of Meromorphic Functions
79. *O. Taussky,* Ternary Quadratic Forms and Norms
80. *S. P. Singh and J. H. Burry,* Nonlinear Analysis and Applications
81. *K. B. Hannsgen, T. L. Herdman, H. W. Stech, and R. L. Wheeler,* Volterra and Functional Differential Equations

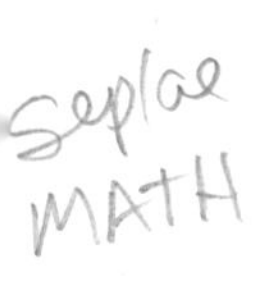

82. *N. L. Johsnon, M. J. Kallaher, and C. T. Long,* Finite Geometries: Proceedings of a Conference in Honor of T. G. Ostrom
83. *G. I. Zapata,* Functional Analysis, Holomorphy, and Approximation Theory
84. *S. Greco and G. Valla,* Commutative Algebra: Proceedings of the Trento Conference
85. *A. V. Fiacco,* Mathematical Programming with Data Perturbations II
86. *J.-B. Hiriart-Urruty, W. Oettli, and J. Stoer,* Optimization: Theory and Algorithms
87. *A. Figa Talamanca and M. A. Picardello,* Harmonic Analysis on Free Groups
88. *M. Harada,* Factor Categories with Applications to Direct Decomposition of Modules
89. *V. I. Istrătescu,* Strict Convexity and Complex Strict Convexity: Theory and Applications
90. *V. Lakshmikantham,* Trends in Theory and Practice of Nonlinear Differential Equations
91. *H. L. Manocha and J. B. Srivastava,* Algebra and Its Applications
92. *D. V. Chudnovsky and G. V. Chudnovsky,* Classical and Quantum Models and Arithmetic Problems

Other Volumes in Preparation

Classical and Quantum Models and Arithmetic Problems

Edited by

David V. Chudnovsky and
Gregory V. Chudnovsky

Department of Mathematics
Columbia University
New York, New York

MARCEL DEKKER, INC. New York and Basel

Library of Congress Cataloging in Publication Data

Main entry under title:

Classical and quantum models and arithmetic problems.

(Lecture notes in pure and applied mathematics ; 92)
Includes index.
1. Numbers, Theory of--Addresses, essays, lectures.
2. Differentiable dynamical systems. 3. Arithmetic functions--Addresses, essays, lectures. 4. Mathematical physics--Addresses, essays, lectures. I. Chudnovsky, D. (David), 1947- . II. Chudnovsky, G. (Gregory), 1952- . III. Series: Lecture notes in pure and applied mathematics ; v. 92)
QA241.C66 1984 512'.7 84-7775
ISBN 0-8247-1825-9

MARCEL DEKKER, INC.
270 Madison Avenue, New York, New York 10016

Current printing (last digit):
10 9 8 7 6 5 4 3 2 1

PRINTED IN THE UNITED STATES OF AMERICA

Dedicated to A. Lavut, N. Meiman, A. Sakharov, and
T. Velikanova - scientists and heroes.

PREFACE

Under the title "Classical and Quantum Models and Arithmetic Problems" we have collected a series of papers devoted to various mathematical aspects of dynamic systems connected with or arising from Number Theory. This volume grew from a seminar of the name "Exact Solutions of Classical and Quantum Models and Parallel Arithmetic Problems," held at Columbia University in 1980-1981 by the editors.

The relationship between the action of the modular group and related number-theoretic objects (continued fraction expansions, class numbers, and modular forms), and dynamic systems of various types connected with them, became one of the key topics of the seminar. Not unexpectedly this subject has its roots in the pioneering works of Artin, Morse, and Hedlund which were later incorporated into modern Symbolic Dynamics. The paper of Gutzwiller traces dynamic systems of this form further back to the original early contribution of Einstein. Gutzwiller's work contains a complete analytic solution of a quantum problem connected with geodesics on a surface of negative curvature. This is one of a very few important cases when a nontrivial quantum problem is completely solved. Other mathematical problems arising from the same action of the modular group on the upper half-plane are studied in the papers of Sheingorn and Cohn. The paper of Sheingorn contains an explicit determination of the growth of modular functions in the neighborhood of a real point. The result is expressed in terms of the continued fraction expansion of a real number. The important contribution of Cohn is devoted to a detailed exposition of a variety of interesting and complex problems of modern Number Theory that can be translated into modular language and are associated with various modular groups. To the same line of research belongs the paper of Schmidt, who studies rational approximations to, and continued fraction expansions of, imaginary quadratic numbers. The corresponding topological objects and transformation groups turn out to have a complicated structure even for the simplest fields. The action of the modular group and Artin's example of the ergodic system connected with continued fractions is reexamined in the Chudnovskys' paper "Note on Eisenstein's System...." Here the authors describe Artin's system by nonlinear ordinary differential equations satisfied by Eisenstein's series.

Dynamic systems that are completely integrable or close to completely integrable constitute the second key subject of papers presented at the seminar. Tabor's paper introduces the reader to the fascinating area of the singularities of solutions of differential equations and to the Painlevé property. In this paper the Painlevé test of complete integrability is applied to various dynamic systems of physical interest. Painlevé expansions and the meromorphity of solutions appear as basic analytic instruments in integrable and nonintegrable cases. The complete integrability property for multidimensional quantum systems and its relationship with S-matrices and star-triangle relations of statistical mechanics are studied in the paper "Some Remarks on θ-Functions and S-Matrices" by the Chudnovskys. This paper can also be used as an introduction to the rapidly developing area of mathematical physics and demonstrates the important role played by multidimensional objects of algebraic geometry. Two other papers on dynamic systems were specially prepared for this volume by their authors.

The paper by Churchill and Lee presents an effective algorithm for reduction of Hamiltonian systems (say, nonlinear oscillators) to their normal form. Anyone who has worked in this field will appreciate the advantages and usefulness of the explicit formulas presented in the paper. The contribution of Barnsley, Geronimo, and Harrington deals with the iterations of polynomial mappings. Beautiful invariant measures and new systems of orthogonal polynomials are associated in this paper with Julia's sets and a potential theory for strange, Cantor-like sets. This gives us a glimpse of the rigorous analytic treatment of complex physical systems beyond modern day numerical and heuristic studies.

Continued fraction expansions and diophantine approximations are the central subject of the Chudnovskys' paper "Recurrences, Padé approximations...." In this paper the authors apply continued fraction expansions of functions in form of Padé approximations to studies of diophantine approximations of numbers. Padé-type approximations that are constructed using the Bäcklund transformation method lead to new sequences of explicit rational approximations to particular numbers such as log 2 and $\pi/\sqrt{3}$.

The interest of the participants in the seminar in exactly solvable models connected with algebra or algebraic geometry is in tune with the recent explosion in "Kortweg-de Vries related" studies. As a tribute to the early and often unrecognized pioneers we include a note, "Travaux de J. Drach (1919)." With the kind permission of A. Rohou of Gauthier-Villars we also reproduce Drach's original note from C.R. Acad. Sci. Paris (1919). Another forgotten treasure is a paper of Naiman (1962) where Burchnall-Chaundy's results on commuting differential operators are extended to the difference case (cf. Toda lattice). We thank Dr. W. Le Veque and the American Mathematical Society for their permission to reproduce the English translation of Naǐman's paper from Sov. Math. Dokl.

We thank very much the participants of the seminar, the Department of Mathematics of Columbia University, the U.S. Air Force Office of Scientific Research, and the National Science Foundation for their support. We thank K. March for typing most of the manuscripts. Our most cordial thanks go to Marcel Dekker, Inc. publishers for their support and help, and especially to V. Kearn for her patience and attention, and to E. Taft and E. Hewitt for advice and encouragement.

David V. Chudnovsky
Gregory V. Chudnovsky

CONTENTS

CONTRIBUTORS

M. F. BARNSLEY, School of Mathematics, Georgia Institute of Technology, Atlanta, Georgia

DAVID V. CHUDNOVSKY, Department of Mathematics, Columbia University, New York, New York

GREGORY V. CHUDNOVSKY, Department of Mathematics, Columbia University, New York, New York

RICHARD C. CHURCHILL, Hunter College, City University of New York, New York, New York

HARVEY COHN, Department of Mathematics, City College of New York, New York, New York

J. S. GERONIMO, School of Mathematics, Georgia Institute of Technology, Atlanta, Georgia

MARTIN C. GUTZWILLER, IBM Thomas J. Watson Research Center, Yorktown Heights, New York

A. N. HARRINGTON, School of Mathematics, Georgia Institute of Technology, Atlanta, Georgia

DAVID LEE*, Hunter College, City University of New York, New York, New York

ASMUS L. SCHMIDT, Matematik Institut, University of Copenhagen, Copenhagen, Denmark

MARK SHEINGORN**, Institute for Advanced Study, Princeton, New Jersey

MICHAEL TABOR***, Center for Studies of Nonlinear Dynamics, La Jolla Institute, La Jolla, California

Current Affiliations:
*Department of Computer Science, Columbia University, New York, New York
**Department of Mathematics, Baruch College, City University of New York, New York, New York
***Department of Applied Physics and Nuclear Engineering, Columbia University, New York, New York

GEOMETRICAL AND ELECTRICAL PROPERTIES OF SOME JULIA SETS

M.F. Barnsley*, J.S. Geronimo**, and A.N. Harrington

School of Mathematics
Georgia Institute of Technology
Atlanta, Georgia

1. Introduction

Let $\mathbb{C}$ denote the complex plane and $\hat{\mathbb{C}} = \mathbb{C} \cup \{\infty\}$. Let $T: \hat{\mathbb{C}} \to \hat{\mathbb{C}}$ denote a polynomial mapping with complex coefficients, $T(z) = z^N + a_1 z^{N-1} + \ldots + a_N$ with $N \geq 2$; and introduce the notation $T^o(z) = z$ and $T^{n+1}(z) = T \circ T^n(z)$ for $n \in \{0,1,2,\ldots\}$. Then the Julia set B for T is the set of points $z \in \mathbb{C}$ where $\{T^n(z)\}$ is not a normal family.

Our main topic is the family of Julia sets B_λ for the quadratic map $T_\lambda(z) = (z-\lambda)^2$ where λ is a real parameter. In Section 2, we examine in detail the correspondence between B_λ and the set of all λ-chains, for $-1/4 < \lambda < \infty$,

$$\lambda \pm \sqrt{(\lambda \pm \sqrt{(\lambda \pm \ldots}}$$

---

*Supported by NSF Grant MCS-8104862
**Supported by NSF Grant MCS-8203325

where all sequences of signs are allowed. The aim is to provide a complete description in terms of λ-chains of some of the phenomena associated with the real map $T_\lambda: \mathbb{R} \to \mathbb{R}$, such as the cascades of bifurcations [Fe] and the sequences of implications which attach to the presence on the real line of itineraries [Gu2]. From this point of view the quadratic map is a paradigm which unites and illustrates the apparently diverse theories which relate to the different aspects of iterated maps.

When T_λ possesses an attractive k-cycle (so that the system is hyperbolic), the correspondence between B_λ and the λ-chains can be made everywhere precise, as is shown in Sec. 2.1 where we also mention the special structure of the Riemann surfaces associated with the iterated inverses of polynomials. In setting up the correspondence, the selection of branch cut, which at first sight would seem to be only a matter of convenience, turns out to be important. To have one-to-oneness in the identification of λ-chains with elements of B_λ, a negative axis cut must be used for real elements, and a positive axis cut otherwise. As λ increases and elements of B_λ become real, one must switch from one cut to the other, which reflects a fundamental change in the geometry of the Julia set, and is explained in Section 2.2. This leads us to show how to calculate the set of real λ-chains which are implied by a given real λ-chain. (A real λ-chain is one which corresponds to a real element of B_λ.) We also investigate the functional equations which are obeyed by the attractive k-cycles, and their relationship with λ-chains.

In Section 2.3 we use the theory of 2.1 and 2.2 to provide a reasonably complete description of the first cascade of period doubling bifurcations from the point of view of λ-chains, the Julia set, and the complex plane as a whole. Some remarks are made concerning the location on the set of various types of k-cycle, the distribution of the invariant measure, and about an attendant sequence of Böttcher equations which we show lead formally to the Feigenbaum functional equation. We also mention the structure of trees and subtrees in relation to B_λ, which are important for the orthogonal polynomials on B_λ [BGH2-4, BGHD].

In Section 3 we consider electrical properties of the Julia set of an arbitrary polynomial. Properties of the equilibrium charge distribution are developed with the aid of the Böttcher equation and Green's star domains. As an illustration, the detailed structure of the Green's function, the lines of force, and the mapping which connects B_0 with B_λ when $\lambda > 2$ are presented. It is also shown how various integrals involving the equilibrium measure can be evaluated explicitly.

2. Geometrical Properties and Sign Structure of Lambda-Chains for the Julia Set of $T_\lambda(z) = (z-\lambda)^2$

Let B_λ denote the Julia set [Ju, Fa, Jo, Br] for the mapping $T_\lambda: \hat{\mathbb{C}} \to \hat{\mathbb{C}}$ defined by $T_\lambda(z) = (z-\lambda)^2$, where $\lambda \in \mathbb{C}$ is a parameter. In this section, we explore the relationship between the λ-chains and the structure of B_λ when $\lambda > -1/4$ and B_λ is hyperbolic. In Section 2.1 we introduce positive-axis and negative-axis λ-chains. In Section 2.2 we analyze the equivalence class structure of the λ-chains. In Section 2.3 we apply the theory to give a description of the first cascade of period doubling bifurcations.

2.1. Lambda Chains

Let R denote the inverse of T_λ. Then the branch points of R are 0 and ∞. Let γ denote any simple continuous path which connects 0 and ∞ on the Riemann sphere $\hat{\mathbb{C}}$. Let $\mathbb{D}$ denote two copies of the Riemann sphere $\hat{\mathbb{C}}$, each slit along the path γ, and joined one to the other at the lips of the slit, see Figure 1. One seam, which we continue to call γ, belongs to one of the spheres while the other, which we call γ', belongs to the other sphere. The end points 0 and ∞

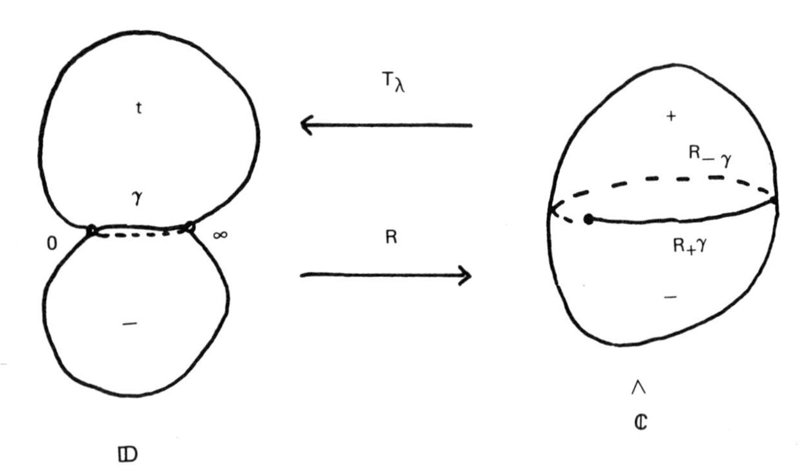

Figure 1. This illustrates the conformal equivalence between $\mathbb{D}$ and $\hat{\mathbb{C}}$ provided by R. $\mathbb{D}$ consists of two spheres slit along γ and joined there. The points 0 and ∞ occur only once, but $\gamma-\{0,\infty\}$ appears twice, once on each sphere. $\hat{\mathbb{C}}$ is divided into two components, labeled + and -, by the simple Jordan curve $R_+\gamma \cup R_-\gamma$.

of γ (and γ') appear only once, being common to both spheres. R maps $\mathbb{D}$ one to one onto $\hat{\mathbb{C}}$, and we use the notation R_+ for this mapping restricted to one of the spheres, and R_- for the restriction to the other sphere. The domain of each mapping can be taken to be $\hat{\mathbb{C}}$, and their ranges divide $\hat{\mathbb{C}}$ into two components separated by the Jordan curve $R_+\gamma \cup R_-\gamma$. R_- is the analytic continuation of R_+ across $\gamma-\{0,\infty\}$ and vice-versa.

Using R_+ and R_- we can build up chains of inverse maps, such as $R_+(R_-(R_-(R_+(R_-(z)))))$, which represent branches of the inverse mappings $R^n(z)$, $n \in \{1,2,3,...\}$. The domain of $R^n(z)$ is $\mathbb{D}^n$, which consists of 2^n copies of $\hat{\mathbb{C}}$ slit and interconnected along paths belonging to the set $\{\gamma, T_\lambda\gamma, ..., T_\lambda^{n-1}\gamma\}$. $T_\lambda^{n-1}\gamma$ appears on two of the spheres, $T_\lambda^{n-2}\gamma$ on four of the spheres,..., and γ on all of them. The finite critical points are: $T_\lambda^{n-1}0$, which occurs once, common to the two spheres which are slit and joined along $T_\lambda^{n-1}\gamma$; $T_\lambda^{n-2}0$, which occurs twice, once common to one pair of spheres which are slit and joined along $T_\lambda^{n-2}\gamma$ and once common to the other pair of spheres which are slit and joined along the same arc; ...; and 0, which occurs 2^{n-1} times being common on each of the 2^{n-1} pairs of spheres which are connected along γ. The point at infinity is common to all of the spheres. The lips of the slits, along which the spheres are joined are the branch cuts of R^n. The domain of a single branch of $R^n(z)$ consists of the projection from $\mathbb{D}^n$ onto $\hat{\mathbb{C}}$ of one of the 2^n copies of $\hat{\mathbb{C}}$, complete with the branch cuts which belong to it. We refer to this domain as the original sheet of the domain of the given branch of $R^n(z)$. By analytically continuing a branch of R^n from its original sheet across a branch cut on that sheet, one arrives at another branch of R^n, complete with its own collection of branch points and cuts. Hence, a branch of $R^n(z)$ can be defined as a holomorphic function on a domain which extends from the original sheet onto other sheets by crossing available cuts.

To represent all of the branches of $R^n(z)$, let Ω denote the set of all half-infinite chains of +1 and -1, so that $\omega \in \Omega$ if and only if

$$\omega = (e_1, e_2, e_3, \ldots)$$

where each $e_i \in \{-1,1\}$. Then we write

$$R^n_\omega(z) = R_{s(e_1)}(R_{s(e_2)}(\ldots R_{s(e_n)}(z)\ldots))$$

where $s(+1) = +$ and $s(-1) = -$. We denote the set of all finite branch points of all inverse branches of T_λ by

$$C = \{T^n_\lambda(0) \mid n \in \{0,1,2,\ldots\}\}$$

Then every one of the functions $R^n_\omega(z)$ is meromorphic in any simply connected domain D of $\mathbb{D}$ such that $P(D) \cap \bar{C} = \phi$, where $P\colon \mathbb{D} \to \hat{\mathbb{C}}$ denotes the projection which identifies elements of $\mathbb{D}$ with the corresponding points in $\hat{\mathbb{C}}$.

The following theorem is readily deduced from Brolin (p. 113, Theorem 6.2, and Lemma 6.3).

<u>THEOREM</u> (due to Cremer 1932). Let $\{R^{n_i}_{\omega_i}(z)\}$ denote any infinite set of inverse branches of T_λ, and let D be any simply connected domain on $\mathbb{D}$ such that $P(D) \cap \bar{C} = \phi$, and such that P(D) contains no accumulation point of successors of a point outside B_λ. Then $\{R^{n_i}_{\omega_i}(z)\}$ is normal in D and every convergent

subsequence of $\{R_{\omega_i}^{n_i}(z)\}$ tends to a constant which belongs to B_λ. Moreover, if $b \in B_\lambda$ then there is a sequence of $\{R_{\omega_j}^{m_j}(z)\}$ which converges to b uniformly on closed subsets of D.

This theorem tells us that we can set up a correspondence between the points of B_λ and infinite sequences of inverse maps. To make such a correspondence we must first specify γ. One important choice is for γ to be the positive axis branch cut extending from 0 to ∞ along the positive axis. Then, for $z = re^{i\theta}$ where $r \geq 0$ and $0 \leq \theta < 2\pi$ we define

$$R_\pm(z) = \lambda \pm \sqrt{r}e^{i\theta/2}$$

For z belonging to the original sheet for $R_+(z)$, either $\operatorname{Im} R_+(z) > 0$, or $\operatorname{Im} R_+(z) = 0$ and $\operatorname{Re} R_+(z) \geq \lambda$, we say that $R_+(z)$ lies in the upper half-plane. Similarly, for z belonging to the original sheet for $R_-(z)$ (which is the second sheet for $R_+(z)$), either $\operatorname{Im} R_-(z) < 0$, or $\operatorname{Im} R_-(z) = 0$ and $\operatorname{Re} R_-(z) \leq \lambda$, and we say that $R_-(z)$ lies in the lower half-plane. Notice that when γ is the positive axis branch cut, the corresponding cuts for R^n also all lie on the positive real axis.

Before giving the next theorem we recall the definition of a k-cycle and of an attractive set. T_λ has a k-cycle when there is a set of distinct bounded points $\{z_1, z_2, \ldots, z_k\} \subset \mathbb{C}$ such that $T_\lambda(z_1) = z_2$, $T_\lambda(z_2) = z_3, \ldots$, $T_\lambda(z_{k-1}) = z_k$, and $T_\lambda(z_k) = z_1$, for some $k \in \{1, 2, 3, \ldots\}$. The k-cycle is attractive, indifferent, or repulsive, according as $\left|\frac{d}{dz} T_\lambda^k(z)\Big]_{z_1}\right|$ is less than unity, equal to unity, or strictly greater than unity, respectively. If T_λ has an attractive k-cycle

$\{z_1, z_2, \ldots, z_k\}$, then the attractive set of the k-cycle is

$$A_k = \{z \in \mathbb{C} \mid \lim_{n\to\infty} T_\lambda^{nk} z \in \{z_1, z_2, \ldots, z_k\}\}$$

T_λ possesses at most one attractive k-cycle because R has only one finite branch point, and there must be at least one such point in the attractive set of the k-cycle, [Ju, Br].

We use Cremer's Theorem to prove the following result.

THEOREM 1. Let γ be the positive axis branch cut. Let $\lambda \in [-1/4,2]$ be such that T_λ admits an attractive k-cycle $\{z_1, z_2, \ldots, z_k\}$. Let Q be any region in the complement of the attractive set, with $\infty \cap Q = \phi$. Then for each $\omega \in \Omega$, $\{R_\omega^n(z)\}_{n=1}^\infty$ converges uniformly on closed subsets of Q to a single element $b(\omega) \in B_\lambda$. Moreover the mapping $b\colon \Omega \to B_\lambda$ is onto.

Proof. Since the branch point λ is attracted to the k-cycle, $\bar{C}$ can be covered by a finite union of disjoint convex open sets F, such that $\bar{F} \subset A_k$. Then $(\mathbb{C}-\bar{F}) \cap \gamma$ is a finite union of disjoint connected components. Let N_i be an open neighborhood of the ith component, such that $N_i \cap N_j = \phi$ for $i \neq j$ and $\bigcup_i N_i \cap \gamma = (\mathbb{C}-\bar{F}) \cap \gamma$, (‡). Then we define a domain D on $\mathbb{D}$ to consist of $\mathbb{C}-\bar{F}$ lifted to one of the spheres of $\mathbb{D}$, together with $\bigcup_i N_i$ lifted to the other sphere. The conditions (‡) ensure D is a simply connected domain on $\mathbb{D}$, and clearly $P(D) \cap \bar{C} = \phi$. Moreover neither the k-cycle nor ∞ belong to D, so D does not contain any accumulation point of successors of a point outside B_λ. Hence Cremer's Theorem applies to $\{R_\omega^n(z)\}_{n=1}^\infty$ over D.

Let ε denote the exterior of a closed disk centered at the origin, containing A_k, and of a radius so large that

$T_\lambda \varepsilon \subset \varepsilon$. Define a region S on $\mathbb{D}$ to consist of $\mathbb{C} - A_k - \varepsilon$ lifted to one of the spheres of $\mathbb{D}$, together with $(\mathbb{C} - A_k - \varepsilon) \cap \gamma$ lifted to the other sphere, such that $S \subset D$. Then Cremer's Theorem applies to $\{R_\omega^n(z)\}_{n=1}^\infty$ over S.

Observe now that $R_\pm S \subset S$. This is true because $\mathbb{C} - A_k$ is totally invariant under T_λ, and $R\,\varepsilon \supset \varepsilon$. It follows that $R_\omega^{n+1}(D) \subset R_\omega^n(D)$, and so $\{R_\omega^n(D)\}_{n=1}^\infty$ converges uniformly to a single constant limit belonging to B_λ. The convergence is uniform because S is closed. The last part of the theorem follows from the last part of Cremer's Theorem. Q.E.D.

We call $b(\omega)$ a positive axis λ-chain, and will use the notation

$$b(\omega) = \lambda + e_1\sqrt{(\lambda + e_2\sqrt{(\lambda + \ldots}}$$

for $\omega = (e_1, e_2, e_3, \ldots) \in \Omega$. Unless otherwise stated we mean that the positive axis cut is to be used for the evaluation of the chain.

When the element of B_λ to be described by a λ-chain lies upon the real axis, it is usually convenient to take γ to be the negative axis branch cut. In this case the branches of the square root function applied to $z = re^{i\theta}$ where $r \geq 0$ and $-\pi < \theta \leq \pi$, are defined by

$$+\sqrt{z} = \sqrt{r}e^{i\theta/2}$$

and

$$-\sqrt{z} = -(+\sqrt{z})$$

(so $+\sqrt{z}$ is the usual principal branch of the square root).

We will use the notation $\tilde{R}^n_\omega(z)$ for the corresponding branches of $R^n(z)$, defined with the aid of the negative axis branch cut.

Any given positive axis λ-chain can be converted into one in which the square roots are to be evaluated using the negative axis branch cut, which we call a <u>negative axis λ-chain</u>, in the following manner, [BGH1]. Let $b(\omega) = \lambda+e_1\surd(\lambda+e_2\surd(\lambda+e_3\ (\ldots$, and let $\tilde{\omega} = (\tilde{e}_1, \tilde{e}_2, \tilde{e}_3,\ldots)$ where $\tilde{e}_1 = e_1e_2$, $\tilde{e}_2 = e_2e_3,\ldots$, $\tilde{e}_j = e_je_{j+1},\ldots$ Then

$$\begin{aligned} b(\omega) &= \lambda+\tilde{e}_1\surd(\lambda+\tilde{e}_2\surd(\lambda+\tilde{e}_3\surd(\ldots\text{[negative cut]} \\ &= \tilde{b}(\tilde{\omega}) \end{aligned}$$

where the tilda on b means that the negative axis cut is to be used in the evaluation of the chain. Thus we have a two-to-one mapping $h\colon \Omega \to \Omega$ defined by $h(\omega) = \tilde{\omega}$. Conversely, any given negative axis λ-chain $\tilde{b}(\tilde{\omega})$ can be converted into a corresponding pair of positive axis chains by choosing e_1 arbitrarily, and then $e_2 = e_1\tilde{e}_1$, $e_3 = e_2\tilde{e}_2,\ldots e_j = e_{j-1}\tilde{e}_{j-1}\ldots$ The mapping $h^{-1}\colon \Omega \to \Omega$ is doubled-valued. The next two theorems provide λ-chain descriptions of B_λ when $\lambda < -1/4$ and $\lambda > 2$.

<u>THEOREM 2</u> [BGH1]. Let γ be the negative axis branch cut, and $2 < \lambda < \infty$. Let $S = [\lambda - 1/2 - \sqrt{\lambda+1/4}, \lambda + 1/2 + \sqrt{\lambda+1/4}]$. Then for each $\tilde{\omega} \in \Omega$, $\{\tilde{R}^n_{\tilde{\omega}}(z)\}_{n=1}^{\infty}$ converges uniformly on S to a single element $\tilde{b}(\tilde{\omega}) \in B_\lambda$. The mapping $\tilde{b}\colon \Omega \to B_\lambda$ is one-to-one and onto.

A proof of this theorem is given in [BGH1]. However, it is easy to see that a related line of argument to the proof of Theorem 1 applies here also. In the present situation the union of the branch cuts $\{T_\lambda^n \gamma \mid n \in \{0,1,2,\ldots\}\}$ consists of the negative real axis together with the positive real axis from λ^2 to ∞. Hence, we can readily find a domain $D \subset \mathbb{C}$ with $D \supset S$, which obeys the conditions of Cremer's Theorem. Now using the fact that $R_\pm S \subset S$, we get the desired convergence of $\{\tilde{R}^n_{\tilde{\omega}}(z)\}_{n=1}^{\infty}$. The one-to-one property of the resulting mapping $\tilde{b}$ of Ω onto B_λ follows from the fact that $S \cap \bigcup_{n=0}^{\infty} T_\lambda^n \gamma = \phi$.

THEOREM 3. Let γ be the positive axis branch cut and $-\infty < \lambda < -1/4$. Let S be any closed bounded simply connected domain. Then for each $\omega \in \Omega$, $\{R^n_\omega(z)\}_{n=1}^{\infty}$ converges uniformly on S to a single element $b(\omega) \in B_\lambda$. The mapping $b\colon \Omega \to B_\lambda$ is one-to-one and onto.

Sketch of the Proof. Without loss of generality we can take S to be a closed disk, centered at the origin of radius so large that $T_\lambda S \supset S$. Let D be an open disk which contains S, such that $T_\lambda D \supset D$. Then there is a finite integer m such that $R^m D \cap \{\bigcup_{n=0}^{\infty} T_\lambda^n \gamma\} = \phi$ (since $B_\lambda \cap \mathbb{R} = \phi$) and $R^m D$ is simply connected. Letting $\tilde{D} = R^m D$, we find that Cremer's Theorem applies to $\{R^n_\omega(z)\}_{n=1}^{\infty}$ over $\tilde{D}$. Moreover, since $R_\pm(R^m S) \subset R^m S$, it now follows that $\{R^n_\omega(z)\}_{n=1}^{\infty}$ converges uniformly to a single constant in $b(\omega)$, for $z \in R^m S$ and consequently for $z \in S$. The one-to-one property follows from $R^m S \cap \{\bigcup_{n=0}^{\infty} T_\lambda^n \gamma\} = \phi$. This completes the sketch of the proof.

The following statements can now be established with the help of Theorems 1, 2, and 3, and the properties of $h\colon \Omega \to \Omega$. They are valid when either $-\infty < \lambda < -1/4$, or $2 < \lambda < \infty$, or $-1/4 < \lambda < 2$ and T_λ has an attractive k-cycle. Both

$b: \Omega \to B_\lambda$ and $\tilde{b}: \Omega \to B_\lambda$ are onto, and b is single-valued. If $z \in \tilde{b}(\omega)$ for some $\omega \in \Omega$, then $\bar{z} \in \tilde{b}(\omega)$. If $b(\omega) \in \mathbb{R}$ for some $\omega \in \Omega$ then there is $\sigma \in \Omega$ with $\sigma \neq \omega$ such that $b(\omega) = b(\sigma)$. When $-\infty < \lambda < -1/4$, $\tilde{b}$ is double-valued and b is one-to-one. When $\lambda > 2$, $\tilde{b}$ is single-valued and one-to-one, while b is two-to-one. As λ increases from less than -1/4 to greater than 2, $\tilde{b}$ changes from double-valued to single-valued, while b changes from one-to-one to two-to-one. These changes mark the progression of B_λ from having the property $B_\lambda \cap \mathbb{R} = \phi$ when $-\infty < \lambda < -1/4$ to having the property $B_\lambda \cap \mathbb{R} = B_\lambda$ when $2 < \lambda < \infty$.

Define the distance between $\omega = (e_1,e_2,e_3,\ldots)$ and $\sigma = (f_1,f_2,f_3\ldots)$ in Ω by $|\omega - \sigma| = |\sum_{i=1}^{\infty} (e_i-f_i)/2^{i+1}|$. Then Ω is a topological space homeomorphic to the real interval [0,1], provided that we identify the elements $(e_1,e_2,\ldots,e_m,1,-1,-1,\ldots)$ and $(e_1,e_2,\ldots,e_m,-1,+1,+1,\ldots)$, whose distance apart is zero.

THEOREM 4. When $-1/4 < \lambda < 2$ and T_λ has an attractive k-cycle, $b: \Omega \to B_\lambda$ is continuous.

Proof. First we show that b is well-defined with respect to the identifications in Ω. Observe that $b(+1,+1,+1,\ldots) = b(-1,-1,-1,\ldots) \in \gamma$, see [BGH1]. Hence $b(+1,-1,-1,-1,\ldots) = b(-1,+1,+1,+1,\ldots)$, which lies on the negative real axis. All preimages of the latter point do not lie on γ, whence $b(e_1,e_2,\ldots, e_m,1,-1,-1,-1,\ldots) = b(e_1,e_2,\ldots,e_m,-1,+1,+1,+1,\ldots)$ and so b is well-defined.

Let $\omega \in \Omega$ and $\varepsilon > 0$. Introduce the projection operator $P_m: \Omega \to \Omega$, defined by $P_m(e_1,e_2,e_3,\ldots) = (e_1,e_2,e_3,\ldots,e_m,$

$-1,-1,-1,\ldots)$. By Theorem 1 there is an integer N such that $|b(\omega) - R^n_\omega(z)| < \varepsilon$ for all $n \geq N$ and $z \in B_\lambda$. Hence $|b(\omega) - b(\sigma)| < \varepsilon$ whenever $P_N\omega = P_N\sigma$, with $\sigma \in \Omega$.

Suppose ω does not terminate in $(+1,+1,+1,\ldots)$ or $(-1,-1,-1,\ldots)$. Then we can choose $\delta > 0$ so that $|\omega-\sigma| < \delta$ implies $P_N\omega = P_N\sigma$, and hence that $|b(\omega)-b(\sigma)| < \varepsilon$.

Suppose ω does terminate in $(+1,+1,+1,\ldots)$ or $(-1,-1,-1,\ldots)$. Then ω possesses two equivalent representations ω and ω', one terminating $(+1,+1,+1,\ldots)$ and the other terminating $(-1,-1,-1,\ldots)$. Note that $b(\omega) = b(\omega')$. Choose the positive integer M so that $|b(\omega')-R^m_{\omega'}(z)| < \varepsilon$ whenever $z \in B_\lambda$ and $m \geq M$. It follows that $|b(\omega)-b(\sigma)| < \varepsilon$ whenever $P_M\omega' = P_M\sigma$. Finally observe that we can pick $\delta > 0$ such that $|\omega-\sigma| < \delta$ implies either $P_N\omega = P_N\sigma$ or $P_M\omega' = P_M\sigma$, in both of which cases $|b(\omega)-b(\sigma)| < \varepsilon$. Q.E.D.

In what follows we assume $b\colon \Omega \to B_\lambda$ is continuous. We then have a useful description of the topology of B_λ in terms of positive axis λ-chains. $b\colon \Omega \to B_\lambda$ is a continuous mapping of a compact topological space onto a Hausdorff space. Hence the identification topology of B_λ which is induced by b is the same as the relative topology of B_λ as a subset of $\mathbb{C}$, [Me]. That is, for any subset $O \subset B_\lambda$ we have that $b^{-1}O$ is open if and only if there is an open subset $Q \subset \mathbb{C}$ such that $O = Q \cap \mathbb{C}$.

Let us consider the construction of some continuous curves lying in B_λ, which join a given pair of points z_1 and z_2. It will be convenient for us to identify each element $\omega = (e_1,e_2,e_3,\ldots)$ of Ω with the corresponding element

$\theta(e_1)\theta(e_2)\theta(e_3)\ldots$ of [0,1] in binary decimal expansion, where $\theta(+1) = 1$ and $\theta(-1) = 0$. Then we refer to [0,1] in place of Ω. Also, when $\delta < \gamma$, we will understand by $[\gamma,\delta]$ the usual closed interval $[\delta,\gamma]$. Let $\alpha \in b^{-1}(z_1)$, $\beta \in b^{-1}(z_2)$, and form $P = [\alpha,\beta_1] \cup [\alpha_2,\beta_2] \cup \ldots \cup [\alpha_{n-1},\beta_{n-1}] \cup [\alpha_n,\beta]$, where $\alpha_{i+1} \in b^{-1}(b(\beta_i))$ and $\beta_i \in [0,1]$ for $i \in \{0,1,2,\ldots, n-1\}$. Then $b(P)$ is a continuous path which lies in B_λ and joins z_1 to z_2. For example, we know that $b(0) = b(1)$, and hence each of $P_1 = [1/3,0] \cup [1,2/3]$ and $P_2 = [1/3,2/3]$ leads to a continuous path which lies in B_λ and joins $b(1/3)$ to $b(2/3)$.

If Γ is a continuous curve in B_λ which connects z_1 to z_2, then its complement $B_\lambda - \Gamma$ is open and $b^{-1}(B_\lambda - \Gamma) = [0,1] - b^{-1}(\Gamma)$ must consist of a countable union of open intervals in [0,1].

b induces an equivalence relation ~ between points in Ω according to $\omega \sim \sigma$ if $\omega \in b^{-1}(b(\sigma))$. When $b\colon \Omega \to B_\lambda$ is continuous we can think of the topology of B_λ as being that of [0,1] "pinched together" or "joined to itself" at equivalent points. For example, when $-1/4 < \lambda < 3/4$ one can show that the only pair of distinct points in [0,1] which are equivalent is $\{0,1\}$, and as a consequence B_λ is a simple Jordan curve. In the next section, we describe the dependence on λ of the equivalence classes of points in Ω.

Not only do the λ-chains codify the topology of B_λ, but also they describe the dynamics of $T_\lambda\colon B_\lambda \to B_\lambda$. Let $T\colon \Omega \to \Omega$ denote the right-shift operator defined by

$$T(e_1,e_2,\ldots,e_n,\ldots) = (e_2,e_3,\ldots,e_{n-1},\ldots)$$

When either $-\infty < \lambda < -1/4$, or $2 < \lambda < \infty$, or $-1/4 < \lambda < 2$

and T_λ has an attractive k-cycle, we have

$$T_\lambda b(\omega) = b(T\omega) \quad \text{for all } \omega \in \Omega$$

Similarly, for the negative axis λ-chains $\tilde{b}(\omega)$, which are single-valued when $\tilde{b}(\omega)$ is real and doubled-valued otherwise, we have

$$\{T_\lambda \tilde{b}(\omega)\} = \{\tilde{b}(T\omega)\} \quad \text{for all } \omega \in \Omega$$

where the parentheses { } denote the set of values of the enclosed set-valued function. These relations are readily proved. For example, when Theorem 1 applies, since $T_\lambda: \mathbb{C} \to \mathbb{C}$ is continuous,

$$T_\lambda b(\omega) = T_\lambda \lim_{n\to\infty} b_n(\omega,z) = \lim_{n\to\infty} T_\lambda b_n(\omega,z)$$

$$= \lim_{n\to\infty} b_n(T\omega,z) = b(T\omega)$$

where $z \in B_\lambda$. The corresponding result for $\tilde{b}$ follows from

$$\{T_\lambda \tilde{b}(\omega)\} = T_\lambda b(\{h^{-1}(\omega)\}) = b(T\{h^{-1}(\omega)\})$$

$$= b(\{h^{-1}(T\omega)\}) = \{\tilde{b}(T\omega)\}$$

Let $\omega = e_1, e_2, e_3, \ldots) \in \Omega$, and introduce the alternative notation $\omega = (s(e_1)s(e_2)s(e_3)\ldots)$ where $s(+1) = +$ and $s(-1) = -$. For example, $(+1,-1,-1,+1,-1,\ldots) = (+--+-\ldots)$. Let $\sigma = (e_1,e_2,\ldots,e_k,f_1,f_2,\ldots,f_\ell,f_1,f_2,\ldots,f_\ell,f_1,f_2,\ldots)$ be an eventually periodic element of Ω. Then we will denote it by

$$\omega = (e_1,e_2,\ldots,e_k \S f_1,f_2,\ldots f_\ell)$$

$$= (s(e_1)s(e_2)\ldots s(e_k) \S s(f_1)s(f_2)\ldots s(f_\ell))$$

For example $(-1,+1,+1,-1,+1,-1,+1,-1,\ldots) = (-1,+1\{+1,-1) =$ $(-+\ \ +-)$. Let $Q = (f_1,f_2,\ldots,f_\ell,f_1,f_2,\ldots,f_\ell,f_1,f_2,\ldots f_\ell,\ldots)$ be a periodic element of Ω. Then we will denote it by $Q = (f_1,f_2,\ldots,f_\ell) = (s(f_1)s(f_2)\ldots s(f_\ell))$. For example $(+1,-1,+1,-1,+1,-1,\ldots) = (+1,-1) = (+-)$.

It is now possible to describe dynamical features of T_λ acting on B_λ in terms of λ-chains. To illustrate this, take $\lambda < -1/4$ so that $b\colon \Omega \to B_\lambda$ is one-to-one. Then the only 2-cycle of T_λ on B_λ must be $\{z_1,z_2\}$ where $z_1 = b(+-)$ and $z_2 = b(-+)$. Any point $z \in B_\lambda$ such that $T_\lambda^n z = z_1$ must be expressible in the form $z = (s_1,s_2,\ldots,s_n\ \ +-)$ where each $s_i \in \{+,-\}$. The only 3-cycles of T_λ on B_λ must be $\{b(++-),$ $b(+-+), b(-++)\}$, and $\{b(--+), b(-+-), b(+--)\}$. The only fixed points (1-cycles) of T_λ on B_λ are $b(+)$ and $b(-)$. Observe that there are exactly 2^n distinct elements $\omega \in \Omega$ such that $\mathcal{T}^n\omega = \omega$. Hence, when either $b\colon \Omega \to B_\lambda$ or $\tilde{b}\colon \Omega \to B_\lambda$ is one-to-one, B_λ contains exactly 2^n distinct points z such that $T_\lambda^n z = z$, and since the polynomial $T_\lambda^n z - z = 0$ possesses at most 2^n distinct rods, we conclude that all k-cycles for all k belong to B_λ. Conversely, when T_λ possesses an attractive k-cycle neither $b\colon \Omega \to B_\lambda$ nor $\tilde{b}\colon \Omega \to B_\lambda$ is one-to-one. Notice that an expression such as $b(++-)$ is not only a symbol for the dynamics of the point in question, but also a prescription for the computation of that point.

When $-1/4 < \lambda < 2$ and T_λ possesses an attractive k-cycle, the representation of cycles belonging to B_λ is more complicated. We have $T_\lambda^n b(\omega) = b(\omega)$ if and only if $b(\mathcal{T}^n\omega) = b(\omega)$, if and only if $\omega \sim \mathcal{T}^n\omega$. Hence there is an interplay between the equivalence class structure of Ω (which, we recall, fixes the topology of B_λ) and the dynamics of T_λ on B_λ.

2.2 Equivalence Classes of Lambda Chains, and the Structures of $B_\lambda \cap \mathbb{R}$

Throughout this section, unless otherwise stated, we suppose that $-1/4 < \lambda < 2$ and that T_λ has an attractive k-cycle. γ denotes the positive axis branch cut $(0,\infty)$. For $m \in \{1,2,3,\ldots\}$, $P_m\colon \Omega \to \Omega$ denotes the projection operator $P_m(e_1,e_2,\ldots) = (e_1,e_2,\ldots,e_{m-1})$, and $P_o\omega = (-)$.

The following theorem specifies which positive axis λ-chains are equivalent, and shows that the equivalence class structure of Ω is completely fixed by $B_\lambda \cap \gamma$.

THEOREM 5. Let $-1/4 < \lambda < 2$, and let T_λ have an attractive k-cycle. Let $z \in B_\lambda$. If $T_\lambda^n(z) \notin \gamma$ for all $n \in \{0,1,2,\ldots\}$ then $\{b^{-1}(z)\}$ consists of a single element. If $T_\lambda^n(z) \in \gamma$ but $T_\lambda^{n-1}(z) \notin \gamma$ for some $n \in \{1,2,3,\ldots\}$, then $\{b^{-1}(z)\} = \{\omega_1,\omega_2\}$ where $\omega_1 \neq \omega_2$, $P_{n-1}\omega_1 = P_{n-1}\omega_2$, and $\{\mathcal{T}^{n-1}\omega_2, \mathcal{T}^{n-1}\omega_2 = \{h^{-1}(h(\mathcal{T}^{n-1}\omega_1))$. If $z \in \gamma$ then $\{b^{-1}(z)\} = \{\omega_1,\omega_2\} = \{h^{-1}(h(\omega_1))\}$.

As an illustration let $-1/4 < \lambda < 3/4$. Then [BGH1] $B_\lambda \cap \gamma$ contains only $z = a = \lambda+1/2+\sqrt{\lambda+1/4}$, and $b^{-1}(a) = \{(+), (-)\}$. Theorem 5 now states that the only elements in Ω whose equivalence classes consist of more than one element are $(+)\sim(-)$, $(+\wr -)\sim(-\wr +)$ and $(s_1s_2\ldots s_n+\wr -)\sim(s_1s_2\ldots s_n-\wr +)$, where each $s_i \in \{+,-\}$. (It was precisely these equivalence classes which permitted the identification of $[0,1]$ with Ω in Theorem 4.)

Proof of Theorem 5. Suppose $T_\lambda^n(z) \notin \gamma$ for all n. Then $T_\lambda^n(z) \notin \mathbb{R}$ for all n, since if $T_\lambda^m z < 0$ for some m then

$T_\lambda^{m+1}(z) \in \gamma$, and $T_\lambda^n(z) \neq 0$ because the branch point 0 is attracted to the k-cycle and so does not belong to B_λ. It follows that, for each n, either $\operatorname{Im} T_\lambda^n z > 0$ or $\operatorname{Im} T_\lambda^n z < 0$. The coefficients in $\omega = (e_1, e_2, \ldots) \in b^{-1}(z)$ are given by $e_n = +1$ if $\operatorname{Im} T_\lambda^n(z) > 0$ and $e_n = -1$ if $\operatorname{Im} T_\lambda^n(z) < 0$, which fixes ω uniquely.

Suppose $T_\lambda^n(z) \in \gamma$ but $T_\lambda^{n-1}(z) \notin \gamma$ for some $n \in \{1,2,3,\ldots\}$. Then $T_\lambda^{n-1}(z) < 0$, and for each $k \in \{0,1,\ldots,n-2\}$ either $\operatorname{Im} T_\lambda^n(z) > 0$ or $\operatorname{Im} T_\lambda^k(z) < 0$, which fixes uniquely the coefficients in $P_{n-1}\omega$ independently of the choice of $\omega \in \{b^{-1}(z)\}$. Since $T_\lambda^m(z)$ for $m \geq n-1$ lies on the real line, it is convenient to consider the associated negative axis λ-chains. Let $\tilde{\omega} = (\tilde{e}_1, \tilde{e}_2, \ldots) \in \{\tilde{b}^{-1}(T_\lambda^{n-1} z)\}$. Then $T_\lambda^{n-1} z < 0$ implies $\tilde{e}_1 = -1$. Moreover $\tilde{e}_j$ for $j \in \{1,2,3,\ldots\}$ is uniquely defined by $\tilde{e}_j = +1$ when $T_\lambda^{n+j-2}(z) > \lambda$ and $\tilde{e}_j = -1$ when $T_\lambda^{n+j-2} z < \lambda$. (Notice that $T_\lambda^n z \neq \lambda$ for any n because $0 \notin B_\lambda$ implies $\lambda = T_\lambda 0 \notin B_\lambda$.) Hence $\tilde{\omega}$ is fixed uniquely, and $\{b^{-1}(T_\lambda^{n-1} z)\} = \{h^{-1}\tilde{\omega}\} = \{h^{-1}(h\mathcal{T}^{n-1}\omega)\}$ consists of exactly two elements, as claimed.

Similarly, if $z \in \gamma$ then $\tilde{b}^{-1}(z)$ has only one element and $\{b^{-1}(z)\} = \{h^{-1}(\tilde{b}^{-1}(z))\}$ consists of two elements $\{\omega_1, \omega_2\} = \{h^{-1}(h(\omega_1))\}$. Q.E.D.

In view of Theorem 5, our next aim is to describe $B_\lambda \cap \gamma$, and to explain how it varies with increasing λ. We find it most convenient to express elements of $B_\lambda \cap \gamma$ in terms of negative axis λ-chains.

Let $\omega = (e_1, e_2, \ldots, e_k)$ be a periodic element of Ω which may contain subperiods, and let λ be such that $\tilde{b}(\omega)$ is defined and real. Then by the cycle $(\tilde{b}(e_1, e_2, \ldots, e_k))$ we mean the set of points $\{\tilde{b}(\omega),\tilde{b}(T\omega), \ldots, \tilde{b}(T^{k-1}\omega)\}$. In general, when we refer to such a cycle it is to be understood that λ is such that the cycle is real. If ω belongs to a k-cycle in $\Omega(T_{\omega}^{j} \neq \omega$ for $j \in \{1,2,\ldots,k-1\})$ then $(\tilde{b}(e_1,e_2,\ldots e_k))$ is a real k-cycle for T_λ. We denote this k-cycle by $\{x_1,x_2,\ldots,x_k\}$ where $x_i = \tilde{b}(T^{i-1})$, so that $T_\lambda x_j = x_{j+1}$ for $j \in \{1,2,\ldots,k-1\}$ and $T_\lambda x_k = x_1$. Since the k-cycle $(\tilde{b}(e_1',e_2',\ldots,e_k'))$ is the same as the k-cycle $(\tilde{b}(e_1,e_2,\ldots,e_k))$ whenever $(e_1',e_2',\ldots,e_k')$ is a cyclic permutation of $(e_1,e_2,\ldots,e_k)$, we can assume without loss of generality that $x_1 < x_j$ for $j \in \{2,3,\ldots,k\}$.

There are two logical orderings of a real k-cycle. Already we have used the notation $\{x_1,x_2,\ldots,x_k\}$ putting the points in iterative order. The points also may be put in increasing order. For example a 4-cycle may have the increasing order $x_1 < x_3 < x_4 < x_2$. We call the combined information the order of visitation. It can be given diagramatically

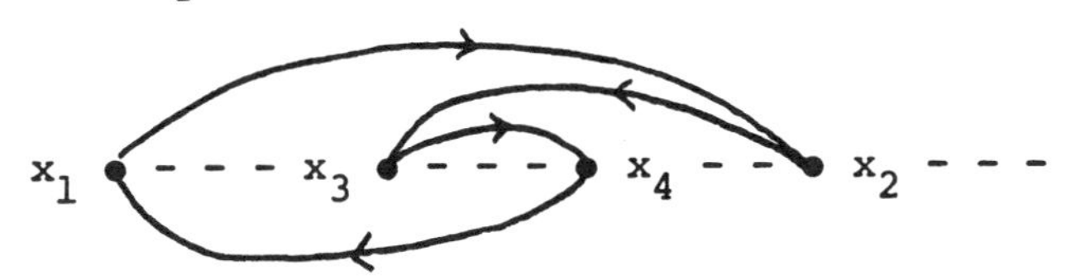

To determine the order of visitation of a k-cycle $(\tilde{b}(e_1,e_2,\ldots,e_k))$ we define a mapping ϕ from $\sigma = (f_1,f_2,f_3,\ldots) \in \Omega$ into [0,1] in binary decimal representation by

$$\phi(\sigma) = 0.\alpha_1\alpha_2\alpha_3 \ldots \alpha_k \ldots$$

where

$$\alpha_i = \begin{cases} 1 & \text{if } f_1f_2\ldots f_i = +1 \\ 0 & \text{if } f_1f_2\ldots f_i = -1 \end{cases}$$

<u>THEOREM 6</u> (A). The order of visitation of the real k-cycle $(\tilde{b}(\omega)) = \{x_1, x_2, \ldots, x_k\}$ is given by the increasing order of the set of real numbers $\{\phi(\omega), \phi(T\omega), \ldots, \phi(T^{k-1}\omega)\}$. If $\omega = (e_1, e_2, \ldots, e_k)$ and $x_1 < x_j$ for $j \in \{2,3,\ldots,k\}$, then $e_i = +1$ when $x_i > x_k$ and $e_i = -1$ when $x_i < x_k$. (B) Conversely, if $(\tilde{e}_1, \tilde{e}_2, \ldots, \tilde{e}_k)$ is a k-cycle in Ω such that $e_i = +1$ when $x_i > x_k$ and $e_i = -1$ when $x_i < x_k$, then the real k-cycle $(\tilde{b}(\tilde{e}_1, \tilde{e}_2, \ldots, \tilde{e}_k))$ has the same order of visitation as $\{x_1, x_2, \ldots, x_k\}$.

In the theorem $\tilde{e}_k$ is not specified. Let us maintain the notation of the theorem and suppose $\tilde{e}_k = -e_k$. If $(\tilde{b}(\tilde{e}_1, \tilde{e}_2, \ldots, \tilde{e}_k)) = \{\tilde{x}_1, \tilde{x}_2, \ldots, \tilde{x}_k\}$ is a real k-cycle, then we say that the two real k-cycles $\{x_1, x_2, \ldots, x_k\}$ and $\{\tilde{x}_1, \tilde{x}_2, \ldots, \tilde{x}_k\}$ are <u>partners</u>. Otherwise we say $\{x_1, x_2, \ldots, x_k\}$ is a loner.

<u>Proof of (A)</u>. It is straightforward to check that if $\tilde{b}(\sigma_1)$ and $\tilde{b}(\sigma_2)$ are real λ-chains then $\tilde{b}(\sigma_1) < \tilde{b}(\sigma_2)$ if and only if $\phi(\sigma_1) < \phi(\sigma_2)$. From this follows the first statement in the theorem.

Since x_1 is the smallest member of the k-cycle, it is closest to zero. Hence x_k is closest to λ. Hence, x_j with $j \neq k$ is greater than λ if and only if x_j is greater than x_k, and x_j is less than λ if and only if x_j is less than x_k. But $x_j > \lambda$ if and only if $e_j = +1$, and $x_j < \lambda$ if and only if $e_j = -1$, which proves the second statement in the theorem.

We defer the proof of (B) until after the proof of Theorem 8. Although a direct combinatorial proof should be available, the only approach we know relies on the structure of the bifurcation diagram for T_λ.

Since some familiarity with computations based on Theorem 6 will be helpful later on, we give some examples. Consider the real 3-cycle $(\tilde{b}(++-)) = \{x_1' = \tilde{b}(++-),\ x_2' = \tilde{b}(+-+),\ x_3' = \tilde{b}(-++)\}$. The order of visitation is given by the ordering of the sequence {.110, .100, .000}, whence $x_3' < x_2' < x_1'$ which can also be expressed

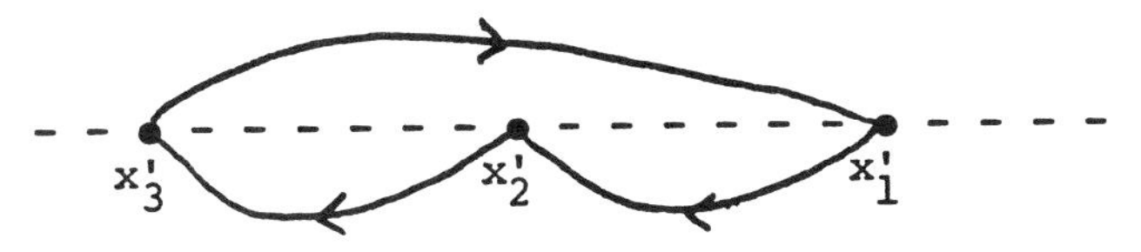

If we relabel the cycle $x_1 = x_3'$, $x_2 = x_1'$, $x_3 = x_2'$, then the order of visitation is $x_1 < x_3 < x_2$ where $x_1 = b(-++)$. Theorem 6 (B) now asserts that the order of visitation for the 3-cycle $(\tilde{b}(-+-)) = \{\tilde{x}_1 = \tilde{b}(-++),\ \tilde{x}_2 = \tilde{b}(++-),\ \tilde{x}_3 = \tilde{b}(+-+)\}$ is $\tilde{x}_1 < \tilde{x}_3 < \tilde{x}_2$, which is readily checked. The two cycles $\{x_1,x_2,x_3\}$ and $\{\tilde{x}_1,\tilde{x}_2,\tilde{x}_3\}$ in this example are <u>partners</u>.

Consider the real 4-cycle $(\tilde{b}(-+--)) = \{x_1 = \tilde{b}(-+--),\ x_2 = \tilde{b}(+---),\ x_3 = \tilde{b}(---+),\ x_4 = \tilde{b}(--+-)\}$. The order of visitation is given by the ordering of the sequence of numbers {.0010, .1010, .0100, .0110}, whence $x_1 < x_3 < x_4 < x_2$ which can also expressed

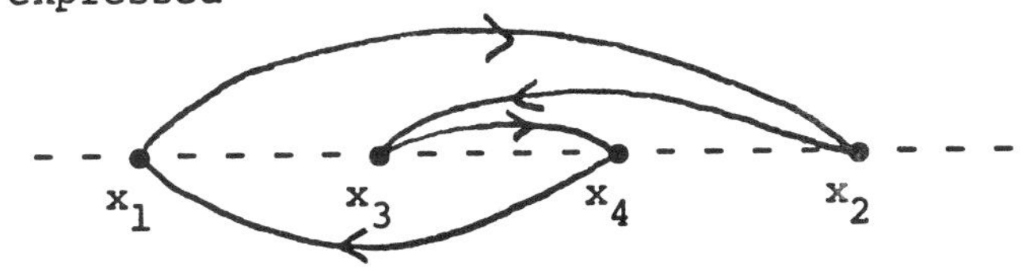

In this case $(\tilde{e}_1,\tilde{e}_2,\tilde{e}_3,\tilde{e}_4) = (-1,+1,-1,+1)$ which is not a 4-cycle in Ω. Hence $\{x_1,x_2,x_3,x_4\}$ is a loner. On the other hand, the two real 4-cycles $(\tilde{b}(-++-))$ and $(\tilde{b}(-+++))$ are partners, each with the order of visitation

Next we consider the continuation of real k-cycles, through decreasing values of λ. Let $\omega = (e_1,e_2,\ldots,e_k)$ be a k-cycle in Ω, and fix $\lambda_o > 2$. For this value of λ, $(\tilde{b}(e_1,e_2,\ldots,e_k))$ is a real k-cycle because $B_{\lambda_o} \subset \mathbb{R}$. We denote this k-cycle by $\{x_1^o,x_2^o,\ldots,x_k^o\}$, where $x_1^o = \tilde{b}(e_1,e_2,\ldots,e_k)$ and we suppose $x_1^o < x_j^o$ for $j \in \{2,3,\ldots,k\}$. We define a function of x and λ by $(e_1,e_2,\ldots,e_k)(x,\lambda) := \lambda+e_1\sqrt{(\lambda+e_2\sqrt{(\lambda+\ldots e_{k-1}\sqrt{(\lambda+e_k\sqrt{x})}\ldots)})}$ where the negative axis branch cut is used, and where the domain is set of real λ and x such that $(e_1,e_2,\ldots,e_k)(x,\lambda)$ is real. Then (x_j^o,λ_o) is a solution of the equation $u_j(x,\lambda) = 0$, for $j \in \{1,2,\ldots,k\}$, where

$$u_j(x,\lambda) := x-(e_j,e_{j+1},\ldots,e_k,e_1,e_2,\ldots,e_{j-1})(x,\lambda)$$

The implicit function theorem states that there is a unique continuation of this solution when $\partial u_j(x_j,\lambda)/\partial x_j$ exists and does not vanish.

We have

$$\frac{\partial u_j}{\partial x_j}(x_j,\lambda) = 1 - \frac{e_1e_2\ldots e_k}{2^k\sqrt{x_1x_2\ldots x_k}} \qquad (*)$$

It follows that there is an open interval I containing λ_o, and a unique set of continuous functions $\{x_1(\lambda), x_2(\lambda),\ldots, x_k(\lambda)\}$ defined for $\lambda \in I$ and obeying $u_j(x_j(\lambda),\lambda) = 0$ for all $\lambda \in I$. Moreover (i) each of $\{x_1(\lambda),x_2(\lambda),\ldots,x_k(\lambda)\}$ is finite and strictly positive for all $\lambda \in I$, and (ii)

$$(1 - \frac{e_1e_2\ldots e_k}{2^k\sqrt{x_1x_2\ldots x_k}}) \neq 0$$

for all $\lambda \in I$. We assume I is the largest open interval such that the above statements are true.

We show that no two memebers of $\{x_1(\lambda), x_2(\lambda), \ldots, x_k(\lambda)\}$ can be equal for $\lambda \in I$. We can have $x_j(\lambda) = x_\ell(\lambda)$ with $j \neq \ell$ only if the polynomial equation $T_\lambda^k(x) - x = 0$ has x_j as a double root. For this it is necessary that $\partial(T_\lambda^k(x_j)-x_j)/\partial x_j = 0$. That is,

$$2^k T_\lambda'(x_1)T_\lambda'(x_2)\ldots T_\lambda'(x_k) - 1 = 0$$

Since $u_j(x_j,\lambda) = 0$ and $x_j > 0$ for $j \in \{1,2,\ldots,k\}$, the last condition can be rewritten

$$e_1e_2\ldots e_k 2^k\sqrt{x_1x_2\ldots x_k} - 1 = 0$$

which is not possible by (ii). We conclude $x_j(\lambda) \neq x_\ell(\lambda)$ for all $\lambda \in I$. It now follows directly from the equation $u_j(x_j(\lambda),\lambda) = 0$ that $\{x_1(\lambda), x_2(\lambda), \ldots, x_k(\lambda)\}$ is a k-cycle for T_λ, with $x_1(\lambda) < x_j(\lambda)$ for $j \neq 1$ and having the same order of visitation as $\{x_1^o,x_2^o,\ldots,x_k^o\}$.

It is important to understand that although $\{x_1^o,x_2^o,\ldots x_k^o\} \subset B_{\lambda_o}$, it may not be true that $\{x_1(\lambda),x_2(\lambda),\ldots,x_k(\lambda)\} \subset B_\lambda$ when $\lambda < 2$. Observe that

$$\left.\frac{\partial}{\partial x} T_\lambda^k(x)\right|_{x=x_1(\lambda)} = e_1e_2\ldots e_k 2^k\sqrt{x_1(\lambda)x_2(\lambda)\ldots x_k(\lambda)}$$

Hence $\{x_1(\lambda),x_2(\lambda),\ldots,x_k(\lambda)\}$ is a repulsive k-cycle and belongs to B_λ when $2^k\sqrt{x_1(\lambda)x_2(\lambda)\ldots x_k(\lambda)} > 1$. $\{x_1(\lambda),x_2(\lambda),\ldots,x_k(\lambda)\}$ is an attractive k-cycle and does not belong to B_λ when $2^k\sqrt{x_1(\lambda)x_2(\lambda)\ldots x_k(\lambda)} < 1$.

Let $I = (\nu,\mu)$. Then $\mu = \infty$. To see this we suppose that μ is finite and consider what can happen to $\{x_1(\lambda), x_2(\lambda), \ldots x_k(\lambda)\}$ as $\lambda \to \mu$. No member of the cycle can approach ∞ as $\lambda \to \mu$, since any continuation of any finite root of $T_\lambda^k(x) - x = 0$ remains finite for finite λ. It follows that we can define $x_j(\mu) = \lim_{\lambda\to\mu} x_j(\lambda)$. We cannot have $x_1(\mu) = 0$ because all k-cycles belong to B_λ and $0 \notin B_\lambda$ for $\lambda > 2$. Finally, we cannot have

$$1 - \frac{e_1 e_2 \ldots e_k}{2^k \sqrt{x_1(\mu) x_2(\mu) \ldots x_k(\mu)}} = 0$$

for if so then $T_\mu^k(x) - x = 0$ has a double root. The latter is not possible because it implies the false assertion that $\tilde{b}\colon \Omega \to B_\mu$ is not one-to-one. It now follows that $\{x_1(\mu), \ldots, x_k(\mu)\}$ has a unique continuation throughout some neighborhood of μ, where it obeys $u_j(x_j(\lambda),\lambda) = 0$. This contradicts our assumption on I.

We must have $\nu \geq -1/4$ because T_λ possesses no real k-cycles when $\lambda < -1/4$; also as above, we can define $x_j(\nu) = \lim_{\lambda\to\nu+} x_j(\lambda) < \infty$. Thus, on the continuation of solutions of $u_j(x_j(\lambda),\lambda) = 0$ we have so far established

THEOREM 7. Let $\omega = (e_1, e_2, \ldots, e_k)$ be a k-cycle in Ω, let $\lambda = \lambda_o > 2$, and let $(\tilde{b}(e_1, e_2, \ldots, e_k)) = \{x_1^o, x_2^o, \ldots x_k^o\}$ be the corresponding real k-cycle in B_{λ_o}. Then there is a unique real continuation $(x_j(\lambda), \lambda)$ for λ in a largest open interval $I = (\nu, \infty)$ containing λ_o, where $\nu \geq -1/4$, such that

$$
\begin{aligned}
u_j(x_j(\lambda),\lambda) &= x_j(\lambda) \\
&\quad -(\lambda+e_j\sqrt{(\lambda+\ldots e_k\sqrt{(\lambda+e_1\sqrt{(\lambda+\ldots+e_{j-1}\sqrt{x_j(\lambda)})\ldots))\ldots)}}} \\
&= 0
\end{aligned}
$$

and $x_j(\lambda_o) = x_j^o$, for $j \in \{1,2,\ldots,k\}$. $\{x_1(\lambda), x_2(\lambda),\ldots,x_k(\lambda)\}$ is a real k-cycle of T_λ, for all $\lambda \in I$.

When we can do so without ambiguity, we will use the notation $(\tilde{b}(e_1,e_2,\ldots,e_k))$ both for the real k-cycle $\{x_1(\lambda),x_2(\lambda),\ldots,x_k(\lambda)\}$ in Theorem 7, and for its continuation through decreasing λ-values as long as it remains a real k-cycle, obeying $T_\lambda^k(x_1(\lambda)) = x_1(\lambda)$, even when it may be that the cycle in question is attractive (thus failing to belong to the Julia set) over a range of λ-values, and the negative axis λ-chain does not converge to the cycle with which it is identified by the notation. Notice that the order of visitation for the k-cycle is independent of λ and hence can be calculated from $\omega = (e_1,e_2,\ldots e_k)$ as in Theorem 6A.

Before we consider what happens at ν, we recall some facts about the bifurcations of real solutions of $T_\lambda^k(x)-x = 0$. Fundamental is the Mandlebrot domain M, which consists of the set of points $\lambda \in \mathbb{C}$ such that B_λ is connected. The following information is drawn from the study of M by Douardy and Hubbard, [DH]. (1) Let $\{x_1(\delta),x_2(\delta),\ldots,x_k(\delta)\}$ be a real k-cycle at $\lambda = \delta$. Then there is a real k-cycle $\{x_1(\lambda),x_2(\lambda),\ldots,x_k(\lambda)\}$ for all $\lambda \geq \delta$, depending continuously on λ, [Su]. (2) Let C denote a component of the interior $\mathring{M}$ of M, such that $C \cap \mathbb{R} \neq \phi$. Then we can write $C \cap \mathbb{R} =$

(α,β) where $-1/4 \leq \alpha < \beta < 2$. For all $\lambda \in (\alpha,\beta)$, T_λ has a unique attractive k-cycle $\{x_1(\lambda),x_2(\lambda),\ldots,x_k(\lambda)\}$. This cycle is real and depends continuously on $\lambda \in (\alpha,\beta)$. Conversely if $\{x_1(\lambda),x_2(\lambda),\ldots,x_k(\lambda)\}$ is a real attractive k-cycle then there is a component C of $\mathring{M}$ with $C \cap \mathbb{R} \neq \phi$ such that $\lambda \in C$. (3) The derivative $\frac{d}{dx} T_\lambda^k(x)\Big|_{x_1(\lambda)}$ is strictly decreasing in $\lambda \in (\alpha,\beta)$, with value +1 at α and -1 at β. At some point $\nu \in (\alpha,\beta)$, called the center of C, $\frac{d}{dx} T_\lambda^k(x)\Big|_{x_1(\lambda)} = 0$, and then the k-cycle is said to be superstable. (4) $\lambda = \alpha$ is a bifurcation point for the real solution $x_1(\lambda)$ of $T_\lambda^k(x)-x = 0$ of one of two types. Either (i) there is, or (ii) there is not another component C' of $\mathring{M}$ with $C' \cap \mathbb{R} \neq \phi$ and $\alpha \in \overline{C'}$. In case (i) C is said to derive from C' by bifurcation at α. This means that C' is associated with a real attractive p-cycle $\{y_1(\lambda),y_2(\lambda),\ldots,y_p(\lambda)\}$ where p divides k and $p \neq k$, such that $\{y_1(\nu),y_2(\nu),\ldots,y_p(\nu)\} = \{x_1(\nu),x_2(\nu),\ldots,x_k(\nu)\}$. In case (ii) we will say (α,β) is a base. (5) There exists a component C" of $\mathring{M}$ such that $C'' \cap C = \phi$, $C'' \cap \mathbb{R} \neq \phi$, and $\beta \in \overline{C''}$. Thus, β is a bifurcation point of type (ii).

Further information about bifurcations of real solutions of $T_\lambda^k(x)-x = 0$ now follows upon combining the above information with [Gul]. The only mechanism by which a real k-cycle of T_λ can appear for increasing λ is via either (i) or pitchfork bifurcation, or (ii) a tangent bifurcation. These correspond exactly to the two types of bifurcation point mentioned above; however, in the description of (i), it must be that k is even and $p = k/2$. Let α be a bifurcation point of type (ii). Then there is an associated real k-cycle $\{x_1(\lambda),x_2(\lambda),\ldots,x_k(\lambda)\}$ for $\lambda > \alpha$, continuously dependent on λ, which does not have a real continuation for $\lambda < \alpha$. This cycle can be continued to $\lambda = \alpha$, and $\{x_1(\alpha), x_2(\alpha), \ldots, x_k(\alpha)\}$ is also a real k-cycle. For $\lambda \geq \alpha$ there exists a real k-cycle $\{\tilde{x}_1(\lambda), \tilde{x}_2(\lambda), \ldots, \tilde{x}_k(\lambda)\}$, continuously dependent upon λ, distinct from $\{x_1(\lambda), \ldots, x_k(\lambda)\}$ for $\lambda > \alpha$ and such that $x_j(\alpha) = \tilde{x}_j(\alpha)$ for $j \in \{1,2,\ldots,k\}$. One of the two k-cycles is attractive and the other is repulsive for $\lambda \in (\alpha, \alpha+\varepsilon)$ for some $\varepsilon > 0$.

From the point of view of the complex plane we see that a tangent bifurcation occurs with increasing λ when two distinct k-cycles, one the complex conjugate of the other, with nonzero imaginary parts, become real at $\lambda = \alpha$ to form a single real k-cycle. As λ increases from $\lambda=\alpha$ to $\lambda > \alpha$ the coalesced pairs of points separate, yielding two real k-cycles one of which is attractive and the other repulsive. On the other hand, a pitchfork bifurcation takes place when a self-conjugate k-cycle, with nonzero imaginary parts, becomes real at $\lambda = \alpha$. In this case, the members of the cycle coalesce in pairs on the real axis to become at $\lambda = \alpha$ an indifferent real (k/2)-cycle. Not only at this value of λ does the cycle merge with itself, but also it coalesces with a second real (k/2)-cycle which was real and stable for $\lambda \in (\alpha-\varepsilon, \alpha)$ for some $\varepsilon > 0$. When λ is increased from $\lambda = \alpha$ to $\lambda > \alpha$ the self-conjugate k-cycle becomes an attractive real k-cycle, and the (k/2)-cycle which was already on the real axis becomes unstable.

We return now to the context of Theorem 7, and consider what happens at ν.

<u>THEOREM 8</u>. Let $(\tilde{b}(e_1, e_2, \ldots e_k)) = \{x_1(\lambda), x_2(\lambda), \ldots, x_k(\lambda)\}$ with $\lambda \in (\nu, \infty)$ be the real k-cycle exhibited in

Theorem 7, and let $x_1(\lambda) < x_j(\lambda)$ for $j \neq 1$. Let $\tilde{e}_i = e_i$ for $i \neq k$, $\tilde{e}_k = -e_k$, and if the k-cycle possesses a partner, denote it by $(\tilde{b}(\tilde{e}_1,\tilde{e}_2,\ldots,\tilde{e}_k)) = \{\tilde{x}_1(\lambda),\tilde{x}_2(\lambda),\ldots,\tilde{x}_k(\lambda)\}$. Assume that $e_1e_2\ldots e_k = -1$. Then $\{x_1(\lambda),x_2(\lambda),\ldots,x_k(\lambda)\}$ is superstable at ν with $x_1(\nu) = 0$, and it possesses a unique real continuation to some largest interval (α,β), containing ν, throughout which it is an attractive real k-cycle. For $\lambda \in (\alpha,\nu)$ it obeys the functional equations

$$\tilde{u}_j(x_j(\lambda),\lambda) = x_j(\lambda) - (\lambda+\tilde{e}_j\sqrt{(\lambda+\ldots+\tilde{e}_k\sqrt{(\lambda+\tilde{e}_1\sqrt{(\lambda+\ldots+\tilde{e}_{j-1}\sqrt{x_j(\lambda)})\ldots)})}}) = 0, \qquad j \in \{1,2,3,\ldots,k\}$$

If the k-cycle is a loner then $\lambda = \alpha$ is a bifurcation point of type (i) at which the cycle takes part in a pitchfork bifurcation. The (k/2)-cycle out of which the real k-cycle appears, and which is attractive over some interval immediately preceeding the bifurcation point is $(\tilde{b}(e_1,e_2,\ldots,e_{k/2}))$, and $e_1e_2\ldots e_{k/2} = -1$. If $\{x_1(\lambda),x_2(\lambda),\ldots,x_k(\lambda)\}$ possesses a partner $\{\tilde{x}_1(\lambda),\tilde{x}_2(\lambda),\ldots,\tilde{x}_k(\lambda)\}$, then the partner obeys $\tilde{u}_j(\tilde{x}_j(\lambda),\lambda) = 0$, $j \in \{1,2,\ldots,k\}$, and is a repulsive k-cycle for all $\lambda \in (\alpha,\infty)$. In this case $\lambda = \alpha$ is a bifurcation point of type (ii) at which the k-cycle and its partner coalesce in a tangent bifurcation. If $e_1e_2\ldots e_k = +1$ then the k-cycle $(\tilde{b}(e_1,e_2,\ldots,e_k))$ possesses a partner, and the roles of $\{x_1(\lambda),x_2(\lambda),\ldots,x_k(\lambda)\}$ and $\{\tilde{x}_1(\lambda),\tilde{x}_2(\lambda),\ldots,\tilde{x}_k(\lambda)\}$ are interchanged.

Proof. Since $e_1 e_2 \ldots e_k = -1$, we cannot have

$$1 - \frac{e_1 e_2 \ldots e_k}{2^k \sqrt{x_1(\nu) x_2(\nu) \ldots x_k(\nu)}} = 0$$

and the only possibility for stopping the continuation of the real solutions $x_j(\lambda)$ of $u_j(x_j(\lambda),\lambda) = 0$ through decreasing λ-values is $x_1(\nu) = 0$, which means that the cycle is superstable and ν is the center of a component of $\mathring{M}$. It follows that the cycle possesses a unique continuation to some largest open interval (α,β), containing ν, throughout which it is an attractive k-cycle. Since the order of visitation is independent of $\lambda \in (\alpha,\infty)$, it follows from Theorem 6A that the cycle must obey either $u_j(x_j(\lambda),\lambda) = 0$ or $\tilde{u}_j(x_j(\lambda),\lambda) = 0$ at each $\lambda \in (\alpha,\nu)$. But the former is not possible, because if it was true then the k-cycle could be continued through decreasing λ-values to a second center, which contradicts the information about the Mandlebrot domain given earlier.

Notice that because $\tilde{e}_1 \tilde{e}_2 \ldots \tilde{e}_k = +1$, the real solution $\{x_1(\lambda), x_2(\lambda), \ldots, x_k(\lambda)\}$ of $\tilde{u}_j(x_j(\lambda),\lambda) = 0$ can be continued through decreasing values of λ until $\lambda = \alpha$ at which we have the bifurcation condition

$$\left.\frac{\partial \tilde{u}_j}{\partial x_j}(x_j,\lambda)\right|_{\lambda=\alpha} = 1 - \frac{\tilde{e}_1 \tilde{e}_2 \ldots \tilde{e}_k}{2^k \sqrt{x_1(\alpha) x_2(\alpha) \ldots x_k(\alpha)}} = 0$$

For $\lambda \in (\alpha,\alpha+\varepsilon)$ where $\varepsilon > 0$ is sufficiently small the equations $\tilde{u}_j(x_j,\lambda) = 0$, $j \in \{1,2,\ldots,k\}$, must possess a second distinct solution which we denote by $\{\tilde{x}_1(\lambda), \tilde{x}_2(\lambda), \ldots, \tilde{x}_k(\lambda)\}$.

This must be either a (k/2)-cycle or a k-cycle according as the bifurcation point of a type (i) or (ii) respectively.

Since for $\lambda \in (\alpha,\alpha+\varepsilon)$ the k-cycle $\{x_1(\lambda),x_2(\lambda),\ldots,x_k(\lambda)\}$ is attractive, all other cycles must belong to B_λ, including in particular $\{\tilde{x}_1(\lambda),\ldots,\tilde{x}_k(\lambda)\}$. It now follows that the latter cycle is given by $(\tilde{b}(\tilde{e}_1\tilde{e}_2,\ldots,\tilde{e}_k))$ which is a k-cycle if and only if $(\tilde{e}_1,\tilde{e}_2,\ldots,\tilde{e}_k)$ is a k-cycle in Ω, which is to say that $\{x_1(\lambda),\ x_2(\lambda),\ldots,x_k(\lambda)\}$ possesses a partner. The only other possibility is that α is of type (i) and $(\tilde{b}(\tilde{e}_1,\tilde{e}_2\ldots,\tilde{e}_k))$ would have to be a k/2-cycle, namely $(\tilde{b}(\tilde{e}_1,\tilde{e}_2,\ldots,\tilde{e}_{k/2})) = (\tilde{b}(e_1,e_2,\ldots,e_k))$, and $\{x_1(\lambda),\ x_2(\lambda),\ldots,x_k(\lambda)\}$ must be a loner. Since the real (k/2)-cycle, out of which the real k-cycle $\{x_1,(\lambda),x_2(\lambda),\ldots,x_k(\lambda)\}$ appeared, must itself be attractive for λ just less than α, we must have $e_1e_2\ldots e_{k/2} = -1$. Q.E.D.

Proof of Theorem 6(B). It is clear from the above that the order of visitation for a real k-cycle is the same as that for its partner, if it has one. Q.E.D.

As an example of Theorem 8 we consider the 4-cycle $(\tilde{b}(-+--))$, for which $x_1 = \tilde{b}(-+--)$ is the least element. Since $(-+-+)$ is not a 4-cycle in Ω, $(\tilde{b}(-+--))$ is a loner, and must have appeared by pitchfork bifurcation from the 2-cycle $(\tilde{b}(-+))$. This 2-cycle is itself a loner and must have appeared by pitchford bifurcation from the 1-cycle $(\tilde{b}(-))$. The latter has the partner $(\tilde{b}(+))$, with which it appeared by tangent bifurcation. This example, including the orders of visitation, is summarized in Figure 2. The dotted portions of the curves denote cycles which do not belong to B_λ and consequently are not represented by λ-chains. These cycles can be indicated by the functional equations which

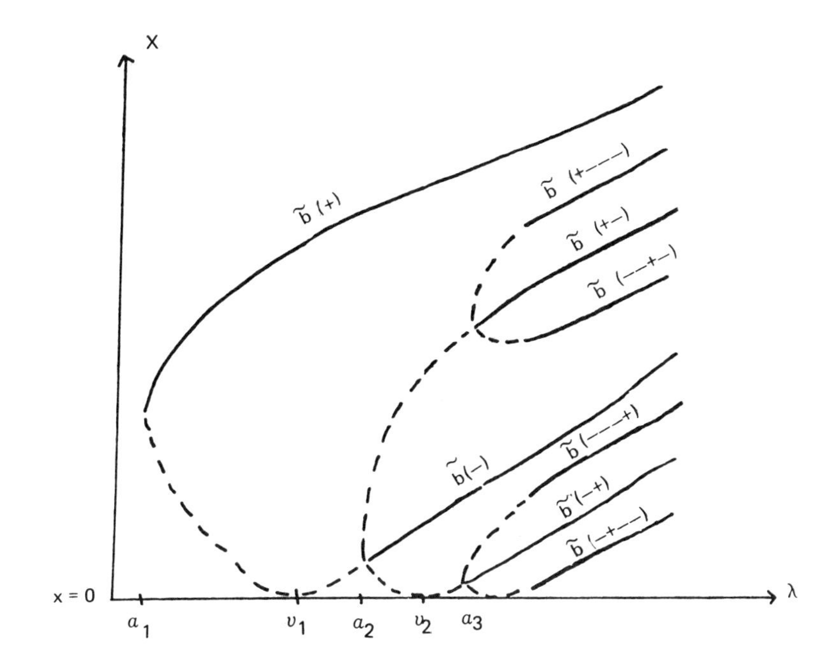

Figure 2. Sketch of the bifurcation diagram associated with the 4-cycle $(\tilde{b}(-+--))$, labelled with the corresponding λ-chains. See text.

they obey. For $\lambda \in (\alpha_1, \nu_1)$ the attractive 1-cycle obeys $x-(\lambda-\sqrt{x}) = 0$, whilst for $\lambda \in (\nu_1, \alpha_2)$ it obeys $x-(\lambda+\sqrt{x}) = 0$. Similarly, Theorem 8 tells us that the attractive 2-cycle $\{x_1, x_2\}$ which exists for $\lambda \in (\alpha_2, \alpha_3)$, obeys $x_1-(\lambda-\sqrt{(\lambda-\sqrt{x_1})}) = 0$, $x_2-(\lambda-\sqrt{(\lambda-\sqrt{x_2})}) = 0$ for $\lambda \in (\alpha_2, \nu_2)$, and $x_1-(\lambda-\sqrt{(\lambda+\sqrt{x_1})}) = 0$, $x_2-(\lambda+\sqrt{(\lambda-\sqrt{x_2})}) = 0$, for $\lambda \in (\nu_2, \alpha_3)$.

To complete our description of $B_\lambda \cap \gamma$ and how it varies with increasing λ, we answer the following question. Let $(\tilde{b}(e_1, e_2, \ldots, e_k))$ be a real k-cycle. Then what is the full set of other λ-chains which <u>necessarily</u> belong to $B_\lambda \cap \gamma$? That is, we continue through decreasing λ-values both the k-cycle and its partner (if it has one) until, at $\lambda = \nu$, one or other is superstable. Then we wish to determine $B_\nu \cap \gamma$. Equivalently, we define $W_\omega = \{\sigma \in \Omega | \tilde{b}(\sigma) \in (0,\infty)$

when $\lambda = \nu\}$, so that $B_\nu \cap \gamma = \tilde{b}(W_\omega)$, and we look for W_ω. Notice that if (α,β) is the interval of stability of $(b(\omega))$ (or its partner) then $\tilde{b}^{-1}(B_\lambda \cap \gamma)$ is independent of $\lambda \in (\alpha,\beta)$. Also $\tilde{b}^{-1}(B_\lambda \cap \gamma)$ is an increasing set-valued function of $\lambda \in \mathbb{R}$.

We describe how to determine W_ω, starting from $\omega \in \Omega$. In this situation, we will call ω the <u>seed</u>. From ω we obtain the order of visitation of $(\tilde{b}(\omega))$ according to Theorem 6. This order is the same both for the cycle and its partner, if it has one; consequently, it is unnecessary to decide whether it is the cycle or its partner which is superstable at $\lambda = \nu$. The outcome of the computation is the same in any case. Indeed, instead of starting with ω we could begin with the attendant order of visitation.

From the order of visitation implied by the seed we can fix certain facts about the real mapping $T_\nu: \mathbb{R} \to \mathbb{R}$. Let the real superstable k-cycle be $\{x_1,x_2,\ldots,x_k\}$ where $x_1 = 0$. Then the set of points $\{(x_1,x_2),(x_2,x_3),\ldots,(x_{k-1},x_k),(x_k,x_1)\}$ must lie on the graph of $T_\nu(x) = (x-\nu)^2$. The graph is a parabola with its minimum on the x-axis at x_k, see Figure 3. Since we do not know ν in general we cannot draw the graph accurately; however, we can make a <u>sketch graph</u> which contains the information we need. To do this we mark on both the x- and y-axes the set of points $\{x_1,x_2,\ldots,x_k\}$ according to their real order $x_1 < x_{\sigma(2)} < x_{\sigma(3)} < \cdots < x_{\sigma(k)}$. We label the intervals defined by these points with the notation $I_o = (-\infty,x_1)$, $I_1 = (x_1,x_{\sigma(2)})$, $I_2 = (x_{\sigma(2)},x_{\sigma(3)})\ldots$, $I_{k-1} = (x_{\sigma(k-1)},x_2)$ and $I_k = (x_2,\infty)$. We also locate the points whose <u>coordinates</u> are $(x_1,x_2),(x_2,x_3),\ldots,(x_{k-1},x_k)$

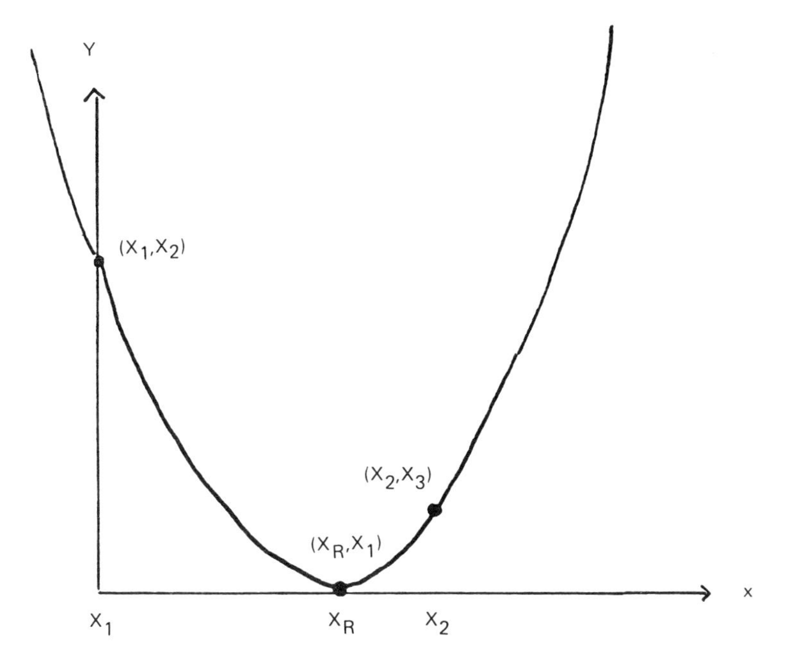

Figure 3. The graph of $T_\nu(x)$, when $\{x_1,x_2,\ldots,x_k\}$ is a superstable k-cycle.

and (x_k,x_1). The sketch graph is completed by joining the neighboring pairs of these points by straight lines, and including both a monotone decreasing straight line through $(x_1,x_{\sigma(2)})$ for $x \in I_o$ and a monotone increasing straight line through $(x_{\sigma(k-1)},x_2)$ for $x \in I_k$.

To illustrate the procedure so far we construct the sketch graph for the seed (+---). Denoting the corresponding 4-cycle (b(+---)) by $\{x_1,x_2,x_3,x_4\}$, its order of visitation is $x_1 < x_3 < x_4 < x_2$, and the intervals are $I_o = (-\infty,x_1)$, $I_1 = (x_1,x_3)$, $I_2 = (x_3,x_4)$, $I_3 = (x_4,x_2)$ and $I_4 = (x_2,\infty)$. The sketch graph is shown in Figure 4.

Next we use the sketch graph to locate where lie the preimages under T_ν of the intervals I_o, $I_1,\ldots,$ I_k. Let $\tilde{R}_\pm$

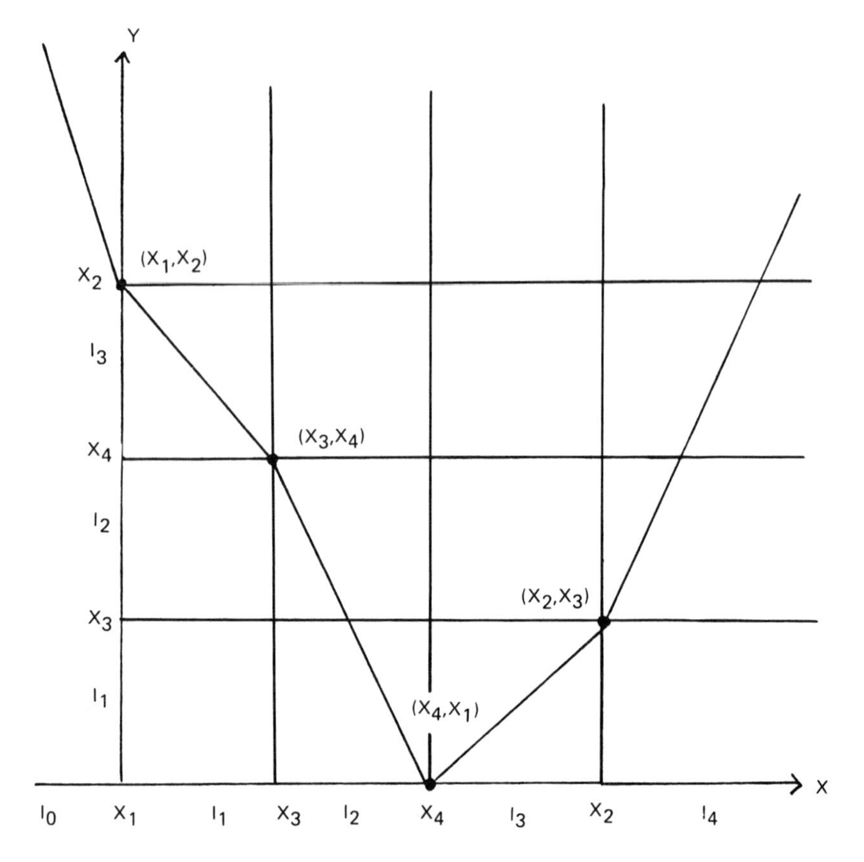

Figure 4. The sketch graph for the seed (+---). It is convenient to mark the intervals I_o, I_1, I_2, I_3, and I_4 on both axes.

denote the two branches of the real inverse of T_ν, where the negative axis cut is chosen, and

$$\tilde{R}_+(x) = x_k + \sqrt{x} \ , \ \tilde{R}_-(x) = x_k - \sqrt{x}$$

Then for $j \in \{1,2,\ldots,k\}$, $\tilde{R}_+(I_j)$ (respectively $\tilde{R}_-(I_j)$) is contained in exactly one of the intervals $\{I_o, I_1, \ldots, I_k\}$ which we denote by I_j+ (respectively I_j-). To find I_j+ (respectively I_j-) one reads off the interval on the x-axis which lies to the right (respectively left) of x_k for which the corresponding portion of the graph has I_j for its set of y-values. Notice that $I_k+ = I_k$ and $I_k- = I_o$, whilst I_o has no real pre-images. Once $I_j\pm$ have been found for $j \in \{1,2,\ldots,k\}$ we construct what we will call the <u>code</u> (from the seed ω) as follows. Label k+1 columns $I_o, I_1, \ldots, I_k$ and for each $j \in \{1,2,\ldots, k\}$ draw two arrows, one labelled ⊕ from the column I_j+ to the column I_j, and one labelled ⊖ from the column I_j- to the column I_j. In place of using columns one can use points if k is not too large.

To illustrate we continue the above example, whose sketch graph is in Figure 4. From the figure, we readily find $\tilde{R}_-(I_1) \subset I_2$ whence $I_1- = I_2$, and $\tilde{R}_+(I_1) \subset I_3$ whence $I_1+ = I_3$. Similarly $I_2- = I_2$, $I_2+ = I_4$, $I_3- = I_1$, $I_3+ = I_4$, $I_4- = I_0$, and $I_4+ = I_4$. This information is represented by the code in Figure 5(a), and also by the code in Figure 5(b) where points are used instead of columns.

In general, once the code for the seed ω has been found, the set W_ω can be calculated as follows. To find a member of W_ω we simply write down from left to right any half infinite sequence of plus and minus signs which is encountered upon following arrows from column to column (or point to point) commencing at I_j with $j \neq 0$. The set of all elements of Ω which are obtainable in this manner is exactly W_ω, and the set of corresponding negative axis λ-chains is $B_\nu \cap \gamma$, as stated formally in the next theorem.

<u>THEOREM 9</u>. Let ω belong to a k-cycle in Ω. Then W_ω is equal to the set of elements of Ω which are obtained by following arrows in the code for ω, as described above.

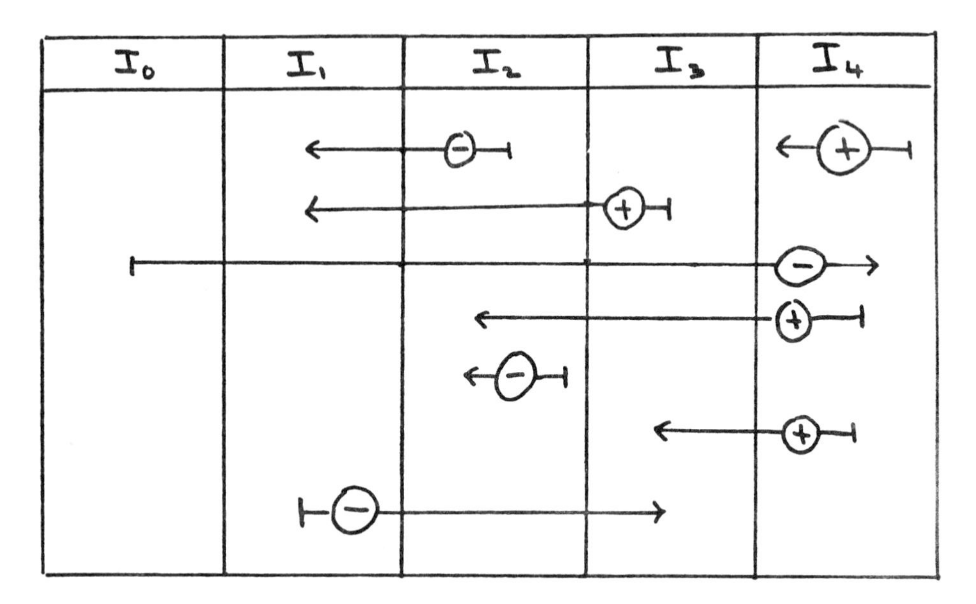

Figure 5(a). The code for the seed (-+++).

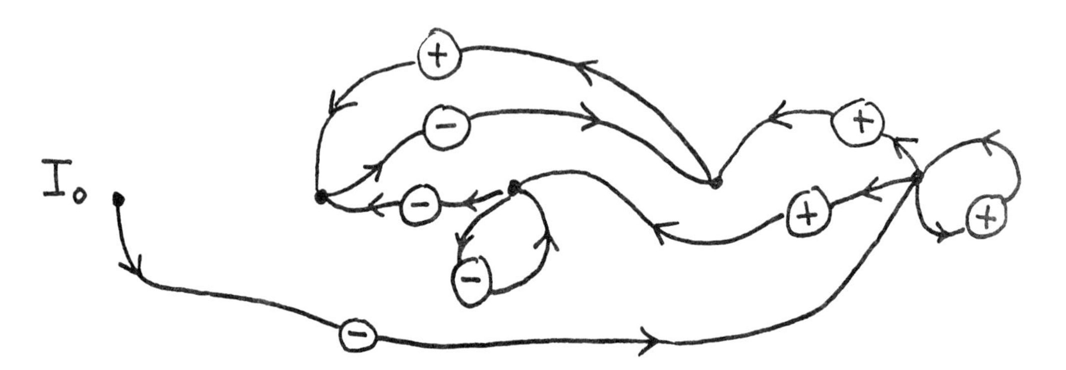

Figure 5(b). The code for the seed (-+++) using points in place of columns.

<u>Proof</u>. Let the continuation through decreasing values of λ of either $(\tilde{b}(\omega))$ or its partner be superstable at $\lambda = \nu$. Then we recall that W_ω is the set of $\sigma \in \Omega$ such that $b(\sigma) \in (0,\infty)$ when $\lambda = \nu$.

Let $\sigma \in \Omega$ be obtained by following arrows in the code for ω, as described above. Then fix $\lambda = \nu$ and consider the sequence of inverse functions $\{\tilde{R}_\sigma^n(x)\}_{n=1}^\infty$ as defined prior to Theorem 2. There must exist two intervals I_j and I_m with $j,m \in \{1,2,\ldots,k\}$, and an infinite subsequence $\{\tilde{R}_\sigma^{n(i)}(x)\}_{i=1}^\infty$ such that

$$\tilde{R}_\sigma^{n(i)}(I_m) \subset I_j \qquad \text{for } i \in \{1,2,\ldots\}$$

(The interval I_j is the one at which commences the chain of arrows giving σ, and I_m is an interval which is visited infinitely many times on following that chain.) Since the k-cycle $(\tilde{b}(\omega))$ or its partner is superstable the set of critical points for all branches of $\tilde{R}^n(z)$ for any n is contained in the k-cycle and does not belong to I_m. Hence the sequence of functions $\{\tilde{R}_\sigma^{n(i)}(z)\}$ is holomorphic in a neighborhood of I_m and hence by [Brolin, Theorem 6.2] possesses a subsequence which converges uniformly on closed subintervals of I_m to an element of B_ν. This element of B_ν lies in $I_j \subset (0,\infty)$. Note that since the sequence of sets $\{\tilde{R}_\sigma^{n(i)}(I_m)\}_{i=1}^\infty$ is nested and decreasing, the convergent subsequence can be taken to be $\{\tilde{R}_\sigma^{n(i)}(z)\}_{i=1}^\infty$, $z \in I_m$. The element to which the subsequence converges can be uniquely represented $\tilde{b}(\tilde{\sigma})$ for some $\tilde{\sigma} \in \Omega$, since every real element of B_ν is given by exactly one negative axis λ-chain. Noting that $T_\nu^n(\tilde{b}(\tilde{\sigma}))$ must lie to the left or right of $x_k = \lambda = \nu$

according as $\sigma_n = -1$ or $+1$ respectively, we conclude that $\sigma = \tilde{\sigma}$, and hence $\sigma \in W_\omega$ as desired.

Conversely, let $\xi \in W_\omega$. Then $\tilde{b}(\xi) \in B_\nu \cap \gamma$ and for any positive integer n there exists $y \in B_\nu \cap \gamma$, namely $y = T_\nu^n(\tilde{b}(\xi))$, such that

$$x = \tilde{R}_\xi^n(y)$$

But this says that from the code for ω by following arrows we must be able to find the sequence of signs belonging to the first n components of ξ. We conclude that ξ can be obtained by following arrows in the code for ω. Q.E.D.

We complete the example of (+---) commenced above. From the corresponding code shown in Figures 5(a) and 5(b) we read off at once that W_ω must contain the two 1-cycles (+) and (-), and also the 2-cycle (+-). In fact in this case we see that W_ω consists of all elements of Ω expressible in the form (+++...+⦚ ---...- ⦚+-). This notation extends that which was introduced earlier, and means all elements of the form (+), or (+++...+⦚ -) where there are finitely many initial plus signs, or (+++...+---...- ⦚+-) where there are finitely many plus signs at the beginning, followed by finitely many minus signs, followed finally by +-+-+-+-... Notice in this example that W_ω does not contain any elements which are not eventually periodic, in contrast to the next example.

As a second illustration we take the seed to be $\omega =$ (+--), for which the order of visitation is

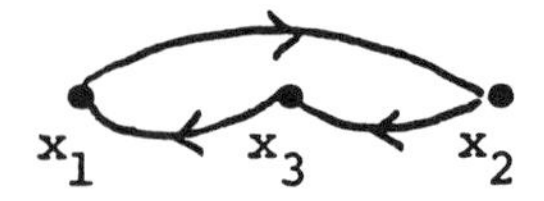

The corresponding sketch graph is given in Figure 6, and

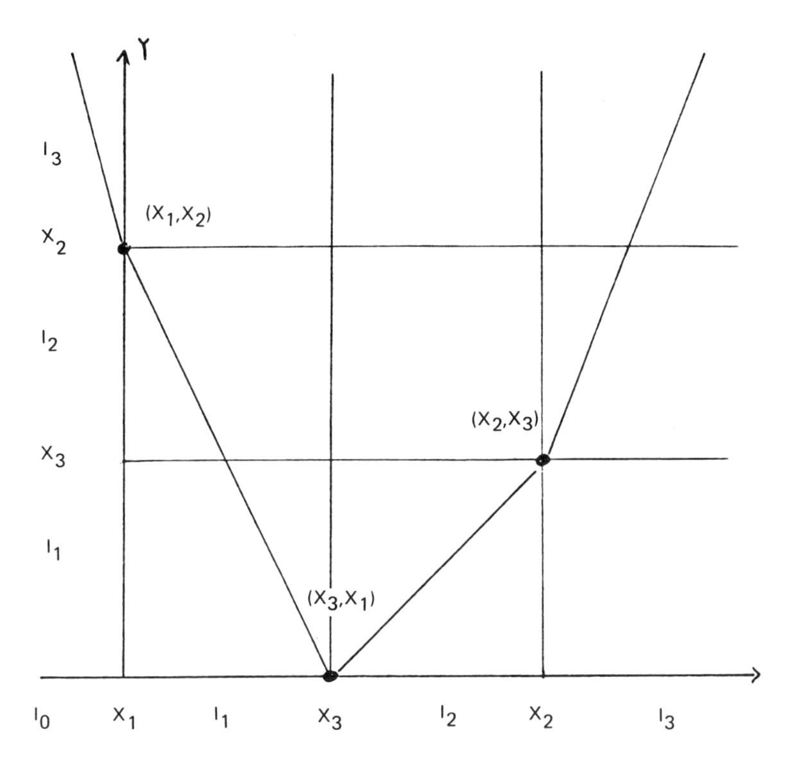

Figure 6. The sketch graph for the seed (-++).

the resulting code is given in Figures 7(a) and 7(b). Notice that although this code is simpler to write down than the one for (+---), it implies that W_ω is much larger than for the previous example. W_ω now consists of all elements of Ω which begin with finitely many plus signs, followed by any sequence of signs from {-, -+}. In particular we see that the existence of the real 3-cycle ($\tilde{b}$(-++)) implies that the Julia set contains real k-cycles of all other orders. Notice that the implied cycles are by no means arbitrary, and their orders of visitation are implicit. For example, the real 4-cycle ($\tilde{b}$(-+++)) is implied but ($\tilde{b}$(-++-)) is not. If ($\tilde{b}$(-++-)) is a real 4-cycle then the real 3-cycle ($\tilde{b}$(-++)) is implied. Thus, although from the more general setting of

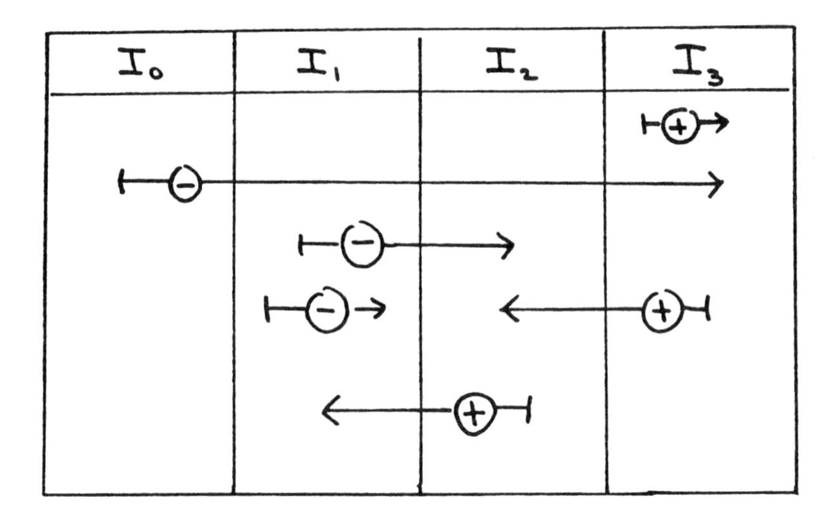

Figure 7(a). The code for the seed (-++).

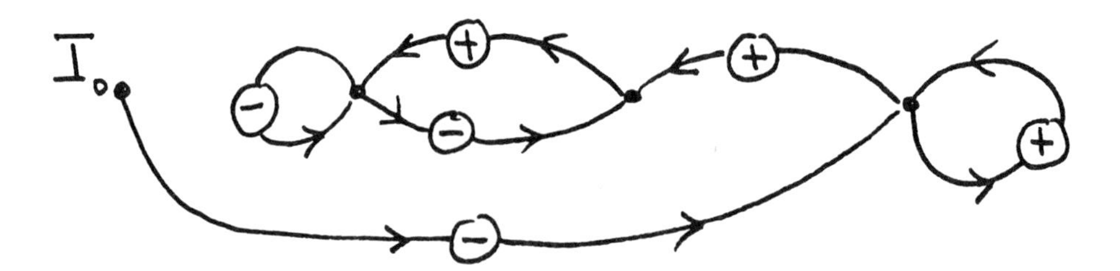

Figure 7(b). The code for the seed (-++) using points instead of columns.

Sharkovskii [Sh] and of Li and Yorke (LY] we have that "real period 3 implies real period 4," we have in our more specialized situation that "period 4 with a certain order of visitation implies period 3," and so on.

We remark that if, in determining the elements of W_ω from the code for ω, we allow the sequence of arrows to start at I_o as well as at $I_1, I_2, \ldots, I_k$, then we obtain the subset of Ω for which the negative axis λ-chains lie on the real axis; that is, we include those members of B_λ which lie on the negative real axis.

We summarize as follows. The seed ω fixes a sketch graph and in turn a code. The code fixes W_ω and hence $B_\nu \cap \gamma$. Once $B_\nu \cap \gamma$ is known, the equivalence class structure for positive axis λ-chains is implied. As described following Theorem 4, the latter fixes the topology of B_ν. Thus, we think of B_ν as having grown from the seed ω, and note that in the obvious sense the topology of B_ν is that of minimal complexity possible, given that T_λ possesses a real k-cycle whose order of visitation is that associated with the seed.

The results here cover territory which overlaps that examined by Guckenheimer [Gu2], and is related to the work of Sharkovskii [Sh] and of Li and Yorke [LY]. Our work refers to a more specialized situation, and is in the context of the complex plane as a whole. Our approach, based upon the interpretation of λ-chains, seems to be more immediate and more unified than any other we know of.

2.3. The First Cascade

Here we apply the theory of the previous sections to describe the progression of Julia sets when λ increases and the real mapping $T_\lambda(x) = (x-\lambda)^2$ yields its first cascade of period doubling bifurcations.

We begin by describing the cascade itself in terms of the real k-cycles involved. The commencement of the cascade was analyzed following Theorem 8, see Figure 2, and it corresponds to the successive occurrence of the sequence of real 2^n-cycles

$(\tilde{b}(-))$	2^0-cycle
$(\tilde{b}(-+))$	2^1-cycle
$(\tilde{b}(-+--))$	2^2-cycle
$(\tilde{b}(-+---+-+))$	2^3-cycle
$(\tilde{b}(-+---+-+-+---+--))$	2^4-cycle

Let the 2^nth cycle in this cascade by $(\tilde{b}(\omega_{2^n}))$. Then $(\omega_{2^{n+1}})$ is obtained by repeating (ω_{2^n}) twice and then reversing the last sign. It is important to observe that the cycles here are represented using the notational shorthand introduced following Theorem 7. The 2^n-cycle is calculated from the stated λ-chain when λ is sufficiently large, and it is obtained by real continuation through decreasing λ-values when the cycle is attractive. In the latter case the 2^n-cycle can be represented by the functional equations which it obeys, these being derived from (ω_{2^n}) as was expressed in Theorems 7 and 8, and illustrated after the proofs following Theorem 8.

One can readily check that (ω_{2^n}) corresponds to the smallest element of $(\tilde{b}(\omega_{2^n}))$, and also that except for the 1-cycle each cycle in the cascade is a loner, derived by pitchfork bifurcation from its predescessor, along the lines of Theorem 8.

Let $(b(\omega_{2^n}))$ be superstable at $\lambda = \nu_n$. Then it follows from Theorem 8 that the ν_n's obey the sequence of functional equations

$$f_o(\nu_0) := \nu_0 = 0$$
$$f_1(\nu_1) := \nu_1 - \sqrt{\nu_1} = 0$$
$$f_2(\nu_2) := \nu_2 - \sqrt{(\nu_2 + \sqrt{(\nu_2 - \sqrt{\nu_2}))}} = 0$$
$$f_3(\nu_3) := \nu_3 - \sqrt{(\nu_3 + \sqrt{(\nu_3 - \sqrt{(\nu_3 - \sqrt{(\nu_3 - \sqrt{(\nu_3 + \sqrt{(\nu_3 - \sqrt{\nu_3}))))))}}}}}} = 0$$

Here the negative axis branch cut is understood. Note that each equation begins and ends with the sequence of signs of its predecessor, and there is one additional sign in between. Numerical computations of these functions suggest that $f_{2^{n+1}}(x)$ is complex valued for $x < \nu_n$ and that their behavior is that sketched in Figure 8. These functions can be used as the basis for numerical methods for the calculation of the "universal constants" such as $\lim_{n\to\infty} \nu_n = \nu_\infty$ (equal to the Myrberg [My] number 1.401155...) and

$$\lim_{n\to\infty} \frac{\nu_n - \nu_{n-1}}{\nu_{n+1} - \nu_n}$$

(equal to the Feigenbaum [Fe] ratio 4.66920...). Since various

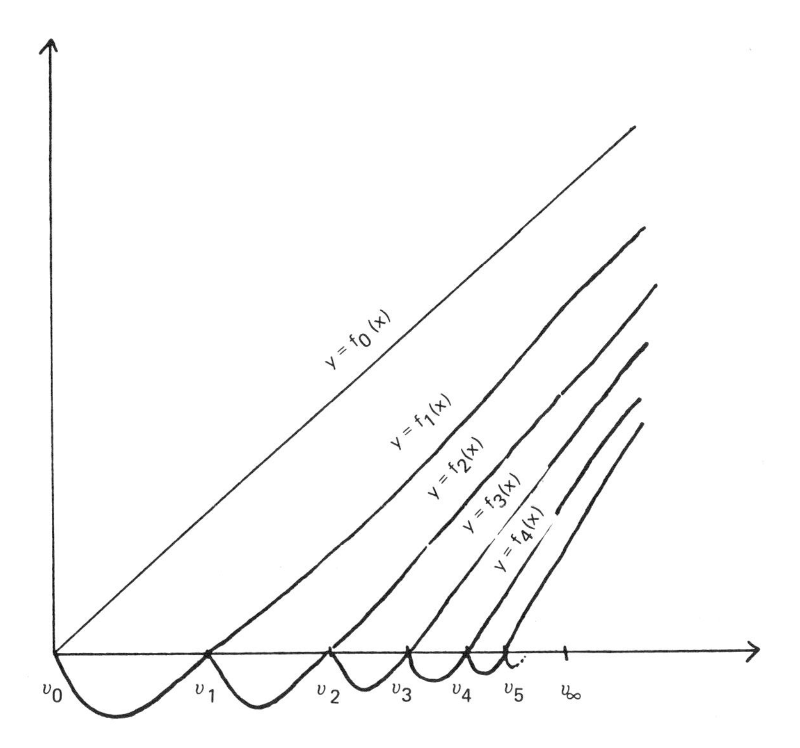

Figure 8. Sketch of the functions $f_0(x)$, $f_1(x)$, $f_2(x)$,...

methods are known for the computation of the numbers we do not dwell on this point. What we do notice is that a similar approach can be adopted for any other cascade. In particular, we see that the analysis of the λ-chains involved allows one to focus attention on exactly the k-cycles one wishes to study without having to consider irrelevant cycles of the same order.

The conclusion of the first cascade is marked by the occurrence of the "∞-cycle" ($\tilde{b}$(-+---+-+-+---+--...)) referred to by [DH] and by [CE].

The occurrence of any real 2^n-cycle in the sequence $\{b(\omega_{2^n})\}_{n=1}^{\infty}$ of course implies that all of its predecessors are real. More generally, with the aid of Theorem 9, we calculate

$$W_{\omega_1} = \{(+)\}$$

$$W_{\omega_2} = \{(+++\ldots+\wr\, -)\}$$

$$W_{\omega_4} = \{(+++\ldots+\wr\, ---\ldots-\wr\, +-)\}$$

$$\text{and } W_{\omega_8} = \{(+++\ldots+\wr\, ---\ldots-\wr\, +-+-\ldots+-\wr\, +---)\}$$

The notation is that introduced near the end of Section 2.1. From these cases it appears that $W_{\omega_{2^{n+1}}}$ can be obtained systematically from $W_{\omega_{2^n}}$ by replacing the tails of elements of the latter by (ω_{2^n}).

We next describe with the aid of Figure 9(a)-(f) the successive structures of B_λ as the cascade proceeds. Here as elsewhere it is most convenient to work in terms of positive axis λ-chains for elements of B_λ with nonzero imaginary parts. Real elements will usually be expressed by their unique negative axis λ-chains and sometimes by their equivalent pair of positive axis λ-chains. The bifurcation point at which the real 2^n-cycle $(\tilde{b}(\omega_{2^n}))$ first appears will be denoted λ_n.

In Figures 9(a)-(f) we use the notation x for the real 1-cycle $(\tilde{b}(-))$ and we write $T_\lambda^{-1}(x) = \{x,y\}$ where $x \neq y$. We use ● to denote elements of k-cycles and □ to denote first predecessors of elements of k-cycles which do not themselves belong to k-cycles. The critical point λ is shown in each figure.

In Figure 9(a) we give a schematic representation of B_λ in the complex plane, when $-1/4 = \lambda_0 < \lambda < \nu_0 = 0$. In this case, as explained in the fourth paragraph after the proof of Theorem 4, B_λ is a simple Jordan curve. The equivalence class structure of Ω follows from W_{ω_1}, and we find that the only elements of Ω whose equivalence classes consist of more than one element are $(+)\sim(-)$, $(+\wr-)\sim(-\wr+)$ and $(s_1s_2\ldots s_n+\wr-)\sim(s_1s_2\ldots s_n-\wr+)$, where each $s_i \in \{+,-\}$. The only real members of B_λ are the repulsive 1-cycle $b(+) = b(-) = \tilde{b}(+)$ and its real preimage $b(+\wr-) = b(-\wr+) = \tilde{b}(-\wr+)$. B_λ separates the complex plane into two components, one of which contains x. This component is not only the attractive set of x but also the <u>immediate attractive set</u> of x. (The immediate attractive set of a k-cycle is the union of the largest connected components of the attractive set of the k-cycle each of which contains an element of the k-cycle, see [Br]). The other component of the complement of B_λ is the immediate attractive set of ∞. x and y obey the functional equations $x = \lambda+\sqrt{x}$ and $y = \lambda-\sqrt{x}$. Also in Figure 9(a) we show the 2-cycle $\{b(+-),b(-+)\}$ together with its first predecessor

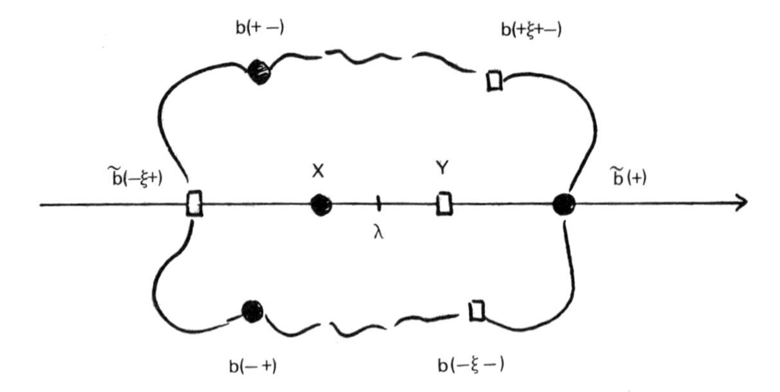

Figure 9(a). Schematic representation of B_λ when $-1/4 = \lambda_0 < \lambda < \nu_0 = 0$. B_λ is a simple Jordan curve.

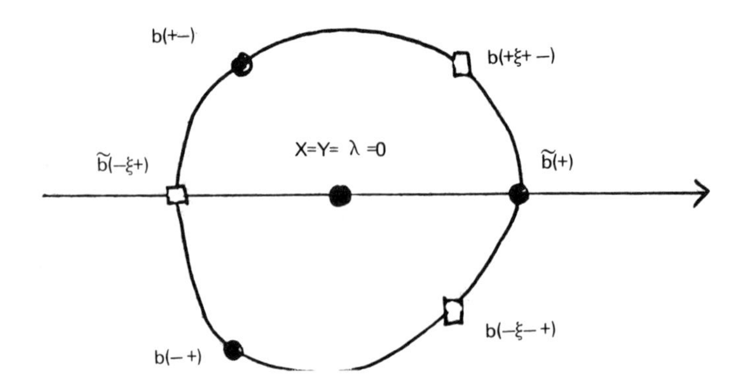

Figure 9(b). Sketch of B_λ when $\lambda = \nu_0 = 0$. B_0 is the unit circle.

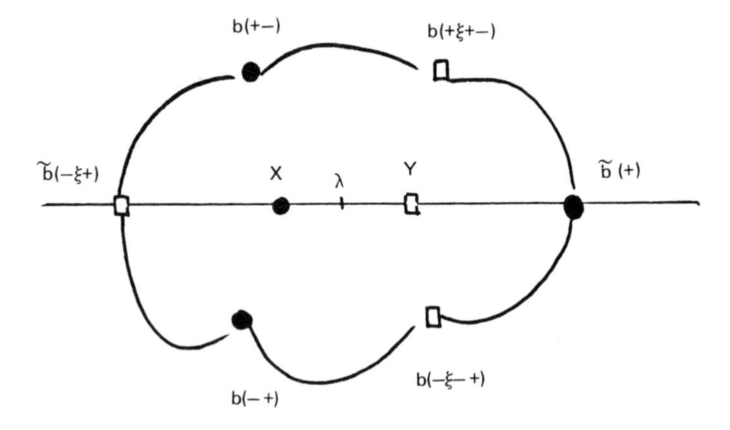

Figure 9(c). Schematic representation of B_λ when $0 = \nu_0 < \lambda < \lambda_1$. B_λ is a simple Jordan curve.

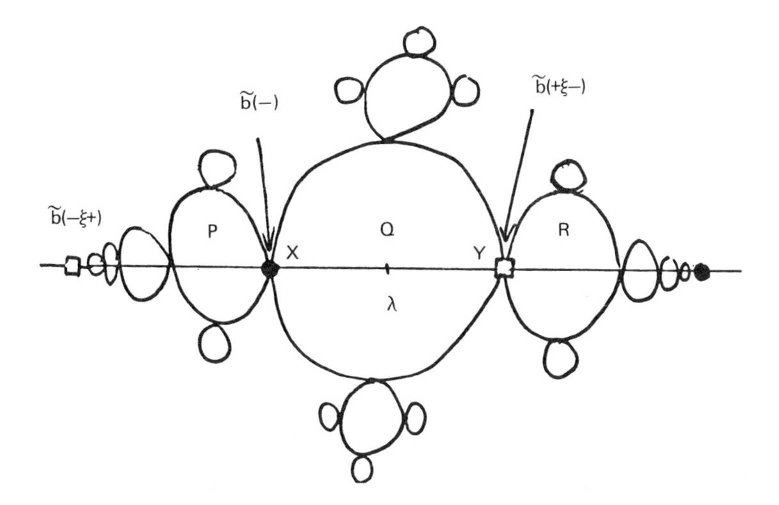

Figure 9(d). Schematic representation of B_λ when $\lambda = \lambda_1$, at which occurs the first pitchfork bifurcation

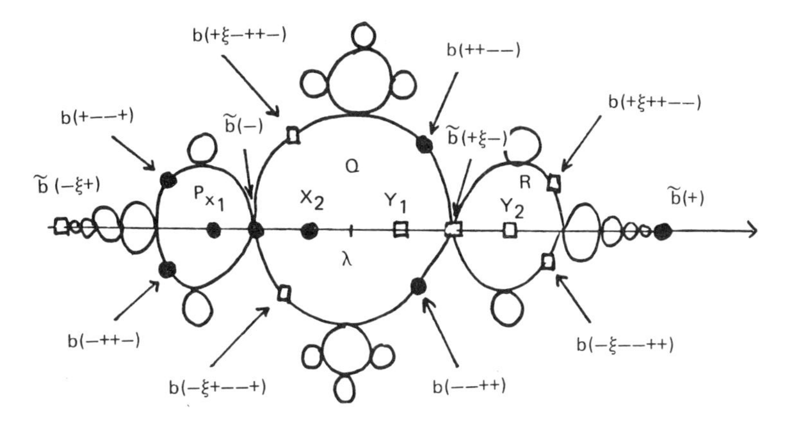

Figure 9(e). Schematic representation of B_λ when $\lambda_1 < \lambda < \nu_1$. T_λ now possesses an attractive 2-cycle.

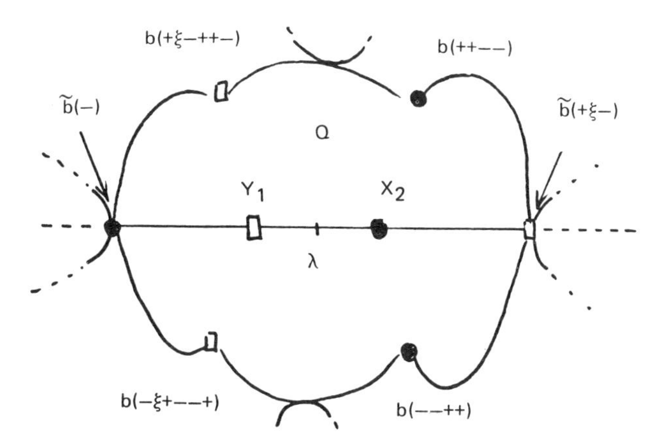

Figure 9(f). Schematic representation of the continuation of the component labelled Q in Figure 9(e). Now $\nu_1 < \lambda < \lambda_2$. Compare with Figure 9(c).

$\{b(+\wr +-), b(-\wr -+)\}$, which also lies on B_λ. The continuation of the 2-cycle, when it first becomes real, will coalesce with real 1-cycle which we denote by $(\tilde{b}(-))$ to yield the first pitchfork bifurcation in the cascade. Note that the 2-cycle $\{b(+-), b(-+)\}$ is the same as the negative axis λ-chain $\tilde{b}(-)$.

In Figure 9(b) $\lambda = \nu_0$ and the attractive 1-cycle $x = (\tilde{b}(-))$ is superstable, being coincident both with the critical point λ and with the set of all predecessors of x. B_0 is in fact the unit circle and is, exceptionally, of finite length (cf. [Br] Theorem 9.1). In Figure 9(c) we have $\nu_0 < \lambda < \lambda_1$, and the main difference from the situation in Figure 9(a) is that the real ordering of the attractive 1-cycle x and its preimage y has been reversed, and for all $\lambda > \nu_0$ we have $x = \lambda - \sqrt{x}$ and $y = \lambda + \sqrt{x}$. As λ increases from ν_0 to λ_1 the complex 2-cycle $\{b(+-), b(-+)\}$ approaches the real attractive 1-cycle x. As it does so the preimage $\{b(-\wr +-), b(+\wr +-)\}$ approaches the preimage y of the 1-cycle. Not shown are the higher order preimages of the 1-cycle and the 2-cycle involved: the predecessors of order n of the 2-cycle lie on the simple Jordan curve B_λ and can be separated into pairs each of which approaches one of the predecessors of order n of x. The latter all lie in the bounded component of the complement of B_λ.

In Figure 9(d) $\lambda = \lambda_1$. B_λ is now pinched together at x, where $b(+-) = b(-+) = (\tilde{b}(-))$. That is, the λ-chain $\tilde{b}(-)$ is now the same as the 1-cycle denoted by $(\tilde{b}(-))$. Prior to λ_1 $\tilde{b}(-)$ actually yields the 2-cycle, which has nonzero imaginary parts, whilst for $\lambda > \lambda_1$ it gives a real 1-cycle. Similarly B_λ is pinched together at y where $b(+\wr +-) =$

$b(-\xi-+) = \tilde{b}(+\xi-)$. The other multiple points in the figure represent a few of the countable infinity of other "pinch-points" at which preimages of higher order of x and of the 2-cycle are coincident. The 1-cycle x is now indifferent rather than attractive, and lies on B_λ. Since there is no attractive k-cycle, not all λ-chains are defined in what we have presented so far. (In fact, all eventually periodic points which are not attracted to the indifferent fixed point can be well-defined in terms of λ-chains.) The meanings of the chains indicated in Figure 9(d) are clear from continuation. The three components in the figure which are labelled P, Q, and R are related by $T_{\lambda_1} P = Q$, $T_{\lambda_1} Q = P$, and $T_{\lambda_1} R = Q$.

In Figure 9(e) B_λ is represented for $\lambda_1 < \lambda < \nu_1$. In this case there is an attractive 2-cycle, namely $\{x_1,x_2\} = (\tilde{b}(-+))$, and the previously attractive 1-cycle denoted by $(\tilde{b}(-))$ now belongs to B_λ. $\{x_1,x_2\}$ has emerged from x, leaving the λ-chain $\tilde{b}(-)$ on B_λ, whilst the preimage $\{y_1,y_2\}$ of the 2-cycle has emerged from y, leaving $\tilde{b}(+\xi-)$ on B_λ. Thus, the 2-cycle which was earlier on B_λ has left it to become an attractive 2-cycle, its stability having been transferred from the previously attractive 1-cycle which has now rejoined B_λ. We see here an example of the interplay between λ-chains and functional equations. When points leave B_λ (not just k-cycles but their preimages as well) they can be represented by the functional equations which they then obey, and the functional equations for the points which rejoin B_λ yield their λ-chain description. We can view the cascade as a sequence of events whereby k-cycles and their preimages, represented by λ-chains, part from B_λ, producing "pinch-points;" and then, represented by functional equations, they return to B_λ to rejoin it at new "pinch-points," the birth places of the next attractive cycle in the cascade and its preimages.

In Figure 9(e) we also show the 4-cycle $\{b(+--+)$, $b(--++)$, $b(-++-)$, $b(++--)\}$ and its first predecessor $\{b(-\wr +--+)$, $b(-\wr --++)$, $b(+\wr -++-)$, $b(+\wr ++--)\}$, which will be involved in the next bifurcation.

The situation for $\nu_1 < \lambda < \lambda_2$ is essentially the same as for $\lambda_1 < \lambda < \nu_1$ except that the real ordering of x_2 and y_1 is interchanged. In Figure 9(f) we represent for $\nu_1 < \lambda < \lambda_2$ the continuation of the component labelled Q in Figure 9(e). In Figures 9(e) and 9(f), P and Q denote the two components of the immediate attractive set of $\{x_1,x_2\}$. The behavior of the boundary of Q, as λ increases from λ_1, is similar to that of the whole of B_λ as λ increases from λ_0. Indeed, if we consider T_λ^2 in place of T_λ we see that $\{b(++--),b(--++)\}$ becomes a 2-cycle instead of part of a 4-cycle, whilst x_2 becomes an attractive 1-cycle. As λ increases $\{b(++--)$, $b(--++)\}$ pinches inwards to join x_2, whilst its predecessors under T_λ^2 on the boundary of Q move to coincidence with the predecessors under T_λ^2 of x_2 in Q. Similar deformations take place with regard to P and to the countable infinite of other components of the attractive set of $\{x_1,x_2\}$.

The equivalence class structure of Ω for $\lambda_1 < \lambda < \lambda_2$ follows from W_{ω_2}, just as when $\lambda_0 < \lambda < \lambda_1$ it followed from W_{ω_1}. The elements of Ω whose equivalence classes consist of more than one element are $(+)\sim(-)$, $(+\ \ -)\sim(-\ \ +)$, and $(s_1s_2\ldots s_n+\wr -)\sim(s_1s_2\ldots s_n-\wr -)$ as when $\lambda_0 < \lambda < \lambda_1$, and in addition $(+++\ldots+\xi-+)\sim(---\ldots-\xi+-)$, and $(s_1s_2\ldots s_n-\xi+++\ldots+\xi-+)\sim(s_1s_2\ldots s_n+\xi---\ldots-\xi+-)$. The images of the latter under b are exactly the "pinch-points" in B_λ referred to above. One of each new equivalent pair terminates with (-+) and the other with (+-); and we recall that {b(+-),b(-+)} was the 2-cycle prior to λ_1.

Let ∂P, ∂Q, and ∂R denote respectively the boundaries of P, Q, and R in Figures 9(e) and (f). One readily shows that ∂P is given by the set of positive axis λ-chains $b(s_1s_1's_2s_2's_3s_3'\ldots s_ns_n'\ldots)$ where each $s_i \in \{+,-\}$ and s_i' is the opposite sign to s_i. Similarly ∂Q is given by $b(s_0s_1s_1's_2s_2's_3s_3'\ldots s_ns_n'\ldots)$ and ∂R is given by $b(s_0s_0s_1s_1's_2s_2'\ldots s_ns_n'\ldots)$. Each of these boundaries is a simple Jordan curve. The boundaries of the countable infinity of other components of the attractive set of the 2-cycle are obtained by taking inverse images of all orders of ∂P, and the set of positive axis λ-chains of which a given one of these boundaries consists can be deduced from the successive branches of the inverse of T_λ which when applied to ∂P yield the desired boundary. Note that, for $\lambda_1 < \lambda < \lambda_2$, B_λ is the closure of the set of all Jordan curves thus obtained. Similar observations apply with regard to the boundaries of the immediate attractive sets which occur as the cascade proceeds.

We can now make some deductions about the locations of cycles during the cascade. Note first that the cycles which participate in the cascade, prior to their becoming real, are given by the sequence of positive axis λ-chains

(2^0) {b(+)} and {b(-)}

(2^1) {b(+-), b(-+)}

(2^2) {b(+--+), b(--++), b(-++-), b(++--)}

(2^3) {b(+--+-++-), b(--+-++-+), ...}

⋮

When the 2^n-cycle in this sequence becomes real, and $(\tilde{b}(\omega_{2^n}))$ is attractive, then for all $j \in \{1,2,3,...\}$ the 2^{n+j}-cycle in the sequence resides upon the boundary of the immediate attractive set of $(\tilde{b}(\omega_{2^n}))$. For example, all of the sequence starting with {b(+-),b(-+)} lie upon B_λ when $(\tilde{b}(-))$ is attractive, and all of the sequence starting with {b(+--+),...} are located upon $\partial P \cup \partial Q$ when $(\tilde{b}(-+))$ is attractive. The general assertion can be proved inductively.

Our second deduction concerns cycles which do <u>not</u> lie either on the boundary of the immediate attractive set of $(\tilde{b}(\omega_{2^n}))$ when this cycle is attractive, or on any of the Jordan curves which are finite order preimages of the boundary of the immediate attractive set. We have already illustrated how one can calculate the set of positive axis λ-chains which make up the boundary of the immediate attractive set of a cycle. Clearly any k-cycle whose positive axis λ-chain is not included, must itself not lie in the boundary of any component of the complement of B_λ which does not contain infinity. For example, when $\lambda_1 < \lambda < \lambda_2$ the 3-cycle {b(++-), b(+-+), b(-++)} does not lie upon the boundary of any of the "bubbles" in Figures 9(e) and (f), because its λ-chains are not included in the ones, described above, which make up these boundaries. Thus the 3-cycle occurs only as an accumulation point of the boundaries in Figures 9(e) and (f). Similarly we discover that when the 3-cycle $(\tilde{b}(++-))$ is attractive, none of the cycles involved in the first cascade are located upon the boundary of the immediate attractive set of the 3-cycle or any of its finite order preimages.

To express our third deduction we introduce the invariant measure μ on B_λ, which is discussed with greater generality in Section 3.2. Let ν denote uniform Borel measure on the real interval [0,1], and identify Ω with [0,1] as was discussed for Theorem 4. Assume that T_λ has an attractive k-cycle. (This assumption is shown to be unnecessary when the more general approach of Section 3.2 is followed.) Then we say that $E \subset B_\lambda$ is μ-measureable if and only if $b^{-1}(E) \subset \Omega \equiv [0,1]$ is a Borel subset of [0,1], and in this case $\mu(E) = \nu(b^{-1}(E))$. μ is invariant under T_λ. What we observe is that the measure of the part of B_λ for $\lambda_1 < \lambda < \lambda_2$ which is actually outlined in Figures 9(e) and (f) is zero. This is because the $\nu(F) = 0$ where F is the Cantor subset of [0,1] in binary decimal representation expressible $.e_1e_1'e_2e_2'e_3e_3'\ldots$ where each $e_i \in \{0,1\}$ and $e_i' = |1-e_i|$. The measure μ does not reside upon any of the "bubbles" one draws, and the only Borel subsets of B_λ which have nonzero μ-measure are those which intersect accumulation points of "bubbles." A similar observation applies to the measure of the "bubbles" which occur at any other stage in the cascade, when $\lambda > \lambda_1$.

We conclude this discussion of the first cascade with some observations concerning an associated sequence of Böttcher functional equations. Recall that $(\tilde{b}(\omega_{2^n}))$ is superstable when $\lambda = \nu_n$, and in this case $x = 0$ belongs to the attractive cycle. Let $\lambda = \nu_n$, let P_n denote the largest connected component of the attractive set of $(\tilde{b}(\omega_{2^n}))$ which contains 0, and let ∂P_n denote the boundary of P_n. Then we construct a conformal mapping $E_n(z)$ of P_n one-to-one onto P_{n+1} as follows.

Let $g_n(z)$ denote the Green's function for P_n with pole at $x = 0$. Then

$$(\ddagger)\quad g_n(T^{2^n}_{\nu_n}(z)) = 2g_n(z) \qquad \text{for } z \in P_n-\{0\}$$

because $T^{2^n}_{\nu_n}(P_n) = P_n$ and $T^{2^n}_{\nu_n}(\partial P_n) = \partial P_n$ so both sides of $(\ddagger)$ vanish as z approaches ∂P_n, and because $g_n(z)$ can be written as $\log\frac{1}{z}$ plus a regular function whilst $T^{2^n}_{\nu_n}(z) = C_n z^2 + O(z^3)$ where C_n is a positive constant.

Let $F_n(z)$ be the unique conformal mapping which takes P_n onto P_0, with $F_n(0) = 0$ and $F_n(z) > 0$ when $z > 0$. Then we must have for $z \in P_n-\{0\}$

$$g_n(z) = g_o(F_n(z)) = \log\left(\frac{1}{F_n(z)}\right)$$

where we have used the fact that P_o is the disk $\{z \in \mathbb{C} \mid |z| < 1\}$ and $g_o(z) = \log(\frac{1}{z})$. It follows that

$$F_n(z) = \exp\{-g_n(z)\} \qquad \text{for } z \in P_n,$$

and from $(\ddagger)$ we now obtain

$$F_n(T^{2^n}_{\nu_n}(z)) = (F_n(z))^2 \qquad \text{for } z \in P_n$$

which is the Böttcher functional equation associated with the superstable fixed point $z = 0$ of $T^{2^n}(z)$. We set

$$E_n(z) = F_{n+1}^{-1}(F_n(z))$$

Then for $z \in P_{n+1}$ we have

$$(*) \quad E_n(T_{\nu_n}^{2^n}(E_n^{-1}(z))) = T_{\nu_{n+1}}^{2^{n+1}}(z)$$

which connects the action of $T_{\nu_n}^{2^n}$ on P_n with that of $T_{\nu_{n+1}}^{2^{n+1}}$ on P_{n+1}.

We now show that if $E_n(z)$ converges to Λz where Λ is a constant, and if certain limiting procedures are justified, then (*) leads to a functional equation of the form of the Cvitanović-Feigenbaum-Coullet-Tresser equation [CT,F]. Our point of view may provide further insights into the analyticity properties found by Epstein and Lascoux [EL]. We stress that our derivation is formal. Let

$$\psi_n(\nu_n, z) = \Lambda^{-n}(T_{\nu_n}^{2^n}(\Lambda^n(z)))$$

so that ψ_n maps $\Lambda^{-n}P_n$ conformally one-to-one onto itself. Taking limits, we suppose that $\psi(z) = \lim_{n\to\infty} \psi_n(\nu_n, z) = \lim_{n\to\infty} \psi_n(\nu_{n+1}, z)$ maps $P = \lim_{n\to\infty} \Lambda^{-n}P_n$ into itself. Then, on rewriting (*) as

$$\Lambda^{-(n+1)}E_n(T_{\nu_n}^{2^n}(E_n^{-1}(\Lambda^{(n+1)}z))) = \Lambda^{-1}(\Lambda^{-n}T_{\nu_{n+1}}^{2^n}(\Lambda^n\Lambda^{-n}T_{\nu_{n+1}}^{2^n}(\Lambda^n\Lambda z)))$$

for $z \in \Lambda^{n+1}P_{n+1}$, and letting $n \to \infty$ we obtain

$$\psi(z) = \Lambda^{-1}\psi(\psi(\Lambda z)) \quad \text{for } z \in P$$

as desired.

Our final observation concerns the tree-like structures introduced in [BGH1] and considered further in [BGH3,5, BGHD]. We recall the following result. Let $0 < \lambda < 2$, $I_o = [0,\lambda+1/2+\sqrt{\lambda+1/4}]$, and $I_j = T_\lambda^{-1}(I_{j-1})$ for $j \in \{1,2,3,\ldots\}$. Then $\{I_j\}_o^\infty$ is an increasing sequence of trees of analytic arcs with $B_\lambda \subset \overline{\bigcup_{j=0}^{\infty} I_j} \subset \{z \in \mathbb{C} | \{T_\lambda^n(z)\}_1^\infty \text{ is bounded}\}$. For infinitely many values of λ with $0 < \lambda < 2$, $B_\lambda = \overline{\bigcup_{j=0}^{\infty} I_j}$ and $\mathbb{C}-B_\lambda$ is connected; in which cases we say B_λ is treelike. The trees $\{I_n\}_{n=1}^\infty$ are of interest even when B_λ is not the closure of their union because of their relationship with the locations of zeros and equioscillation points of certain orthogonal polynomials associated with B_λ.

Here we note the existence of subtrees which lie within the components of the attractive set of $(\tilde{b}(\omega_{2^n}))$ when that cycle is attractive. For example, there is a subtree located in the closure of the component P in Figure 9(e). This is constructed as follows. Let $\tilde{I}_1 = \overline{P \cap \mathbb{R}}$ and $\tilde{I}_{n+1} = \bar{T}_\lambda^{-2}(\tilde{I}_n)$ where $\bar{T}_\lambda^{-2}$ denotes a restricted inverse of T_λ^{-2}. The domain of $\bar{T}_\lambda^{-2}$ consists of P repeated on those sheets of the domain of T_λ^{-2} which are reachable on following curves within P. The range of $\bar{T}_\lambda^{-2}$ is of course P itself. It is easy to see that $\{\tilde{I}_j\}_1^\infty$ is an increasing sequence of trees of analytic arcs with $\partial P \subset \overline{\bigcup_{j=1}^{\infty} \tilde{I}_j} \subset \bar{P}$. The tree-like structure $\overline{\bigcup_{j=0}^{\infty} I_j}$

is the closure of the set of all preimages of $\overline{\bigcup_{j=1}^{\infty} \tilde{I}_j}$. Each preimage of finite order of $\overline{\bigcup_{j=1}^{\infty} \tilde{I}_j}$ is itself a subtree, and is located inside one of the "bubbles" in Figure 9(e). In a similar manner we can associate subtrees with the components of the attractive set of $(b(\omega_{2^n}))$.

Let $\{\tilde{I}_j^m\}_{j=1}^{\infty}$ denote the increasing family of subtrees which is associated with the component P_m of the attractive set of $(b(\omega_{2^m}))$. Then we find that for $n \in \{1,2,3,...\}$

$$E_n(\overline{\bigcup_{j=1}^{\infty} \tilde{I}_j^n}) = \overline{\bigcup_{j=1}^{\infty} \tilde{I}_j^{n+1}}$$

which shows the topological equivalence of subtrees obtained as the cascade proceeds.

3. Electrical Properties of Julia Sets

In this chapter we develop properties of the equilibrium charge distribution on the Julia set for a polynomial, from the point of view of the Böttcher equation and Green's star domains. We use Julia sets for $T(z) = (z-\lambda)^2$ as examples and relate the electrical properties to the geometry of the Julia set.

3.1. The Böttcher Equation

Let B be the Julia set for $T(z) = z^N + a_1 z^{N-1} + ... + a_N$. Let B_0 be the unit circle, and let D_0 and D be the components of the complements of B_0 and B respectively containing ∞. Fatou [Fa] has shown that there is a unique function analytic at ∞, normalized so that $F(z) = z+O(1)$, which obeys the Böttcher functional equation

$$F(T(z)) = (F(z))^N \tag{1}$$

F may be extended to be a well-defined analytic function in any simply connected subregion of D.

Let $g(z)$ be the Green's function with pole at ∞ for D. That is, $g(z)$ is a multiple-valued analytic function on D except at ∞ where $g(z)-\log(z)$ is analytic, and $\lim_{z\to B} \mathrm{Re}(g(z)) = 0$.

First we establish

$$g(z) = \frac{1}{N} g(T(z)) \tag{2}$$

and then recover the Böttcher functional equation from this. Since the Green's function is unique we prove (1) by showing $\frac{1}{N} g(T(z))$ satisfies the conditions for the Green's function. Since $T(D) = D$ and $T^{-1}(\infty) = \infty$, $g(T(z))$ is analytic on D except for a logarithmic singularity at ∞. Furthermore, since $T(z) = z^N(1+O(\frac{1}{z}))$ we may expand $\frac{1}{N} g(T(z))-\log(z)$ at ∞ to show it is analytic. Finally, since $TB = B$, the boundary condition is satisfied.

Now

$$F(z) = e^{g(z)} \tag{3}$$

since

$$(F(z))^N = e^{Ng(z)} = e^{g(T(z))} = F(T(z))$$

Let $G(z) = \log(z)$, the Green's function for D_o with pole at ∞. Then, equivalent to (3), we may write

$$G(F(z)) = g(z) \tag{4}$$

so F gives a correspondence between Green's functions. Note that if $H(z) = F^{-1}(z)$ then (1) can be restated as

$$H(z^N) = T(H(z)) \tag{5}$$

3.2. Green's Star Domains and Equilibrium Measures

For any domain D contained in the extended complex plane $\hat{\mathbb{C}}$, having Green's function $g(z)$ with pole at ∞, we may set $F(z) = e^{g(z)}$ and $H = F^{-1}$. For each $\theta \in [0,2\pi)$ we define $t_o(\theta)$ to be the minimum number ≥ 1 such that H may be analytically continued from ∞ along $R_\theta = \{te^{i\theta} | t > t_o\}$. We find that $t_o(\theta) > 1$ for at most a countable set of values of θ which are arguments of branch points of H. The set $\underset{\theta}{\cup} H(R_\theta)$ is called the Green's star region for D, see for example [SN]. Let μ be the equilibrium distribution of a unit charge on the boundary of D. μ is a Borel measure [Ts]. Then the Green's star domain may be thought of physically as the lines of force emanating from the boundary of D which do not branch.

We are mainly concerned with boundary behavior. If $t_o(\theta) = 1$, let Γ_θ be the set of limit points of the closure of $H(R_\theta)$ not in $H(R_\theta)$, so Γ_θ is a singleton or a connected set contained in the boundary of D.

When D is as in Section 3.1, the boundary of D is the Julia set B. We illustrate the situation for $T(z) = (z-\lambda)^2$ and $B = B_\lambda$, when

(i) $-1/4 < \lambda < 2$ and T has an attractive k-cycle,

(ii) $\lambda \geq 2$.

(i) We have discussed the geometry in this case. In particular, B_λ is connected. Using (5), the fact that λ-chains are well-defined, and the mapping $b: [0,1] \to B_\lambda$ discussed in Theorem 4, one can easily show that $\Gamma(2\pi\alpha) = \{b(\alpha)\}$ for $0 \leq \alpha \leq 1$. Thus, we can extend H continuously to B_o so $H(e^{2\pi i\alpha}) = b(\alpha)$. In other words, if $b(\alpha) = z \in B_\lambda$, then the imaginary part of the Green's function on a line of force terminating at z must be $2\pi\alpha \bmod 2\pi$.

(ii) When $\lambda = 2$, $B_2 = [0,4]$ and the extension of $H(z) = z + \frac{1}{z} + 2$ to B_o becomes two-to-one at all points of B_2 except the endpoints. Points on (0,4) all have conjugate lines of force emanating from them. As is well known, see for example [BGH1], a dramatic change takes place as λ increases past 2: B_λ stays on the real line but breaks up to become a generalized Cantor set. We review the construction. T^{-1}: $[0,a] \to [0,a]$ where $a = \tilde{b}(+)$ is the fixed point which obeys $a = \lambda+\sqrt{a}$. In particular $[0,a]$ is mapped onto $[\lambda,a]$ by the branch $\lambda+\sqrt{z}$, and $[0,a]$ is mapped onto $[\lambda-\sqrt{a},\lambda] \subset [0,\lambda]$ under the branch $\lambda-\sqrt{z}$.

Let

$$I_o = T^{-1}([0,a]) = [\lambda-\sqrt{a},a]$$

$$I_1 = T^{-1}(I_o) = I_o \backslash \theta_1$$

where $\theta_1 = T^{-1}([0,\lambda-\sqrt{a})) = (\lambda-\sqrt{\lambda-\sqrt{a}},\ \lambda+\sqrt{\lambda+\sqrt{a}})$,

$I_j = T^{-1}(I_{j-1})$,

and $\theta_j = I_j \backslash I_{j-1} = T^{-1}(\theta_{j-1})$ for $j = 2,3,\ldots$

Then $B = \cap I_j = I_o - \cup\theta_j$.

θ_j consists of disjoint open intervals θ_j^k, $k = 1,2,\ldots,$ 2^{j-1}, each containing a predecessor of 0.

From (2) we have

$$g(z) = \frac{1}{2} g(T(z)),$$
$$g'(z) = \frac{1}{2} g'(T(z))T'(z) \tag{6}$$

Using (6), or the symmetry of B_λ, we see that $g'(\lambda) = 0$ and vertical lines of force from ∞ join at λ, the center of θ_1, then split and go sideways to the two endpoints of θ_1 on B_λ. For $j > 1$, $k = 1,2,\ldots,2^{j-1}$, there are two conjugate lines of force from ∞ joining at the single point in $T^{-j}(0) \cap O_j^k$, where they split and terminate at the ends of O_j^k in B. See Figure 10.

We see that the Green's star domain D' has boundary $B_\lambda' = B_\lambda \cup \cup\theta_j = I_o$ and $D_o' = F(D')$ consists of D_o minus radial

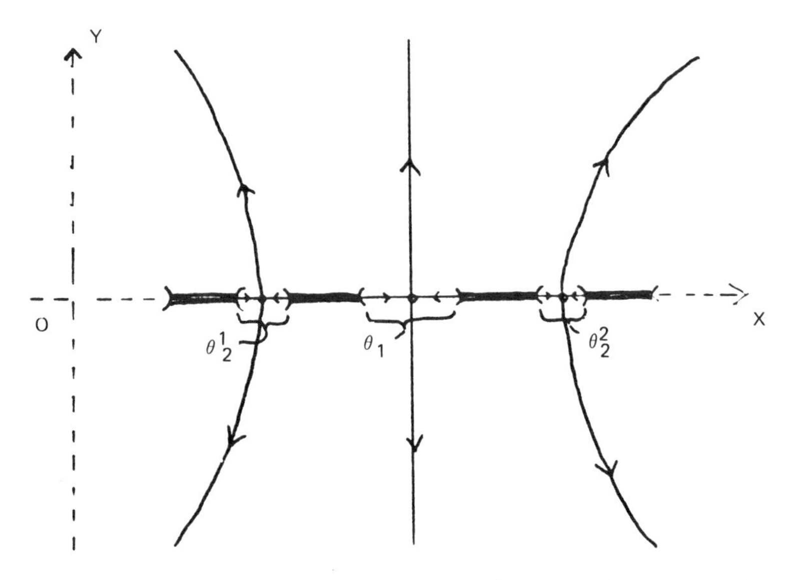

Figure 10. Lines of force associated with θ_1 and θ_2.

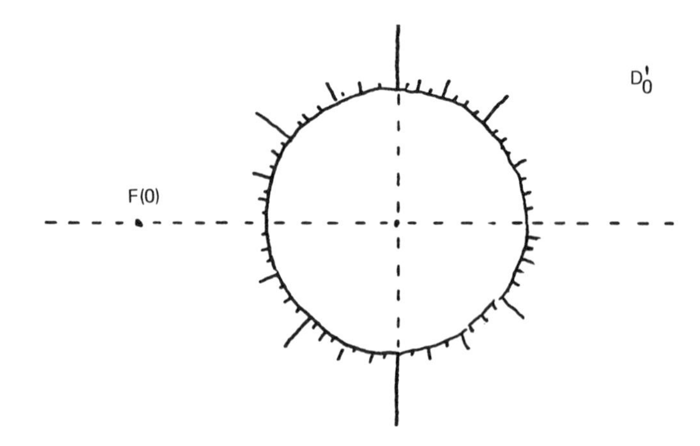

Figure 11. The image $D_o' = F(D')$ of the Green's star domain D'.

segments out of all 2^jth roots of $F(0) = e^{g(0)}$, $j = 1,2,3,\ldots$ Note that, from symmetry, $F(0) < -1$. See Figure 11.

If $z \in B_\lambda$ is not an endpoint of any θ_j^k, then there are two complex conjugate lines of force terminating at z.

When $\lambda = 2$ we find that $\tilde{b}(-\xi+) = 0$. The inverse image points are $\tilde{b}(+-\xi+)$ and $\tilde{b}(--\xi+)$, both of which equal λ. As λ increases these points separate to become the endpoints of θ_1, and their predecessors separate to become the endpoints of the other intervals θ_j^k omitted from the Cantor set.

3.3. Isomorphisms, Balanced Measures, and Integrals

For a region D bounded by analytic curves the equilibrium measure on an arc Γ in the boundary is $\mu(\Gamma) = \int_\Gamma \mathrm{Im}\, dg$. Since Γ_θ is a singleton for almost all θ, we can extend H to B_o almost everywhere by setting $\{H(e^{i\theta})\} = \Gamma_\theta$. If $C_{\alpha\beta} = \{e^{i\theta}: \alpha \leq \theta \leq \beta\}$ and $\Gamma = H(C_{\alpha\beta})$, then $\mu(\Gamma) = \mu(H(C_{\alpha\beta})) = \mu_o(C_{\alpha\beta})$, where μ_o is normalized uniform Lebesgue measure on B_o.

With one assumption a similar relationship holds when D is the unbounded component of the complement of a Julia set B for a polynomial T, except that one point on the boundary of B may be approached from several sides and lie on Γ_θ for more than one θ. To deal with this problem we make an intermediate step and introduce the subset of $\mathbb{C} \times \mathbb{C}$,

$$B^* = \{(z, e^{i\theta}): \quad z \in \Gamma_\theta\}$$

Let $\Pi: B^* \to B$ be the projection $\Pi(z, e^{i\theta}) = z$. In order to define a Borel measure on B^* and later set up an isomorphism, we shall need the following condition on B:

$$(\dagger) \quad \mu_o(E) = 0$$

where $E = \{e^{i\theta} | \Gamma_\theta$ is undefined or not a single point$\}$. Recall that Γ_θ is defined for all but a countable set of points.

Condition ($\dagger$) is satisfied in particular if B is connected, totally disconnected, or is contained in $\mathbb{R}$. In the first case it follows from the theory of boundary correspondence for conformal mapping [Go]. In the second case all components are single points. In the third case E contains at most a countable number of points since there can be only a countable number of disjoint continua contained in $\mathbb{R}$.

Using ($\dagger$) we can define an extension of H to B_o a.e. by $\{H(e^{i\theta})\} = \Gamma_\theta$. Then we can introduce a Borel measure μ^* on B^* determined by

$$\mu^*(\{(H(z), z): \ z \in C_{\alpha,\beta}\}) = \mu_o(C_{\alpha\beta}) = \frac{\beta - \alpha}{2\pi}$$

where $C_{\alpha\beta} = \{e^{i\theta} | \alpha \leq \theta \leq \beta\}$. The equilibrium measure μ satisfies

$$\mu(S) = \mu^*(\Pi^{-1}(S)) = \mu_0(H^{-1}(S)) \tag{7}$$

when S is a Borel set in B. If we define

$$H^*(e^{i\theta}) = (H(e^{i\theta}), e^{i\theta})$$

and

$$T^*(H(e^{i\theta}), e^{i\theta}) = (T(H(e^{i\theta})), e^{iN\theta})$$

for $0 \leq \theta \leq 2\pi$, then using (5) we can prove

<u>THEOREM 10</u>. When (†) is true, H^* provides an isomorphism in the sense of Billingsley [Bi] between the systems (B_0, Borel sets, μ_0, z^N) and (B^*, Borel sets, μ^*, T^*).

Equations (1) and (5) can be reformulated for branches R_j of $\sqrt[N]{z}$ and corresponding branches T_j^{-1} of T^{-1}, $j \in \{1,2,\ldots, N\}$:

$$F(T_j^{-1}(z)) = R_j(F(z)) \tag{8}$$

$$H(R_j(w)) = T_j^{-1}(H(w)) \tag{9}$$

Then from (7) we obtain

$$\mu(T_j^{-1}(S)) = \frac{1}{N}\mu(S)$$

whenever S is a Borel set in B. Thus μ is balanced [BGH5].

Also from (7) we get, for $f \in L^1(B,\mu)$,

$$\int_B f(z)d\mu(z) = \int_{B_0} f(H(w))d\mu_0(w)$$

It follows from the isomorphism and the fact that z^N is mixing on B_o, that T is mixing on B.

Thus, the important properties of the equilibrium measure may be derived from the Böttcher equation when (†) is true, without reference to Brolin's sequence of singular measures μ_n converging weakly to μ. Furthermore, the isomorphism allows us to reexpress integrals with respect to μ in terms of H:

$$\int_B f(z)d\mu(z) = \int_0^{2\pi} f(H(e^{i\theta}))\,\frac{d\theta}{2\pi} = \int_{B_o} \frac{f(H(w))}{2\pi i w}\,dw \qquad (10)$$

when $f \in L'(B,\mu)$.

Let D' be the Green's star region, B' its boundary, $D'_o = F(B')$, and B'_o the boundary of D'_o. Then B'_o consists of B_o and radial branch cuts from B_o to the branch points of H. Each branch cut in B' is traversed once in each direction as B' is followed all the way round the boundary of D'. Suppose f is analytic on B'. The contributions to

$$\int_{B'} \frac{f(z)dF(z)}{2\pi i F(z)}$$

from B'-B cancel. Then using (10) and the substitutions $w = F(z)$, $z = H(w)$, we obtain

$$\int_B f(z)d\mu(z) = \int_{B'} \frac{f(z)dF(z)}{2\pi i F(z)} = \int_{B'_o} \frac{f(H(w))dw}{2\pi i w}$$

and we may deform the contours out into D' and D'_o and make residue calculations. Some examples are:

(a) $$\int_B \frac{d\mu(z)}{(\phi-z)^n} = \begin{cases} \frac{1}{n!}\frac{d^n}{dz^n}\log F(z)\Big|_{z=\phi} & \text{if } \phi \in D \\ 0 & \text{if } \phi \notin D \cup B \end{cases}$$

(b) Let $H(w) = w + \sum_{j=0}^{\infty} \frac{a_j}{w^j}$. Then

$$\int_B z^2 d\mu(z) = \frac{1}{2\pi i}\oint \frac{(w+a_o+\frac{a_1}{w}+\ldots)^2}{w}\,dw$$

$$= a_o^2 + a_1$$

(c) If B is connected the Laurent series for H about ∞ converges for $|z| > 1$ and

$$\int_B |z|^2 d\mu(z) = \int_{B_o} \frac{|H(w)|^2 dw}{2\pi i w} = 1 + \sum_{j=0}^{\infty} |a_j|^2$$

Note that the coefficients a_j for $H(w)$ can be calculated recursively from the Böttcher equation, as is done in [BGH1] when $T(z) = (z-\lambda)^2$. The method in examples (a) and (b) can then be used to calculate explicity $\int_B R(z)d\mu(z)$ for any rational function with no finite poles in $D \cup B$.

A Conjecture Concerning Assumption (†)

Condition (†) is satisfied for connected or totally disconnected Julia sets. The set I of singleton components of B, the set C of continua in B, and the set $X \subset C$ of all sets Γ_θ which are not singleton are invariant under T. Hence, from the ergodicity of T on B, each set must have measure zero or one. We conjecture that unless B is connected, $\mu(C) = 0$, and hence $\mu(X) = 0$, so (†) will always be true.

References

[BGH1] M. F. Barnsley, J. S. Geronimo, A. N. Harrington, "On the invariant sets of a family of quadratic maps," Original Aug. 1981. Revised Dec. 1981. Shortened version submitted to Comm. Math. Phys. (1982).

[BGH2] ——— "Orthogonal Polynomials associated with invariant measures on Julia sets," Bulletin A.M.S. 7 (1982), 381-384.

[BGH3] ——— "Some treelike Julia sets and Padé approximants," submitted to Letters in Mathematical Physics.

[BGH4] ——— "Infinite dimensional Jacobi matrices associated with Julia sets," submitted to Proc. A.M.S. (1982).

[BGH5] ——— "Geometry, electrostatic measure and orthogonal polynomials on Julia sets for polynomials," in preparation.

[BGHD] M. F. Barnsley, J. S. Geronimo, A. N. Harrington, and L. D. Drager, "Approximation theory on a snowflake," to appear in Multivariate Approximation Theory, Vol. 2, ISNM Series, Birkhäuser Verlag, Basel-Boston-Stuttgart, edited by W. Schempp, September/October 1982.

[Bi] P. Billingsley, Ergodic Theory and Information, Wiley, (New York), 1965.

[Br] H. Brolin, "Invariant sets under iteration of rational functions," Arkiv för Matematik 6, 103-144, (1965).

[CE] P. Collet, J. Eckmann, Iterated Maps on the Interval as Dynamical Systems, Birkhauser, (Basel, Boston), 1980.

[CT] P. Coulet, C. Tresser, "Itérations d'endomorphismes et groupe de renormalisation," C.R. Acad. Sc. Paris, 287A (1978) pp. 577-580.

[DH] A. Douady, J. H. Hubbard, "Itération des polynômes quadratiques complexes," C.R. Acad. Sc. Paris, Dec. 1981.

[Fa] M. P. Fatou, "Sur les equations fonctionnelles," Bulletin de Societé Mathématique de France 47 (1919) 161-271; ibid. 48, 33-94; ibid. 48, 208-314.

[EL] H. Epstein and J. Lascoux, "Analyticity properties of the Feigenbaum Function," Comm. Math. Phys. 81 (1981) 437-453.

[Fe] M. Feigenbaum, "Quantitative universality for a class of nonlinear transformations," J. of Stat. Phys. 19 (1978), 25-52.

[Go] G. M. Goluzin, Geometric Theory of Functions of a Complex Variable, Translations of Mathematical Monographs Vol. 26, American Mathematical Society (Providence, Rhode Island), 1969.

[Gu1] J. Guckenheimer, "On the bifurcation of maps of the interval," Inventiones Math. 39 (1977), 165-178.

[Gu2] J. Guckenheimer, "Sensitive dependence to initial conditions for one-dimensional maps," Comm. Math. Phys. 70 (1979) 133-160.

[Ju] G. Julia, "Mémoire sur L'iteration des fonctions rationelles," Journal de Mathématique Pures et Appliqueés, 4 (1918) 47-245.

[Jo] R.-J. de Jonkere, "Convergence de L'iteration des fonctions rationelles," (1963) Thesis: Université Catholique de Louvain, Faculté des Sciences Appliqueés,

[LY] T. Li and J. A. Yorke, "Period three implies chaos," Amer. Math. Monthly 82, (1975) 985-992.

[Me] B. Mendelson, Introduction to Topology, Blackie and Son, Ltd. (London) 1963.

[My] P. J. Myrberg, "Sur L'iteration des polynomes réels quadratiques," J. de Math. Pures et Appliqueés Ser. 9, 41 (1962) 339-351.

[Sh] A. N. Sharkiovskii, "Coexistence of cycles of a continuous map of a line into itself," Ukr. Mat. Z 16 (1964) 61-71.

[Su] D. Sullivan, private communication (1982).

[SN] L. Sario, M. Nakai, Classification Theory of Riemann Surfaces, Springer-Verlag (New York, Heidelberg, Berlin) 1970.

[Ts] M. Tsuji, Potential Theory in Modern Function Theory, Maruzen (Tokyo) 1959.

MATHEMATICAL MICROCOSM OF GEODESICS, FREE GROUPS, AND MARKOFF FORMS

Harvey Cohn

Department of Mathematics
City College of New York
New York, New York

Here we are concerned with a development which began as a problem in number theory even before Markoff, but which has progressed recently to new results with implications ranging over many fields, epitomized by the use of geodesics to parallel purely arithmetic results.

In the interests of clarification, it is noted that the classic work was performed by A. A. Markoff (1856-1922), the Russian number theorist, who is perhaps better known for his later statistical work in random variables. (In the latter connection, he latinized his name as 'Markov' in accordance with changing orthography.) While there is little chance to confuse him with his younger brother, W. A. Markoff (1871-1897), also a number theorist, there is a more understandable confusion with his son, A. A. Markov (1903-) the logician and algebraist.

The initial problem was the study of minimal properties of real quadratic forms,

---

Research sponsored by NSF Grant MCS 7903060

$$aX^2 + bXY + cY^2 = f(X,Y)$$

with coefficients a,b,c and variables X,Y all integers (in Z). The object is to find forms f(X,Y) for which the minimum is large as compared with the discriminant.

This led Markoff to a special set of forms, later known as Markoff forms about 1879. Almost simultaneously, however, Fricke discovered modular functions and (later) certain Fuchsian groups whose behavior bore a curious resemblance to the identities for Markoff forms. This was first noticed in 1955 by the speaker and led to connections with such topics as

integral matrix identities
geodesics on a torus
primitive words on a free group
invariants of a ternary form
complex multiplication
matrix inequalities

all of which is based ultimately on geodesics in the upper half complex plane or a torus (in negative constant curvature).

Most of these aspects were expounded in individual papers of the speaker, but here an attempt is made for the first time to present the results sequentially. It should still be pointed out that many of the topics mentioned here will bear further investigation.

1. MARKOFF'S MINIMAL FORMS

We consider the indefinite forms with integral coefficients, a,b,c and variables X,Y

$$f(X,Y) = aX^2 + bXY + cY^2, \quad d = b^2 - 4ac > 0 \tag{1.1}$$

with d the discriminant. We consider forms which are not zero forms (irrational roots), so f = 0 only when X = Y = 0. This leads to the invariant ratio

$$l(f) = \min(f)/d^{1/2} \tag{1.2a}$$

for min(f) the minimum nonzero absolute value represented by f(X,Y) for integral X,Y. Clearly, l(f) is dimensionless and it avoids the elimination of common divisors of a,b,c. The trick is clearly to make l(f) as large as possible. (It is trivial to make l(f) arbitrarily small by making a = 1 = min(f), b = 0, and -c an arbitrarily large integer not a perfect square.)

Markoff discovered an infinity of forms for which

$$l(f) > 1/3 \tag{1.2b}$$

The largest possible value for l(f) turns out to be

$$l(f) = 1/5^{1/2} \tag{1.3a}$$

associated with the well known illustration

$$f = X^2 + XY - Y^2, \quad d = 5, \ \min(f) = 1 \tag{1.3b}$$

Also, for any lim > 1/3, it will happen that l(f) > lim for only a finite number of values of l(f).

The reciprocals, 1/l(f), constitute the so-called _Markoff spectrum_, (see Bumby). It is discrete at the low end with $5^{1/2}$ = 2.236... as its minimum and 3 as its first limit point.

Some small values of l(f) > 1/3 occur below with the forms abbreviated by (a,b,c), together with periods which will be explained later.

TABLE 1

(a,b,c)	d	min(f)	period
(1,1,-1)	5	1	(1,1)
(2,4,-2)	32	2	(2,2)
(5,11,-5)	221	5	(1,1,2,2)
(13,29,-13)	1517	13	(1,1,1,1,2,2)

(Some forms are written with a common factor of 2 for reasons which emerge later on.)

Markoff's construction uses the continued fraction expansions of the roots of the form $f(X,Y)$, namely r and s defined by

$$f(X,Y) = a(X - rY)(X - sY), \qquad r > s \tag{1.4}$$

Either root will have a continued fraction which is periodic (purely periodic for now) so we write

$$\text{period} = (a_1, a_2, \ldots, a_p)$$

$$r = a_1 + \frac{1}{a_2 +} \frac{1}{a_3 + \ldots} \frac{1}{a_p + \frac{1}{r}} = t(r) \tag{1.5a}$$

The period gives the linear fractional transformation

$$t(r) = (Ar + B)/(Cr + D) \tag{1.5b}$$

For this transformation, r (or s) is the fixed point. We also call r (or s) fixed points of the period. The period is always chosen as even so that $t(r)$ is always strictly unimodular (determinant 1). This can be done by doubling if necessary. According to rules going back to Gauss, the roots can be chosen so that $r > 1$ and $-1/s > 1$. If we start with an equivalent form whose roots are not purely periodic, we get the same period except for possible rotations and reversals.

We wish to avoid becoming emmeshed in the complicated interrelationships involving the quantities A, B, C, D of $t(r)$ and the period and coefficients a,b,c. It is important to note that if a happens to be the minimum, $l(f) = 1/(r-s)$, for example, so clearly the periods must consist only of 1's and 2's. Beyond that, the construction of the periods is far from trivial. We describe it in the next section, but for now, we note only that only very special combinations of 1's and 2's (even when doubled) will give Markoff forms. For instance take

$$\text{period} = (1,1,1,1,2,2,2,2) \tag{1.6a}$$

$$r = (181r + 75)/(111r + 46), 111r^2 - 135r - 75 = 0 \tag{1.6b}$$

$$f(X,Y) = 111X^2 - 135XY - 75Y^2 \tag{1.6c}$$

$$\min(f) = 75,\ d = 51525,\ l(f) = .3304\ldots < 1/3 \tag{1.6d}$$

The most interesting observation for now was made by Markoff that the progression of Markoff forms is not linear, but, in a sense, two-dimensional, in that the forms really come in pairs. In fact they come in triples with an interrelationship. Indeed, a so-called <u>Markoff number</u> is defined for each Markoff form so that

$$m = \min(f), \quad d = 9m^2 - 4 \tag{1.7}$$

This clearly shows why $l(f) > 1/3$ each time. For three forms (f_1, f_2, f_3) which are makde into a triple, $m_i = \min(f_i)$, satisfy

$$m_1^2 + m_2^2 + m_3^2 = 3m_1m_2m_3 \tag{1.8}$$

Furthermore, any member of a triple satisfying (1.8) (<u>Markoff's equation</u>) actually corresponds to a Markoff form. Indeed all of the triples are interrelated by permutations or by the relation

$$m_0 = 3m_1m_2 - m_3 \ (> 0) \tag{1.9}$$

where (m_1, m_2, m_0) replaces (m_1, m_2, m_3).

Thus the first triple of Markoff numbers is (1,1,1). From this we successively build (1,1,2) from (1.9) followed by (1,2,1), (1,2,5), (1,5,2), (1,5,13), etc. (see Table I).

We have taken certain liberties with terminology in referring to a Markoff form as though it is the special form for which min(f) (or m) appears as (say) the coefficient a. Obviously, we can use an equivalent form and have the same $l(f)$ although, the coefficients of f would be different and the period would not appear immediately in the continued fraction expansion for r and s. The construction is discussed in Markoff [1] as well as texts like that of Cassels and Lekkerkerer. (Also compare Nicholls.)

2. THE MATRIX IDENTITIES

We now use the symbols T,U,V,... to denote two by two real matrices of determinant 1. Let [T,U] denote the commutator of two matrices and trU the trace, so by well known relations

$$[T,U] = TUT^{-1}U^{-1}$$

$$\mathrm{tr}[T,U] = \mathrm{tr}[U,T] = \mathrm{tr}[U^{-1},T] = \mathrm{tr}[U,T^{-1}] = \text{etc.}$$

$$\mathrm{tr}(UT) = \mathrm{tr}(TU), \quad \mathrm{tr}(UTU^{-1}) = \mathrm{tr}T$$

Now, Fricke [3] discovered the identities

$$(\mathrm{tr}T)^2 + (\mathrm{tr}U)^2 + (\mathrm{tr}(TU))^2 = \mathrm{tr}T\ \mathrm{tr}U\ \mathrm{tr}(TU) + \mathrm{tr}[T,U] + 2 \qquad (2.1)$$

$$\mathrm{tr}(UTU) + \mathrm{tr}T = \mathrm{tr}U\ \mathrm{tr}(TU) \qquad (2.2)$$

as part of a trace calculus (also see Horowitz), with the purpose of (essentially) characterizing the traces lying in a matrix group. We now restrict ourselves to integral matrices and situations where the commutator has trace -2. Then if 1/3 of each trace is viewed as a Markoff number, formulas (2.1) and (2.2) become Markoff's equation (1.8) and the recursion relation (1.9).

For example, let us write the periods (1,1) and (2,2) (see (1.5ab) as linear transformations t(r) to form matrices T and U

$$\text{period} = (1,1) \qquad r = 1 + \cfrac{1}{1 + \cfrac{1}{r}} \qquad r = \frac{2r+1}{r+1} \qquad T = \begin{pmatrix} 2 & 1 \\ 1 & 1 \end{pmatrix} \tag{2.3a}$$

$$\text{period} = (2,2) \qquad r = 2 + \cfrac{1}{2 + \cfrac{1}{r}} \qquad r = \frac{5r+2}{2r+1} \qquad U = \begin{pmatrix} 5 & 2 \\ 2 & 1 \end{pmatrix} \tag{2.3b}$$

This gives us $\operatorname{tr}[T,U] = -2$, since

$$TU^{-1}T^{-1}U = U^{-1}[U,T]U = \begin{pmatrix} -1 & 0 \\ -6 & -1 \end{pmatrix} \tag{2.3c}$$

The important fact is that this situation is part of a self-perpetuating chain. For instance, the pair (T,U) can be replaced by any one of the pairs

$$(T,U),\ (U,T),\ (TU,U),\ (UT,U),\ (T,TU),\ (T,UT),\ \text{etc.} \tag{2.4}$$

and the commutator will still have the same trace. (Note in (2.2), (T,U) is replaced by (U,TU)). Actually the operations indicated in (2.4) are a semigroup generating _equivalent bases_ of the _free group_ $\langle T,U \rangle$ generated by T and U. Thus we have the definitive result:

Consider all periods which are concatenations of double 1's (T = (1,1)) _and double_ 2's (U = (2,2)), _but are generated_ *as word pairs under the operations in* (2.4). *These periods and only these periods belong to Markoff forms. All Markoff triples belong to (one-third) the traces of the* T, U, *and* TU *which come out of these operations, together with* (1.9).

If we consider the free group $\langle T,U\rangle$ there will be many equivalent bases, e.g., $\langle T',U'\rangle$ which might have negative exponents. For instance, the inverses of T or U could replace them. Any word of such a pair leads conversely to a linear transformation whose roots belong to a Markoff form. This, in effect, constitutes a test for whether or not an arbitrary word T' of the free group $\langle T,U\rangle$ is *primitive*, (i.e., whether a word U' exists so that $\langle T',U'\rangle$ is the same free group as $\langle T,U\rangle$).

An explicit analytic formula can be written for the most general Markoff period as a concatenation of T (= (1,1)) and U (= (2,2)) by what we call a *step-word*. For given integers $a > 0$ and $b > 0$ we define

$$e(t) = [tb/a] - [(t-1)b/a], \quad t = 1,2,\ldots,a$$

(Note these are the jumps in the step-function for the line $y = bx/a$.) Then we designate

$$W = (T,U)^{(a,b)} = TU^{e(1)}TU^{e(2)} \ldots TU^{e(a)} \tag{2.5}$$

Markoff [1] has derived a much more complicated set of rules to determine which sequences of exponents are admissible, but all of this is really contained in the concept of premitivity of W. In fact, the second word W' (for which $\langle W,W'\rangle = \langle T,U\rangle$) is

$$W' = (T,U)^{(a',b')}, \quad ab' - a'b = 1 \tag{2.6}$$

for suitably chosen integers a',b'.

If $b < a$ many of the exponents are 0 so we have in effect powers of T separated by individual symbols U. In fact

$$(T,U)^{(a,b)}(U^{-1},T^{-1})^{(b,a)} = 1 \tag{2.7}$$

In the extreme case for $(a,b) = (1,0)$, clearly $W = T$, and for $(a,b) = (0,1)$, by (2.7) we see that W should be U. If we return to the case (1.6), we see that the word TTUU could not be the period of a Markoff form since this can not be a primitive word.

If the word W, given above, were abelianized (as additive), we should write

$$W = aT + bU \tag{2.8}$$

In fact, there is a unique class of Markoff forms corresponding to any abelianization (a,b) for a and b nonnegative relatively prime integers. It is necessary, but certainly not sufficient for primitivity, that W have such an abelianization. (This easily excludes TTUU in (1.6).) If (a,b) has a common factor, however, the definition (2.5) still is valid as such, by repetition.

According to (2.5) and (2.6), we can now write a general Markoff triple as follows:

periods: (W,W',WW')

Markoff triple: $(\text{tr}W/3,\text{tr}W'/3,\text{tr}(WW')/3)$

(The Markoff forms are found by using as roots the fixed points of the periods, e.g., $r = t(r)$ in the notation of (1.5ab).) Thus in reference to Table 1 (above) we have these two Markoff triples and periods, namely,

(1,2,5) (T,U,TU)

(1,5,13) (T,TU,TTU)

There are currently many disturbingly elementary questions which can be asked about the Markoff numbers, and which have not yet been answered. For example, are there more than one (essential) period which produce a Markoff number? More simply, are there more than one triple with the same maximum Markoff number? Markoff carried out the computations for all numbers up to 1,000,000, but more extensive calculations have created even more uncertainty. We are really asking, analytically for a discrete spectrum, but values in the spectrum are very close. For example, $m = 195025$ for TU^7 and $m = 196418$ for $T^{12}U$, although the gaps in the Markoff numbers are essentially exponential, (see Table 2 and (8.6) below).

Some further details are discussed in Cohn [3], [6], [7] and in Magnus. There is a difficult problem in parametrizing all the words (negative exponents included) which lead to the same Markoff forms. Such matters can be better understood when the problem is next interpreted geometrically.

3. THE GEODESIC MODEL

We now return to the problem of describing $l(f)$ in terms of the roots, r,s of the quadratic form

$$f(X,Y) = a(X - rY)(X - sY), \quad r > s \tag{3.1}$$

Both $\min(f)$ and (discriminant d) are independent of unimodular changes of variable, called the special linear group,

$$SL(2,Z): \qquad \begin{array}{ll} X = AX' + BY' & A,B,C,D \text{ in } Z \\ Y = CX' + DY' & AD - BC = 1 \end{array} \tag{3.2}$$

The roots are clearly transformed as follows:

$$f(X,Y) = a'(X' - r'Y')(X' - s'Y') \tag{3.3}$$

$$r' = (rD - B)/(-rC + A), \quad s' = (sD - B)/(-sC + A) \tag{3.4}$$

by the inverse of (3.2). Note $r' > s'$, also.

We can therefore identify with each form $f(X,Y)$ the aggregate of root-pairs (r,s) and we find that over these pairs,

$$1/l(f) = \sup(r - s) \tag{3.5}$$

This is easily seen, since $d^{1/2} = a(r - s)$ always, and the relation $a = \min(f)$ is valid when $r - s$ takes its sup value, by Gaussian reduction.

Now we consider the root pair (r,s) to lie on the real axis of H, the upper half $z = x + iy$ plane. This pair is further associated with a semicircle in H perpendicular to the real axis at r and s. We then have the curious result that these semicircles all lie below the line

$$y = \operatorname{Im} z = 3/2 \tag{3.6}$$

if and only if the form f to which the roots belong is a Markoff form. Indeed, by the discreteness of the spectrum, all of the semicircles lie below $y = \lim$ for some $\lim < 3/2$ for only a finite number of Markoff forms.

Yet this is not a satisfactory condition, since the one form $f(X,Y)$ is associated with an infinity of semicircles. In order to be able to associate a form with one object, we eliminate the replication under $SL(2,Z)$ in (3.4). We take the upper half plane H under $PSL(2,Z)$, which associates with each z the aggregate of z' where

$$z' = (Ax + B)/(Cz + D), \quad H: y = \text{Im } z > 0 \tag{3.7}$$

This is the projective form of SL(2,Z). It preserves H so there is a fundamental domain, F, in H under this group. Symbolically,

$$H/PSL(2,Z) = F \tag{3.8a}$$

which is well-known to be

$$F: x^2 + y^2 > 1, \quad -1/2 < x < 1/2, \quad (z = x + iy) \tag{3.8b}$$

with the boundaries identified according to

$$z' = z + 1, \quad z' = -1/z \tag{3.8c}$$

We can then restrict the semicircles to just those arcs lying in F. Such a portion exists, e.g., when $r > 1$ and $-1/s > 1$ (by Gaussian reduction). See Figure 1.

We next introduce the concept of the Poincare geodesic corresponding to the noneuclidean metric in H

$$ds^2 = (dx^2 + dy^2)/y^2 \tag{3.9}$$

These are the circles orthogonal to the real axis with the special case of vertical lines. In this way the root-pairs (r,s) all correspond to a single geodesic drawn in F and reflected from one boundary to another by (3.8c). Actually, there are only a finite number of arcs in F because of the periodicity of the continued fraction.

Note that if we wanted to have the geodesic bounce off each side of F like a billiard ball, it would be necessary to introduce into the group PSL(2,Z) some reflection operations,

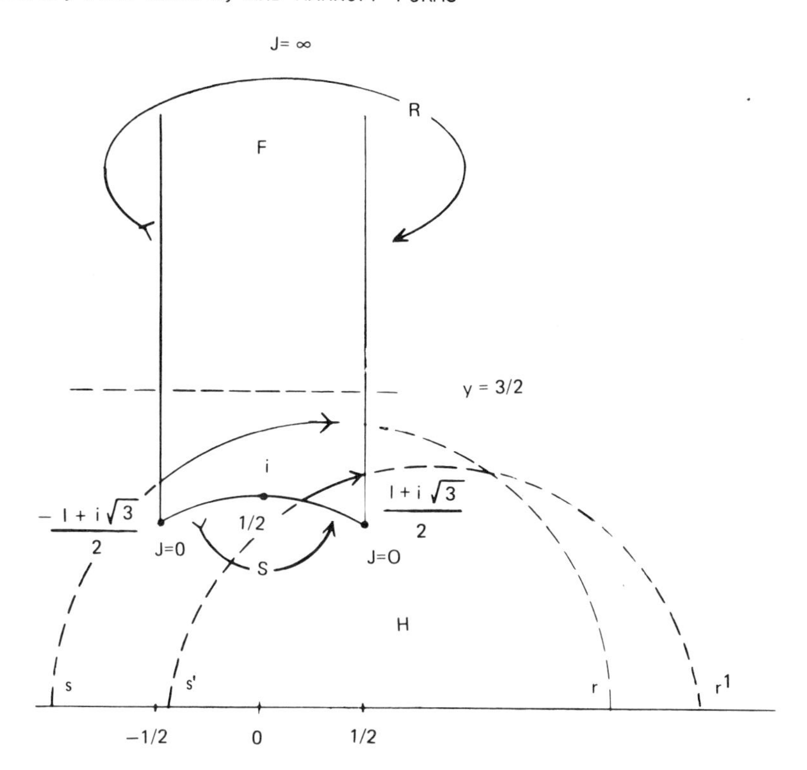

Figure 1. Fundamental domain F in the upper half plane H for G, the modular group PSL(2,X), showing paths of geodesics connecting root-paris for a form f(X,Y). Matching sides of F are identified according to (3.8c), or (4.1).

but this is not advantageous later on. So we have translations rather than reflections at the boundary.

Again, a Markoff form is one for which the roots yield a geodesic which remains below y = 3/2 within F. We call such a curve (or images of it) a <u>Markoff geodesic</u>. The problem with the Markoff geodesics is that they do not look like straight lines! We could remove the interruptions at the boundaries by using the mapping w = J(z) where J(z) is the <u>Klein modular function</u>, which yields a one-to-one map of F onto the complex w-plane. Yet in the w-plane the Markoff geodesic will still loop the special points w = 0 and w = 1 in a very circuitous manner, which we do not explore further. What we have done is to combine the work of Klein, who constructed the root representations and Poincaré, who introduced the geodesic concept. This is not yet enough.

4. THE COMMUTATOR MAPPING

We have just considered the modular group, PSL(2,Z), also denoted by G, for which F (in Figure 1) is the fundamental domain in H, the upper half plane. If we introduce the matrices R,S,Q defined as

$$R = \begin{pmatrix} 1 & 1 \\ 0 & 1 \end{pmatrix}, \quad S = \begin{pmatrix} 0 & -1 \\ 1 & 0 \end{pmatrix}, \quad Q = RS = \begin{pmatrix} 1 & -1 \\ 1 & 0 \end{pmatrix} \tag{4.1}$$

we can then describe this group in terms of its word structure in R and S (the generators of F), or more conveniently, in terms of R and Q. We have

$$G = PSL(2,Z) = \langle S,Q : S^2 = Q^3 = 1\rangle \tag{4.2}$$

Note +M and -M denote the same transformation of PSL(2,Z), or

$$SL(2,Z)/(+1,-1) = PSL(2,Z) \tag{4.3}$$

It is easily seen how T and U enter (compare (2.3ab)),

$$[S,Q] = T^{-1}, \quad [S,Q^2] = T^{-1}UT \tag{4.4}$$

Actually, the group G' = ⟨T,U⟩ as used in Section 2 is the commutator subgroup of PSL(2,Z) (= G). Thus, modulo the second commutator of G, in (2.8), we get the abelianization ZT + ZU, an abelian Z-module of rank 2.

The commutator subgroup G' = ⟨T,U⟩ has index 6 in G. Thus it has a fundamental domain F' (see Figure 2) with 6 replicas of F. (Two models of F' are shown.)

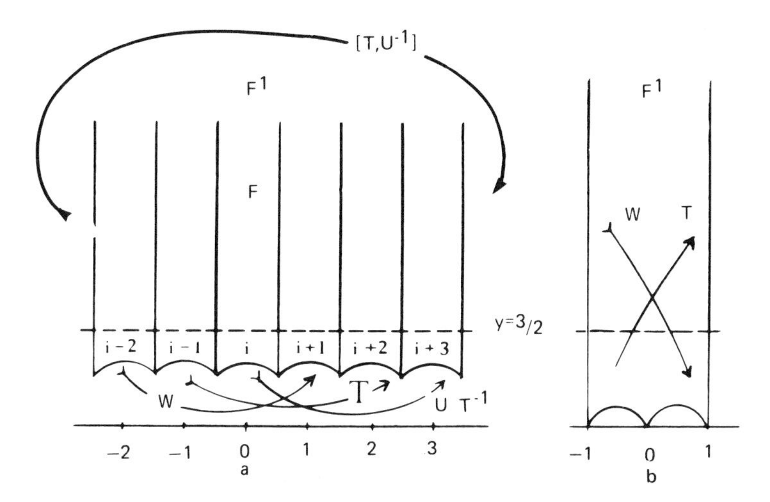

Figure 2. Two representations of F', the torus-type fundamental domain of G' the commutator group of G, the modular group. In (a) six replicas of F are shown; in (b) they are present but rearranged. The identifying mappings are shown using T and U (see (2.3ab)) and $W = (z+1)/(z+2) = TU^{-1}T$. Also $[T,U^{-1}]$ is the translation by 6 identifying the vertical sides.

The main difference between F and F' is topological. Essentially, F (= H/G) is of <u>genus zero</u>, equivalent to a <u>sphere</u>, and F' (= H/G') is of <u>genus one</u>, equivalent to a <u>torus</u>. More precisely, the sphere or torus is <u>punctured</u>, through the absence of a point at infinity. The genus makes no immediate difference when a geodesic is viewed in H (in the z-plane), but it makes a tremendous difference when the mapping is made onto a sphere or onto a torus.

We imbed F' = H/G' into the u-plane with a periodic structure derived from elliptic functions. Thus,

$$1 + p'(u)^2 = 4p(u)^3 = J(z) \tag{4.5a}$$

$$du = \text{const. } dJ/J^{2/3}(J-1)^{1/2} \tag{4.5b}$$

for J(z) the Klein modular function. Note that by symmetries the u-image of infinity determining the lattice of periods has the symmetry of equalateral triangles, (see Figure 3). We introduce the cube root of unity

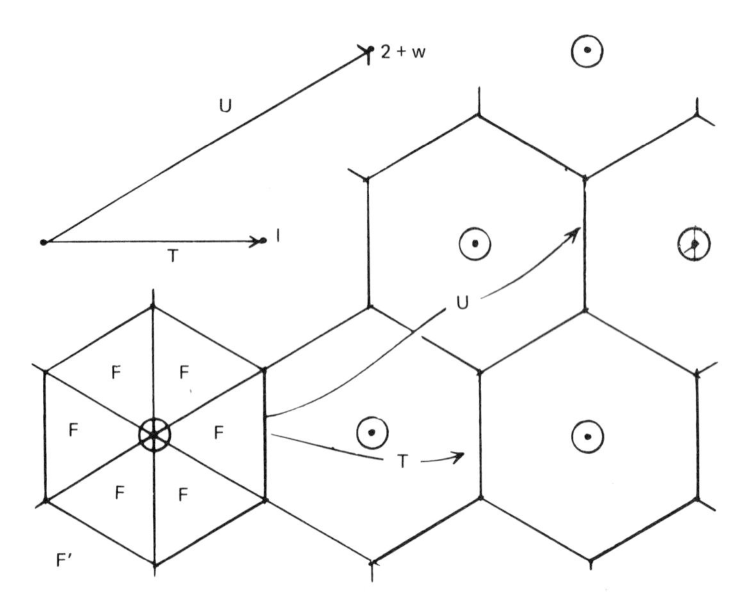

Figure 3. The u-plane determined by (4.1) showing the Markoff geodesics corresponding to periods T (=(1,1)) and U (=(2,2)). Note for Table II, how aT + bU follows the vector a + (2+w)b.

$$\omega = (-1 + i3^{1/2})/2 \tag{4.6}$$

So $Z[\omega]$ becomes the (Eisenstein) ring of complex multipliers of the period lattice of p(u).

The whole point to this transformation is that the Markoff geodesics under the Poincare metric (3.9) (as transferred by (4.5ab)) become the approximately straight geodesics in the u-plane.

The use of the word "straight" requires some clarification: If we refer to Figure 3, we see that T and U are periodic and a period segment has extremities which form vectors determining the period lattice. If we transfer the u-plane neighborhoods of infinity determined by $y > 3/2$ in Figure 1, we then have a lattice arrangement of holes in the u-plane arranged about the period lattice points. Then the Markoff geodesics may be characterized as precisely those which are bounded away from the holes. Even more important, the Markoff geodesics are characterized by a slope b/a in the step-word formula (2.5) for W. <u>Thus these geodesics not only avoid the holes but they weave their way through the lattice as though they were straight lines</u>. (See Cohn [2].)

If we introduce a slope parameter

$$2 + t = b/a \tag{4.7a}$$

we find that the Markoff geodesic corresponding to $aT + bU$ (see (2.8)) is the same if t is changed by the group

$$t,\ (1 - t),\ 1/t,\ (t - 1)/t,\ 1/(1 - t),\ t/(t - 1) \tag{4.7b}$$

(This represents the 3-dihedral group whose fundamental domain on the real axis is the interval $t > 2$, hence (4.7a)).

It should be noted that the historical roots of this commutator mapping are very strange. Fricke [1], [2] first used it to construct a noncongruence subgroup of G (one which is not characterized by congruence conditions on the matrix elements). The one he found was the subgroup of G' for which (in (2.8)) the coefficients a and b were multiples of 3. Yet Fricke used neither topology (as such) nor even the concept of a commutator subgroup (although he knew the practical techniques in either case). A more grievous omission was the concept of a variable basis for the period lattice. If he had done so, he would have discovered the Markoff forms.

In Cohn [3], [4] the invariances and symmetries are further discussed in reference to the free group and its automorphisms. For now we note the characterization of Markoff forms by the complex multiplier determined by its coordinates. If $W = aT + bU$, the corresponding multiplier is $(a + 2b) + b\omega$ whose norm and coordinates are next given in Table 2. This is followed by the Markoff number ($= \mathrm{tr}W/3$). The Markoff number is imbedded in a triple of which it is the maximum. The

TABLE 2

norm	a	b	Markoff triples		
1	1	0	1	1	1
3	0	1	2	1	1
7	1	1	5	2	1
13	2	1	13	5	1
19	1	2	29	5	2
21	3	1	34	13	1
31	4	1	89	34	1
37	1	3	169	29	2
39	3	2	194	13	5
43	5	1	233	89	1
49	2	3	433	29	5
57	6	1	610	233	1
61	1	4	985	169	2
67	5	2	1325	34	13
73	7	1	1597	610	1
79	4	3	2897	194	5
91	8	1	4181	1597	1
91	1	5	5741	985	2
93	3	4	6466	433	5
97	5	3	7561	194	13
103	7	2	9077	89	34
111	9	1	10946	4181	1
109	2	5	14701	169	29
133	10	1	28657	10946	1
127	1	6	33461	5741	2
129	3	5	37666	433	29
133	5	4	43261	2897	5
139	7	3	51641	1325	13
147	9	2	62210	233	89
157	11	1	75025	28657	1
151	4	5	96557	6466	5
163	8	3	135137	1325	34
169	1	7	195025	33461	2
183	12	1	196418	75025	1
181	7	4	294685	7561	13
199	11	2	426389	610	233
193	2	7	499393	985	169
211	13	1	514229	196418	1
201	6	5	646018	43261	5
217	10	3	925765	9077	34

Markoff Tribles for exponents (a,b) written as complex multiplications by $(a + 2b) + b\omega$ (for ω a cube root unity).

coordinates of the smaller elements of the triple, of course, have occurred previously.

5. HOMOTOPY CLASSES AND LIMIT GEODESICS

A word in T and U which is infinitely long in both directions describes an arbitrary geodesic which does not end in a singularity (lattice point of periods or point at infinity). This is not trivial to see as it depends on a continued fraction discussion (Cohn [2]). The periodic geodesics, particularly the Markoff geodesics, are clearly viewed as repeating in both directions, as a special case.

A periodic Markoff geodesic W can be paired with a W' so that $\langle T,U\rangle = \langle W,W'\rangle$ indicating an equivalent basis of the free group. The free homotopy classes such as W and W' correspond to Markoff periods, and every equivalent basis pair, W and W', represents an automorphism of the free group. More simply, any Markoff period is transformable into every other Markoff period by an automorphism of the free group which is the homotopy group of the punctured torus.

An interesting example of these concepts is provided by the doubly infinite word

$$W = \ldots TU^{e(-1)}TU^{e(0)}TU^{e(1)}TU^{e(2)}\ldots \tag{5.1a}$$

$$e(t) = [kt] - [k(t-1)] \tag{5.1b}$$

Here the slope k is a real quadratic irrational. Then it is an easy result that an automorphism exists in the free group which transforms the infinite geodesic into itself. (The method is to use the linear transformation for which the totally positive unit is an eigenvalue.) For example, for

$$k = (1 + 5^{1/2})/2 \tag{5.2a}$$

we have the word (starting with $t = 0$),

$$W = \ldots TU, TUU, TU, TUU, TUU, TU, TUU, TU, TUU, TUU, TU, \ldots \quad (5.2b)$$

(with commas placed to emphasize the form only). Now, if we replace T by TU and U by TUU, we obtain the same word. This might seem very long to check, but we are merely observing the fact that the corresponding basis change of $\langle T,U \rangle$, abelianized, has the unit k^2 as eigenvalue. Specifically, the matrix is

$$\begin{matrix} T \to T + U: \\ U \to T + 2U: \end{matrix} \quad \begin{pmatrix} 1 & 1 \\ 1 & 2 \end{pmatrix}$$

Some examples of infinite invariant words are given with more details in Cohn [4]. Such examples correspond to the limit point 3 in the Markoff spectrum. Geometrically they become arbitrarily close to the holes around the lattice points in the u-plane, even touching one, but not entering the holes. Strangely, Markoff [2] was also interested in sequences like (5.1b), attributed to Bernoulli, but he made no connection with his periods. There are many unanswered questions concerning nonquadratic irrationalities and, generally, limiting geodesics. (See also Stolarsky.)

6. VECTOR BUNDLES AND TERNARY FORMS

If all the periods are interchangeable under automorphisms of the free group, then why do they lead to inequivalent forms? More positively, is there a single item represented by the aggregate of Markoff periods?

There is an answer: all Markoff forms are represented in a single <u>vector bundle</u>. Consider two dimensional vector spaces associated with each point of a punctured torus with consistency relations in each neighborhood, thus forming a bundle. Along one period the vector spaces match by a matrix T, along the other period by a matrix U, (here T and U are taken in $SL(2,Z)$ for convenience). Around the hole (or puncture) the matching is by $[T,U]$. Then since the choice of basis for the periods of the torus is arbitrary, any other matching system would give a pair (say) W and W' for which $[W,W']$ has the same trace as $[T,U]$. It is this bundle which constitutes the single entity from which all Markoff forms are derived.

But is there a single form to which all Markoff forms are related? The answer is a ternary form of the following nature: Let A and B be matrices in $SL(2,Z)$ and set

$$g(X,Y,Z) = tr((X + YA + ZB)(X + YA^{-1} + ZB^{-1}))/2 \qquad (6.1a)$$

$$g(X,Y,Z) = X^2 + Y^2 + Z^2 + XY\, trA + XZ\, trB + YZ\, tr(AB^{-1}) \qquad (6.1b)$$

Here X,Y,Z are integers, and it can be verified that the change of basis operations of the free group $\langle A,B\rangle$ lead to unimodular transformations of $SL(2,Z)$. So the equivalence class of form (6.1ab) represents the whole vector bundle of A and B.

For Markoff forms, we are fortunate enough to have some kind of uniqueness. The determinant of the form $g(X,Y,Z)$ is 1 if and only if A,B,AB are a Markoff triple. Then we can choose the traces in (6.1b) as (say) all 3 (for the Markoff triple $(1,1,1)$) and we find the Markoff triple is characterized by the equivalence of $g(X,Y,Z)$ to

$$-Z^2 + Y(2X + Y + Z) \qquad (6.2)$$

It would be interesting to be able to deduce the Markoff minima directly from (6.2) or (6.1b). (See Cohn [5].)

7. THE GENERAL FUCHSIAN GROUP AND SQUARE LATTICES

In the speaker's work, the results of Markoff (periodic) and Dickson (nonperiodic) on continued fractions were taken as gospel for the purpose of primarily establishing connections with groups and geometric objects. Actually, all these results can be independently verified. This was done by Asmus Schmidt in an even broader setting, generalizing the Fuchsian group $PSL(2,Z)$ to one with two generators from $PSL(2,R)$ with only the assumption of trace = -2. His papers are too comprehensive to even summarize, but it would be of special interest to sketch one special case, where the torus is conformally equivalent to a square (and not a 60-degree rhombus). This is of interest because we are using the theory of free groups in either case, yet many properties of free groups seem tied to the 3-dihedral group (see Cohn [3]).

We define new quantities for the old symbols

$$T = \begin{pmatrix} 2 & 1 \\ 1 & 2 \end{pmatrix}, \quad U \begin{pmatrix} -1 & -\sqrt{2} \\ -3\sqrt{2} & 5 \end{pmatrix} \tag{7.1}$$

$$W = (T,U)^{(a,b)}, \quad W' = (T,U)^{(a',b')}, \quad ab'-a'b = 1 \tag{7.2}$$

$$(\mathrm{tr}W)^2 + (\mathrm{tr}W')^2 + (\mathrm{tr}(WW'))^2 = \mathrm{tr}W\ \mathrm{tr}W'\ \mathrm{tr}(WW'). \tag{7.3}$$

In these cases, exactly two traces are divisible by 2, so there is an interesting diophantine variant of Markoff's equation coming from (7.3). (See Schmidt.) To continue, there is a group analogous to the modular group here

$$G = \langle S,Q : S^2 = Q^4 = 1\rangle \tag{7.4}$$

$$R = \begin{pmatrix} 1 & \sqrt{2} \\ 0 & 1 \end{pmatrix}, \quad S = \begin{pmatrix} 0 & -1 \\ 1 & 0 \end{pmatrix}, \tag{7.5}$$

$$Q = SR = \begin{pmatrix} 0 & -1 \\ 1 & \sqrt{2} \end{pmatrix}$$

Figure 4 shows how to identify the opposite sides of the fundamental domain F = H/G into a (punctured) sphere. Now

$$T = SQ^2, \quad U = QSQSQ^2. \tag{7.6}$$

Actually, $\langle T,U \rangle = G^*$ is not the commutator group (G') of G. It is a subgroup of index 4 which contains G' (which is of index 8), in fact,

$$T^2 = [S,Q^2], \quad U = Q[S,Q]Q^{-1}$$

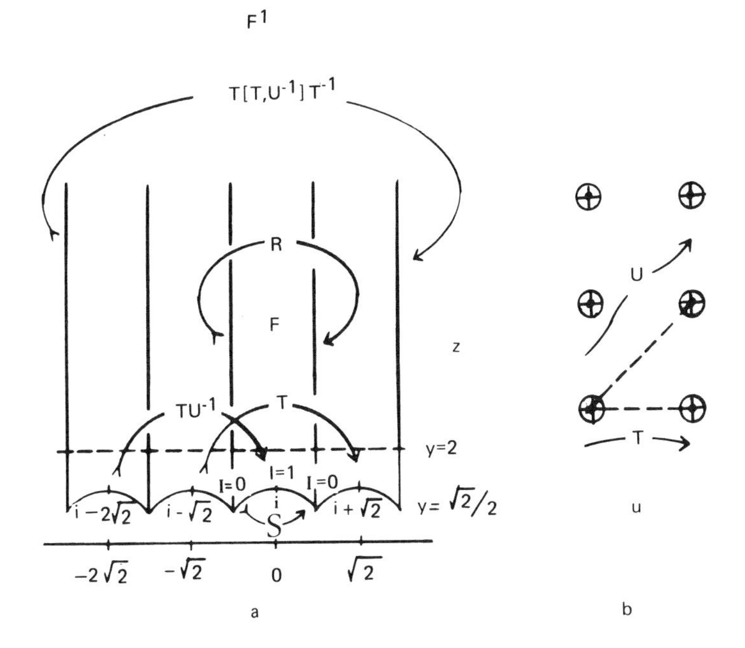

Figure 4. Actual numerical example of singularity structure in the t-plane showing double spirals of alternating regular (o) and irregular (x) poles about a regular pole.

If we refer to the fundamental domains $F = H/G$ and $F^* = H/G^*$, we have the analogue of (3.8b)

$$F: x^2 + y^2 < 1, \quad x^2 < 1/2 \tag{7.8}$$

and F^* consists of four replicas of F (making for the square). The mapping $I(z)$, for $I(z)$ a *Hecke modular function*, takes F on to the I-plane, but to obtain a square, we set

$$du = \text{const } dI/I^{1/2}(I - 1)^{1/2} \tag{7.9}$$

Now F^* is finally mapped onto the square period of the u-plane. We have the same step-formula for analogous minimal geodesics as in (2.5). Like the Markoff geodesics, they avoid the holes in the u-plane by a finite amount. They also lie below $\text{Im } z = 2$ (not 3/2) in the z-plane. Most features could be well imagined from the earlier case, but we refer, again to Schmidt for finer details.

8. MATRICES AND MINIMAL GEODESICS

Many seemingly simple questions have led to unsolved problems or to unexplored areas of unexpected depth. We conclude with such a problem (far from resolution) in the form of a matrix trace inequality which is also a minimality statement for geodesics. (This is the subject of Cohn [7].)

As before, let T and U be matrices generating a free group $\langle T,U\rangle$ in (say) $PSL(2,R)$ (so as to include both (2.3ab) and (7.1)). These matrices are to satisfy

$$\text{tr}[T,U] = -2, \ \text{tr } T > 2, \ \text{tr } U > 2 \tag{8.1}$$

Let V be a variable word in the group $\langle T,U\rangle$ with a fixed abelianization, i.e., fixed positive constants a and b where

$$V = T^a U^b K, \quad a > 0, \quad b > 0 \tag{8.2a}$$

for K in L the commutator of $\langle T,U\rangle$. Additively expressed,

$$V = aT + bU \pmod{L} \tag{8.2b}$$

(We always choose +V or -V to make tr $V > 0$.) Then under (8.2ab),

$$\min \operatorname{tr} V = \operatorname{tr}(T,U)^{(a,b)} \tag{8.3}$$

and <u>this minimum is achieved only by a</u> V <u>conjugate to a step-word</u> W, ($V = XWX^{-1}$, X in $\langle T,U\rangle$). As a special case, for instance,

$$\operatorname{tr}(T^a U^b) > \operatorname{tr}(T,U)^{(a,b)} \tag{8.4}$$

(unless a = 1 or b = 1, whence (8.3) follows). We shall now assume (a,b) = 1 for convenience (although the above inequalities hold regardless, and even are valid for negative a and b).

Now we are actually talking about the lengths of arcs of geodesics. Let g_W be the (infinite) geodesic, i.e., the semicircle in H connecting the fixed points r and s of some transformation of PSL(2,R). Also let d be the noneuclidean distance from z to Wz, its image. Then for z on g_W,

$$\operatorname{tr} W = 2 \cosh d/2 \tag{8.5a}$$

If z is not on g_W, however,

$$\operatorname{tr} W < 2 \cosh d/2 \tag{8.5b}$$

Thus our trace inequality (8.4) is a proposition about geodetic distances: if W is a (primitive) step-word in

$\langle T,U\rangle$, subject to (8.1), then among all elements V of $\langle T,U\rangle$ with the same abelianized coordinates (8.2ab), only these conjugate to W will move an arbitrary point the minimum non-euclidean distance (from z to Wz), which is achieved only along the geodesic g_W.

We may wonder why straightness is so much more accessible than shortness (a purely analytic property). Part of the answer lies in a principle communicated by W. Frenchel (see his work with J. Nielsen): Every group theoretic homomorphism of the free group on two elements is geometric, i.e., induced by a homeomorphism of the surface (for which the free group is the homotopy group). Here straightness is algebraically contained in the step-word and geometrically contained in the property of the absence of self-intersections of the geodesic. There seems to be no such correspondence involving minimality of lengths.

It turns out that like most optimal situations, the deviation from neighboring ones is of second order (as though a derivative vanishes). This fact can be used to prove that the inequality (8.4) is remarkably close to equality, and from this m_N the N-th Markoff number has the asymptotic behavior

$$\log m_N = \text{const } N^{1/2} \quad \text{(approximately)} \tag{8.6}$$

Very little can be done to improve such estimates (see Gurwood). For instance, in Table II, we can verify that the Markoff numbers do not order themselves linearly by means of the norm (which essentially represents the euclidean length of the vector describing the geodesic in the u-plane, see Figure 3).

To return to the intuitive remarks about the main result (8.3) in terms of geodesics, we reconstruct the fundamental domain of $\langle T,U\rangle$ in terms of equivalent generators $\langle W,W'\rangle$ where, for the given W, some W' exists such that

$$\langle T,U\rangle = \langle W,W'\rangle, \quad tr[T,U] = tr[W,W'] \tag{8.7a}$$

In fact, by minor changes (equivalences) we can have

$$[W,W'] = [T,U] = \begin{pmatrix} -1 & k \\ 0 & -1 \end{pmatrix} \tag{8.7b}$$

For any such W' we have a fundamental domain like that of Figure 2b, (with four sides identified by opposites), representing a (punctured) torus. According to a precedure of A. Schmidt, we can choose W' specially so that a point z_* exists on g_W from which a vertical line hits Z_o on the lower boundary circle. Thus in Figure 5a, the left-hand (shaded) area can be moved by W to the right. This is done more plainly in the u-plane to produce a rectangle D_W (with honest right angles in each corner and at g_W (connecting z_* with Wz_*) on the median.

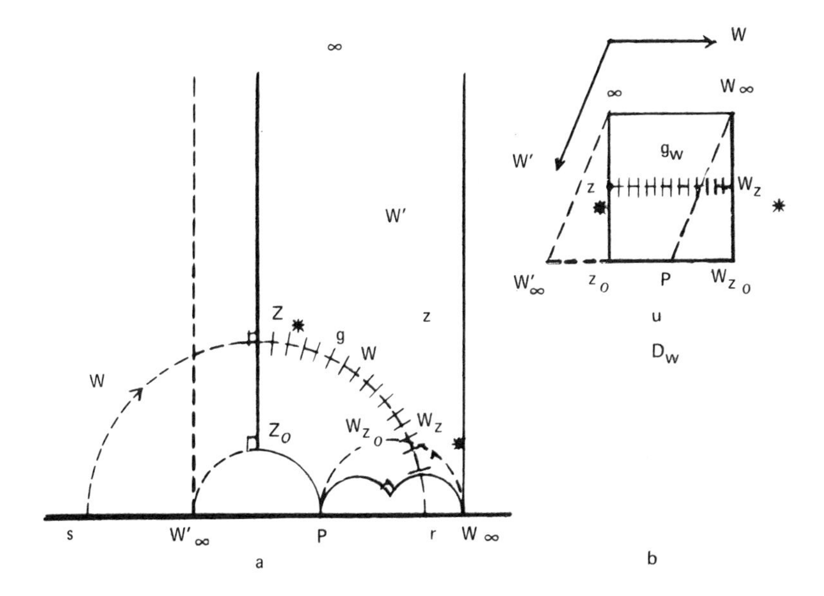

Figure 5. Special torus-like fundamental domains (a) in the z-plane and (b) in the u-plane. Note WW' and W'W each act on infinity to produce the same point P, (ss (8.7)). Also note the translation of the shaded triangle to make the figure in the z-plane into D_W the rectangular region of the u-plane where the geodesics perpendicular to the axis $(z*,Wz*)$ are a field.

The geodesics perpendicular to this median form a Jacobi-field filling out D_W uniquely.

Because of the negative curvature, we can say that the median connecting z_* and Wz_* is shorter than any other geodesic which connects the right and left hand sides of D_W, provided that the geodesics are restricted to the field. Now it would seem obvious that any geodesic wandering outside the field must be all the longer, yet this is the missing link in what would be otherwise a geometric proof of the matrix or geodesic theorem. We would, ideally, hope for a purely algebraic or elementary analytic proof of (8.3) under any conditions, but the involvement of far-flung techniques, essentially of differential topology, is typical of the interrelations arising from Markoff theory.

REFERENCES

[1] R.T. Bumby, The Markov spectrum, Diophantine Approximations and its Applications, Academic Press, 1973, 25-58.

[2] J.W.S. Cassels, An Introduction to Diophantine Approximations, Cambridge University Press, 1957.

[3] H. Cohn,

1. Approach to Markoff's minimal forms through modular functions, Annals of Math. 61 (1955) 1-12.
2. Representation of Markoff's binary quadratic forms by geodesics on a perforated torus, Acta. Arith. 18 (1971) 125-136.
3. Markoff forms and primitive words, Math. Annalen, 196 (1972) 8-22.
4. Some direct limits of primitive homotopy words and of Markoff geodesics, Conf. Disc. Groups and Riemann Surf., Annals Math. Stuides, vol. 79, Princeton, 1974, 81-98.
5. Ternary forms as invariants for Markoff forms and other SL2(Z) bundles, Journ. Lin. Alg. and Applic., 21 (1978) 3-12.
6. Growth types of Fibonacci and Markoff, Fibonacci Quart. 17 (1979) 178-183.

7. Minimal geodesics on Fricke's torus-covering, in Riemann Surfaces and Related Topics, Proc. 1978 Stony Brook Conf., Princeton 1980, 73-85.

[4] L.E. Dickson, Studies in the theory of numbers, Chicago, 1930, 79-107.

[5] W. Fenchel and J. Nielsen, Discontinuous Groups (to appear).

[6] R. Fricke,

1. Über die Substitutionsgruppe, welche zu den aus dem Legendre'schen Integralmodul gezogen Wurzeln gehören, Math. Ann. 28 (1887), 99-119

2. Die Congruenzgruppen der sechsten Stufe, Math. Ann. 29 (1887), 97-123

3. Über die Theorie der automorphen Modulgruppen, Gott. Nach., 1896. 91-101.

[7] C. Gurwood, Diophantine approximation and the Markov chain, Diss. N.Y.U., 1976.

[8] R.D. Horowitz, Characters of free groups represented in the two dimensional special linear group, Comm. Pure Appl. Math., 25 (1972) 635-649.

[9] C.G. Lekkerkerker, Geometry of Numbers, Wiley Interscience, 1969.

[10] W. Magnus, Noneuclidean Tesselations and their groups, Acad. Press, 1974.

[11] A.A. Markoff,

1. Sur les formes binaires indéfinies, I, Math. Ann. 15 (1879), 381-409; II, Math. Ann. 17 (188), 379-400.

2. Sur une question de Jean Bernoulli, Math. Ann. 19 (1882) 27-3

[12] P.J. Nicholls, Diophantine approximations via the modular group, J. London Math. Soc. 17 (1978) 11-17.

[13] A.L. Schmidt, Minimum of quadratic forms with respect to Fuchsian groups, I, J. reine und angew. Math. 286/287 (1976) 341-368, II, J. reine und angew. Math. 292 (1977) 109-114.

[14] K.B. Stolarsky, Beatty sequences, continued fractions, and certain shift operators, Canad. Math. Bull. 19 (1976) 473-482.

NOTE ON EISENSTEIN'S SYSTEM OF DIFFERENTIAL EQUATIONS: AN EXAMPLE OF "EXACTLY SOLVABLE BUT NOT COMPLETELY INTEGRABLE SYSTEM OF DIFFERENTIAL EQUATIONS"

David V. Chudnovsky and Gregory V. Chudnovsky

Department of Mathematics
Columbia University
New York, New York

§1.

In this paper we would like to attract attention to the phenomenon that "there are exactly solvable systems that are not completely integrable". The explanation of this ambiguous statement lies in the definition of complete integrability as a property entirely opposite to ergodicity. We call a Hamiltonian system completely integrable if there is a holomorphic transformation that reduces the system to action-angle variables or, in the case of compact phase space, the phase space of the system is decomposed into tori of different dimensions. Such systems having n degrees of freedom admit n involutive and independent first integrals. Moreover in the case, when these n integrals are algebraic, the solutions are

---

This work was supported in part by grants from the U.S. Air Force, NSF and ONR

given according to Liouville's theorem [1] in terms of algebraic quadratares. In this sense "completely integrable systems" are solved explicitly in terms of functions of classical analysis: Riemann θ-functions or Abelian integrals. Of course, such systems, in addition to the explicit expressions for their solutions, have strong stability properties and absence of strong mixing [1].

On the other hand, there is a large class of systems of nonlinear differential equations of physical interest that exhibit different kinds of ergodic behavior: transition from "completely integrable"to ergodic (cf. [2]); ergodic with strange attractors of different Hausdorff dimensions [3], [4], classical systems with strong ergodicity properties like axiom A [5] etc. Usually, for a given system of nonlinear differential equations the proof of the erodicity property is extremely difficult and had been obtained in one sense or another for a few types of system, cf. [1] . Here we exhibit a large class of nonlinear autonomous differential equations of the third order (or three coupled autonomous nonlinear differential equations of the first order) that describe ergodic flow, have a natural mechanical interpretation and, in addition, have all their solutions explicitly expressed in terms of classical automorphic functions [6] and, in the most interesting cases, in terms of Jacobi θ-functions.
The particular class of systems we exhibit here is a special subclass of a general family of "exactly solvable systems", which is in a sense, parallel to a class of "completely integrable systems". In the same way as classical "completely integrable system" are associated with Jacobian varieties $J(\Gamma)$ of algebraic curves Γ of positive genus [7], [8], the class of nonlinear differential equations under consideration is naturally associated with hyperbolic spaces H^3/G for a Fuchsian group G. One would like to conjecture that all physically interesting and meaningful systems can be reduced or

properly approximated by "exactly solvable" ones. As to the definition of "exactly" expression we mean now only "automorphic" functions among which we include Riemann θ-functions, classical, Hilbert and Siegel automorphic functions, and the still mysterious "matrix" θ-functions corresponding to vector bundles over Riemann surfaces.

§2.

We follow pattern of constructing nonlinear "completely integrable" systems of differential equations from an Abelian flow on a Jacobian of an algebraic curve of genus g, which is a linear flow on a g-dimensional torus. Let Γ be a curve given by an equation $P(x,y) = 0$, where P is a polynomial in x and y of degrees n and m, respectively, and let $J(\Gamma)$ be the Jacobian of Γ. We can choose any linear flow on $J(\Gamma)$ given by a fixed direction $\vec{A}$ on a torus $J(\Gamma)$. To this linear flow on $J(\Gamma)$ there corresponds a system of nonlinear differential equations, whose solutions are expressed in terms of derivatives of $\frac{d}{dx} \log \theta(\vec{A}x)$ for a Riemann g-dimensional θ-function $\theta(\vec{x})$ corresponding to Γ. This system of differential equations is always written in terms of differential operators as in [7] (cf. [9]).

$$P(L_1, L_2) = 0$$

for two linear differential operators $L_1 = \Sigma_{j=0}^{n} u_j \frac{\partial^j}{\partial x^j}$, $L_2 = \Sigma_{i=0}^{n} v_i \frac{\partial^i}{\partial x^i}$. All known classical "completely integrable systems" including the Euler equations of motion of the top Jacobi case of geodesics on ellipsoids, the stationary Korteweg-de Vries equation [10] and the periodic Toda lattice [11], [12] etc. all fall into this category (cf. [13] for an early difference version of Burchnall-Chaundy results [7]). The sense of all these equations is very simple. They are translations of linear flows on a Jacobian into Hamiltonian systems, achieved by the inverse Abel map. Complete integrability of such systems is a consequence of the existence of g independent linear flows on the Jacobian J commuting with each other. In a sense the "completely integrable" flows we are considering are composed from trivial motions along straight lines on tori $\mathbb{C}^g$ modulo the lattice of periods $L(\simeq \mathbb{Z}^{2g})$ where $J(\Gamma) \simeq \mathbb{C}^g/L$.

Our first remark is that similar simple linear flows associated with manifolds of different topological structure are natural examples of arbitrary dynamic behavior: classical quasiperiodic, ergodic and ergodic with complete mixing.

Linear flows considered by us are defined in elliptic, hyperbolic and parabolic cases as natural linear flows induced from g-dimensional complex space $\mathbb{C}^g$ (g-dimensional complex disk D^g or g-dimensional Siegel domain S) to a g-dimensional manifold

$$\mathbb{C}^g/G, \quad D^g/G, \quad H^g/G \quad \text{or} \quad S/G$$

where G is a group of linear fractional transformation acting discontinuously on $\mathbb{C}^g$, D^g or S, respectively. We consider only those groups G for which a fundamental domain F exists. For such groups, a linear flow on g-dimensional complex space induces a billiard problem in F, where boundary behavior and reflections from the boundary are prescribed by the action of the group G. The simplest example is the elliptic curve with $g = 1$ and the action of G on $\mathbb{C}$ generated by two translations (this is a one-dimensional completely integrable system).

In the topological description of the class of dynamical systems under consideration, it should be remembered that these dynamical systems, easily topologically described, are not defined by differential equations. Moreover, a flow topologically defined as a "billiard system" in the fundamental domain cannot possibly be described through a solution of an algebraic system of nonlinear ordinary differential equations. Hence nonlinear differential equations with the same topological properties as "billiard" flows on F appear only after proper transformations analogous to the inverse Abel map from a curve to its Jacobian.

We claim that this translation from topological description of a dynamical system to a system of algebraic ordinary differential equations is carried by a transformation from complex coordinates in F to function automorphic with respect to G and defined on F.

§3.

Our main attention is devoted to systems of exhibiting ergodic, rather than completely integrable, behavior. Among classical systems with complete mixing and ergodic behavior the best known are those arising from hyperbolic spaces of constant negative curvature. For them ergodicity had been rigorously proved by Morse and Hedlund [14], [15]. From the point of view of differential equations, such systems can correspond to an autonomous sytem having $1\frac{1}{2}$-degrees of freedom: 3 nonlinear ordinary differential equations of the first order. Indeed, this is the first dimension for which ergodicity is possible. The nonautonomous systems of the first order, including e.g. Painlevé transcendents,fall into the same category as those autonomous of the third order.

Equations of the third order associated with hyperbolic spaces H^3/G, with G a Fuchsian group of the first kind, can be easily derived using Schwarzian derivatives. It should be recalled that $H^3 = SL(2,\mathbb{R})/SL(2,\mathbb{Z})$ and G is a discrete subgroup of $SL(2,\mathbb{Z})$, where the fact that G is Fuchsian group of the first kind means that there is a circumference invariant under all transformations of G and every element on the circumference is a limit point of G. We assume G to act on H^+ (upper half-plane). In this case geodesic flows on H^+/G and on H^3/G are ergodic in a strong case [14]-[15]. According to the Poincaré uniformization theorem, functions automorphic with respect to the group G arise from an inversion map generated by solutions of Fuchsian linear differential equations with monodromy group generated by G.

THEOREM 1. Let G be a Fuchsian group. Then there is an algebraic function $R(w)$ such that the third order nonlinear differential equation

$$\frac{\dddot{w}}{\dot{w}^3} - \frac{3}{2}\frac{\ddot{w}^2}{\dot{w}^4} = R(w)$$

describes a geodesic flow on H^3/G.

The algebraic function $R(w)$ can be defined in terms of accessory parameters associated with the group G. If there is a univalent function on G, then $R(w)$ is a rational function of w.

Proof. (See [6], [16].) The explicit construction of $R(w)$ follows from the uniformization theorem for an algebraic function on a Riemann surface H^+/G. However the existence of $R(w)$ can be easily proved. Namely, let $w = w(z)$ be a function automorphic with respect to G. Then, since H^+/G is a Riemann surface, any other automorphic function for G is algebraically dependent on $R(w)$. For this most important property, we need G only to be a discontinuous group with fundamental domain F.

Now for an automorphic function w, simple computations show that w' is the form of weight -2. For an arbitrary form f of weight $-k$ it is clear that $kf\cdot f'' - (k+1)f'^2$ is the form of weight -2k - 4. Hence $\dddot{w}/\dot{w}^3 - 3/2\ \ddot{w}^2/\dot{w}^4$ is the form of weight zero, i.e. an automorphic function. So it is algebraically dependent with w. This proves that $w = w(z)$ satisfies an algebraic differential equation of the third order.

The fact that this differential equation describes the topological structure of the flow on H^3/G follows from the remark that all the solutions of the differential equation above are expressed as $w_0(\frac{az+b}{cz+d})$ for $\binom{a\ b}{c\ d} \in SL(2)$ and a fixed automorphic function $w_0(z)$. In other words, time evolution according to this system of nonlinear differential equations for w is equivalent to the evolution of initial data $\binom{a\ b}{c\ d}$ according to the geodesic flow on $H^3/G = SL(2,\mathbb{R})/G$.

The general theorem established above describes a large class of systems exhibiting various ergodic, transitional or "completely integrable" behavior. It is not easy, however, to construct explicitly equations with a given kind of behavior, since, first of all one needs a Riemann surface with given topological properties of the geodesic flow and, second, one should construct the uniformization of this Riemann surface in terms of Fuchsian automorphic functions.

In this note we present examples of systems described by the theorem corresponding to the case in which R(w) is a rational function with two poles (and so uniformization is achieved by Schwarzian or hypergeometric functions). Moreover we naturally decompose third order linear differential equations into coupled system of three first order linear differential equations. The transformation from a single third order differential equation into a coupled system of first order equations is achieved by substituting functions automorphic with respect to G into functions modular with respect to G of the lowerest weights. The relationship between automorphic and modular functions is based on a graded structure of a ring of modular functions. For a given group G and arbitrary m let us denote by $M_m(G)$ the space of functions f(z) modular with respect to G of the weight m. If G is realized as a group of conformal transformations of H^+, then the $f(z) \in M_m(G)$ means that $f(\frac{az+b}{cz+d}) = (cz+d)^m f(z)$ for $\binom{a\ b}{c\ d} \in G$. Usually one takes only those modular forms that are holomorphic in H^+ including infinity; the space they generate is a subspace of $M_m(G)$ denoted by $M_m^+(G)$. It is easy to determine the degree of transcendence of the whole graded algebra $\oplus_{m\in\mathbb{Z}} M_m(G)$. Obviously, the ratio of two modular forms of the same weight is an automorphic form. Also the space of all automorphic forms is naturally a function field on a Riemann surface so it has transcendence degree one. Hence the ring of modular forms $\oplus_{m\in\mathbb{Z}} M_m(G)$ has the dimension of at most two. In the case of a finite group G the dimension is one, otherwise it is two. Derivatives of modular forms are, unfortunately, not modular forms. In order to obtain differentially closed field one needs an algebraically closed field of degree of transcendence three. An additional third algebraically independent generator of this differentially closed field is not multiplicatively automorphic but rather is a multiplicative-additive automorphic function and is transformed like integrals of the second kind. Such a function can be obtained by taking the first derivative f'(z) of a modular function (e.g. automorphic function) f(z). Higher dirivatives do not extend our ring because, e.g. a nonlinear differential operator $(k+2)f''(z) - kf'(z)^2$ maps $M_k(G)$ into $M_{k+2}(G)$. We refer the reader to the discussion by Rankin [16] and Resnikoff [17] on this and similar nonlinear differential operators transforming modular forms into modular forms.

Now the decoupling of the third order differential equation described in Theorem 1 is obtained in the following way. We take a modular (parabolic) form from $M_m^+(G)$ of the smallest positive integer weight m (usually, m = 4) and a modular form algebraically independent with this one of the next smallest positive weight. With these two modular forms denoted by y and z one constructs a multiplicative-additive automorphic function x (of modular "weight" two) such that the ring $\mathbb{C}[x,y,z]$ is closed under differentiation. Previous arguments with the computation of dimension show that such an x actually exists.

Our next example corresponds to the case $G = \Gamma(1) = SL(2,\mathbb{Z})$, in which we speak of ordinary modular functions and x,y,z are normalized Eisenstein series of respective weights 2, 4 and 6.

§4.

In this section we describe geodesic motion on $H^3 = SL(2,\mathbb{R})/SL(2,\mathbb{Z})$ using differential equations satisfied by Eisenstein series $E_k(z)$. These equations were derived by Eisenstein himself [18]. For complete and nice discussion of these equations and recurrence formulae for Eisenstein series see Weil [19]. Later these equations were rewritten by Halphen [20] and then by Ramanujan [21], who applied them to congruences of modular forms. A suprising remark of Hardy states: "I have not seen (it) anywhere except in Ramanujan's work" [22, p. 166]. Later Van der Pol [23] reviewed these results. Differential equations of Eisenstein are written for the following three functions

$$x = 1 - 24 \Sigma_{n=1}^{\infty} \sigma_1(n)q^n$$

$$y = 1 + 240 \Sigma_{n=1}^{\infty} \sigma_3(n)q^n$$

$$z = 1 - 504 \Sigma_{n=1}^{\infty} \sigma_5(n)q^n$$

with $q = e^{2\pi i\tau}$ and $\sigma_k(n) = \Sigma_{d|n} d^k$. These functions are the first in the sequence of the normalized Eisenstein series

$$E_k = 1 - \frac{2k}{B_k} \Sigma_{n=1}^{\infty} \sigma_{k-1}(n)q^n$$

where B_k is k-th Bernoulli number.

LEMMA 2. For a time variable $t = -2\pi i\tau$ we have three ordinary nonlinear differential equations:

$$\frac{\partial}{\partial t}x = \frac{1}{12}(y - x^2)$$

$$\text{(E)} \quad \frac{\partial}{\partial t}y = \frac{1}{3}(z - xy)$$

$$\frac{\partial}{\partial t}z = \frac{1}{2}(y^2 - xz)$$

The proof of this lemma is simple; it is based on the following remark.

LEMMA 3. The three functions x, y, z satisfy modular equations:

$$x(g\tau) = (c\tau + d)^2 x(\tau) + \frac{6c}{\pi i}(c\tau + d)$$

$$y(g\tau) = (c\tau + d)^4 y(\tau)$$

$$z(g\tau) = (c\tau + d)^6 z(\tau)$$

for $g = \begin{pmatrix} a & b \\ c & d \end{pmatrix} \in SL(2;\mathbb{Z})$. In particular $\partial_k = \frac{\partial}{\partial t} + k \cdot x$ is a differentiation of $M_k(\Gamma(1))$ that maps it into $M_{k+2}(\Gamma(1))$.

The proof of Lemma 2 follows from Lemma 3 and the well-known fact that $M_k^+(\Gamma(1))$ is generated polynomially for $k \geq 4$ by y and z.

Moreover all solutions of the system of equations (E) can be explicitly described in terms of the three Eisenstein series E_2, E_4, E_6:

COROLLARY 4. All solutions of the system (E) depend on four (homogeneous) parameters a, b, c, d and have the form

$$x(t) = \frac{\delta}{(c\tau + d)^2} E_2\left(\frac{a\tau + b}{c\tau + d}\right) - \frac{2c}{c\tau + d}$$

$$y(t) = \frac{\delta^2}{(c\tau + d)^4} E_4\left(\frac{a\tau + b}{c\tau + d}\right)$$

$$z(t) = \frac{\delta^3}{(c\tau + d)^6} E_6\left(\frac{a\tau + b}{c\tau + d}\right)$$

for $\delta = ad - bc$ and $t = -2\pi i\tau$.

In these formulae one can pass to nonhomogeneous coordinates a, b, c, d, if one assumes that $\Delta = ad - bc$. After this transformation, all the solutions of the system (E) are parametrized by three independent parameters (just as there are three independent initial conditions). We arrange these parameters into a single matrix $\gamma = \begin{pmatrix} a & b \\ c & d \end{pmatrix} \in SL(2)$. However, the space of all solutions of system (E) is parametrized not by $SL(2)$ but rather by $SL(2)/SL(2;\mathbb{Z})$. Indeed, we have a consequence of Corollary 4 and the formulae of Lemma 3:

COROLLARY 5. Two solutions $(x_\gamma(t), y_\gamma(t), z_\gamma(t))$ and $(x_{\gamma'}(t), y_{\gamma'}(t), z_{\gamma'}(t))$ of (E) corresponding to two matrices $\gamma = \begin{pmatrix} a & b \\ c & d \end{pmatrix}$ and $\gamma' = \begin{pmatrix} a' & b' \\ c' & d' \end{pmatrix}$ are identical if and only if

$$\gamma = M\gamma' \text{ for some } M \; 2 \; SL(2;\mathbb{Z})$$

In other words the dynamical system defined by (E) is naturally associated with the three dimensional hyperbolic space $H^3 = SL(2;\mathbb{R})/SL(2;\mathbb{Z})$ [24]. Note that, instead of arbitrary a, b, c, d, we took only real ones; in general $SL(2;\mathbb{C})/SL(2;\mathbb{Z})$ can be considered. However much more can be said: not only phase spaces of two systems (the geodesic flow on H^3 and that of (E)) are identical, but the geodesic flow on H^3 coincides with the time evolution induced by (E).

In order to prove this we remark that time evolution on H^3 is induced by the right action of $g_t = \begin{pmatrix} 1 & t \\ 0 & 1 \end{pmatrix} \in SL(2;\mathbb{R})$. Let us look on the evolution of an arbitrary solution $(x_\gamma(t),$ $y_\gamma(t),\ z_\gamma(t))$ of (E) corresponding to $\gamma \in SL(2;\mathbb{R})$. Let us look at what happens to this solution after time T passes. The explicit formulae of Corollary 4 show that after time T the initial solution is transformed into a new solution

$$(x_{\gamma_T}(t), y_{\gamma_T}(t), z_{\gamma_T}(t)) = (x_\gamma(t+T), y_\gamma(t+T), z_\gamma(t+T))$$

where

$$\gamma_T = \gamma \cdot g_{-2\pi iT} = \gamma \begin{pmatrix} 1 & -2\pi iT \\ 0 & 1 \end{pmatrix}$$

Consequently (up to a constant factor) the evolution according to the geodesic flow on H^3 and according to equations (E) are identical.

The system (E) is, as noted above, a system with $1\frac{1}{2}$ degrees of freedom. However, system (E) can be embedded in a Hamiltonian system with two degrees of freedom. For this we use a method analogous to a method of transformations of Lagrangian systems and Lagrangian manifolds into Hamiltonians systems and symplectic manifolds. For this we perform a slight change of variables in the system (E) changing it into an equivalent system in Halphen's [20] notation. Let $x_1 = 1/6\ x$, $y_1 = 1/3\ y$, $z_1 = 1/27\ z$. Then the system (E) takes the form:

$$\dot{x}_1 = \frac{1}{2}x_1^2 - \frac{1}{24}y_1$$

$$(E_1)\quad \dot{y}_1 = 2x_1y_1 - 3z_1$$

$$\dot{z}_1 = 3x_1z_1 - \frac{1}{6}y_1^2$$

This system of equations arises from the following Hamiltonian

$$H = \frac{1}{2}q_1^2p_1 - \frac{1}{48}p_1^2p_2^8 + 3q_1q_2p_2.$$

If one writes $q_1 = x_1$, $q_2 = z_1$, $p_1 = \lambda y_1$, $p_2 = \lambda$, then the Hamiltonian equations corresponding to H have the following form:

$$\dot{x}_1 = \frac{1}{2}x_1^2 - \frac{\lambda^9}{24}y_1$$

$$\lambda\dot{y}_1 = 2\lambda x_1y_1 - 3\lambda z_1$$

$$\dot{z}_1 = -\frac{1}{6}\lambda^9 y_1 + 3x_1z_1$$

$$\dot{\lambda} = -3\lambda x_1$$

which is identical to the equations (E_1) if $\lambda = 1$. Hence the system (E_1) is a restriction of the Hamiltonian H to $\lambda = 1$. One of the most important properties of the system (E) is its ergodicity. Another important property of this system is the existence of a large class of closed trajectories. While the determination of all initial conditions leading to the closed trajectories is an easy task, the determination of the topological structure of all closed trajectories is not so trivial. First of all, from Corollary 5 and our definition of time evolution, it follows that a solution of (E) corresponding to initial condition $\gamma \in SL(2)$ is closed if and only if for some $T \neq 0$ (the period of trajectory) $\gamma g_T \gamma^{-1} \in SL(2;\mathbb{Z})$. If $\gamma = \begin{pmatrix} a & b \\ c & d \end{pmatrix}$, then this condition turns out to be equivalent to the following equations:

$$a = -c\frac{k}{(1-\ell)} \qquad nk = -(1-\ell)^2 \qquad Ta = nb + kc$$

where n, k, ℓ are integers, $\ell \neq 1$.

We want to consider again a "billiard" flow on H^+/G using arithmetical algorithms similar to that of continued fractions (which correspond to the case $G = SL(2;\mathbb{Z})$.

Let us start with the best known example of a modular figure or Lobachevsky plane. We take an arbitrary geodesic on H^+, which is either a half circle with a center on a boundary (real axis) or an open half line $\{x + iy: y > 0\}$. These circles are mapped by fractional transformations from $\Gamma(1) = SL(2;\mathbb{Z})$ into the fundamental domain F of $\Gamma(1)$: $F = \{z \in H^+: |z| > 1, |\mathrm{Re}\ x| \leq \frac{1}{2}\}$. This mapping gives the trajectories of a "billiard" flow in F. Depending on the location of the center of the geodesic, its image in F is a continuous curve, which is composed from finitely or infinitely many parts of circles reflected from the boundary of F. The precise description of this trajectory can be given in terms of symbolic dynamics for an arbitrary G and F. In our case these trajectories can be determined by the formalism of continued fraction expansion of the initial positions of geodesics on the boundary.

Let us take the case of a half-line $\ell = \{x_0 + iy: y > 0\}$. In this case the sequence of circles in F reflected from the boundary and corresponding to the limit point x_0 can be explicitly expressed in terms of the continued fraction expansion of x_0. Namely, let $x_0 = [a_0; a_1, a_2, \ldots]$ be the continued fraction expansion of x_0 with $a_i \in \mathbb{Z}$, $a_i > 0$: $i = 1,2,\ldots$.

Let p_n/q_n be the n-th principal convergent of this continued fraction expansion: n = 0,1,... . There exists an infinite sequence of transformations $\gamma_0, \gamma_1, \ldots$ and a partition $\ell = \cup'^{\infty}_{n=0} \ell_n$ of ℓ, where $\ell_n = \{x_0 + iy: \xi_n > y \geq \xi_{n+1}\}$ for $\infty = \xi_0 > \xi_1 > \ldots$ and $\lim_{n\to\infty} \xi_n = 0$ such that γ_n transforms ℓ_n into a circle in F:

$$\gamma_n(\ell_n) \subset F$$

with $\gamma_n(x_0 + i\xi_{n+1}) = \gamma_{n+1}(x_0 + i\xi_{n+1})$ lying on the boundary of F. The points ξ_n and the matrices γ_n are easy to determine. For example,

$$\gamma_0 = \begin{pmatrix} 1 & a_0 \\ 0 & 1 \end{pmatrix} = \begin{pmatrix} p_{-1} & p_0 \\ q_{-1} & q_0 \end{pmatrix}$$

In general, we have

$$\gamma_n = (-1)^{n+1} \begin{pmatrix} p_{n-1} & p_n \\ q_{n-1} & q_n \end{pmatrix}$$

where $\gamma_n \in SL(2;\mathbb{Z})$.

REFERENCES

[1] V. I. Arnold, Mathematical methods of classical mechanics, Springer, N.Y. 1978.

[2] R.C. Churchill, On proving the nonintegrability of a Hamiltonian system, Lecture Notes Math. v. 925, Springer, 1982, 103-123.

[3] J. Guckenheimer, Bifurcations of dynamical systems, Progress in mathematics, v. 8, Birkhauser, Boston, 1980, 115-232.

[4] C. Tresser, Modèles simples de transitions vers la turbulence, Thèse, Nice, 1981.
P. Coullet, Quelques aspects de la transition vers le chaos pour des systemes dynamiques dissipatifs, These, Nice, 1980.

[5] D.V. Anosov, Geodesic flows and closed Riemann manifolds with negative curvature, Proc. Steklov Inst. Math. 90 (1967).

[6] L.R. Ford, Automorphic functions, 2ed., Chelsea, N.Y., 1951.

[7] J. Burchnall, T. Chaundy, Commutative ordinary differential operators, Proc. London Math. Soc. 21 (1922), 420-440; Proc. Royal Sol. London, 118 (1928), 557-583; 132 (1931), 471-485.

[8] D. Mumford, An algebraico-geometric construction of commuting operators and of solutions to the Toda lattice equation, Korteweg-de Vries equation and related nonlinear equations, Proc. Intern. Symp. Algebraic Geometry Kyoto, 1977, 115-153.

[9] E.L. Ince, Ordinary differential equations, Dover, N.Y. 1944.

[10] J. Drach, Détermination de cas de réduction de l'equation differentielle $d^2y/dx^2 = [\varphi(x)+h]y$, C.R. Acad. Sci. Paris, 168 (1919), 47-50; Sur l'intégration par quadratures de l'équation $d^2y/dx^2 = [\varphi(x)+h]y$, C.R. Acad. Sci. Paris, v. 168 (1919), 337-340, erratum p. 532.

[11] H. Flaschka, The Toda lattice, Phys. Rev. 9 (1974), 1924-1925; Prog. Theor. Phys. 51 (1974), 703-716.

[12] M. Gutzwiller, see this volume.

[13] P.B. Naǐman, On the theory of periodic and limit-periodic Jacobian matrices, Soviet Math. Dokl, v. 3 (1962), 383-385 = Dokl. Akad. Sci. USSR v. 143 (1966), 277-279 (Russian), see this volume.

[14] M. Morse, Symbolic dynamics, Lectures by M. Morse 1937-1938, The Institute for Advances Study, Princeton, New Jersey, Notes by Rufus Oldenburg, reprinted by University Microfilms, Ann Arbor, Michigan, 1974.

[15] M. Morse, G.A. Hedlund, Symbolic dynamics II. Amer. J. Math. 62 (1940), 1-42.

[16] R. A. Rankin, The construction of automorphic forms from the derivatives of a given form, J. Indian Math. Soc. (N.S.) 20 (1956), 103-116.

[17] H.L. Resnikoff, A differential equation for an Eisenstein series of genus two, Proc. Natl. Acad. Sci. USA 65 (1970), 495-496.

[18] F.G.M. Eisenstein, Mathematische Werke, N.Y., Chelsea, 1975, 2v.

[19] A. Weil, Elliptic functions according to Eisenstein and Kronecker, Springer, N.Y., 1976.

[20] G.H. Halphen, Ouevres, v. III, Gauthier-Villars, 1900.

[21] A.S. Ramanujan, The collected papers, ed. by G.H. Herdy, P.V. Seshu Aiyer, and B.M. Wilson, Cambridge Univ. Press, 1927.

[22] G.H. Hardy, Ramanujan, Cambridge, 1940.

[23] B. Van der Pol, On a non-linear partial differential equation satisfied by the logarithm of the Jacobian theta-functions with arithmetical applications. I, II. Nederl, Akad. Wetensch. Proc. Ser. A. 54 = Indagationes Math. 13 (1951), 261-271, 272-284.

[24] I.M. Gelfand and S.V. Fomin, Geodesic flows on manifolds of constant negative curvature, Uspekhi. Math. Nauk 7 (1952), 118-137; = Amer. Math. Soc. Translations (2) 1 (1955), 49-65.

SOME REMARKS ON THETA FUNCTIONS AND S-MATRICES

David V. Chudnovsky and Gregory V. Chudnovsky

Department of Mathematics
Columbia University
New York, New York

INTRODUCTION

One of the most attractive problems in the investigation of field theories with strong internal symmetries is the possibility of the exact evaluation of their S-matrix and its spectral properties as the result of the existence of large family of classical and quantum conservation laws [3], [4], [9].[1)] Moreover, the complete knowledge of S-matrix enable us to find a canonical Hamiltonian structure and higher conservation laws [2], [3], [4], [6]. At least in the two dimensional case one knows that complete integrability of a field theory is closely connected with the factorization of S-matrix [2], [3], [4], [6], [9]. It is reasonable to speculate that for higher dimensional cases as well "complete integrability" implies or is implied by similar conditions for S-matrix meaning the decomposition of

---

1) As it is shown in [21], sometimes it is enough to know only a few higher conservation laws

an arbitrary interactions into a sequence of simple ones. Equations describing the condition of factorization of an S-matrix are called themselves factorization equations. It is more or less clear that every completely integrable system gives rise to a certain solution of factorization equations and every solution of factorization equations generates a class of families of completely integrable systems [3], [4], [11], [18]. Factorization equations, as authors understand them now, are certain homological conditions that describe symplectic structure of completely integrable quantum systems. It is the most natural analyzing them, to use the knowledge of quantization according to Heisenberg and Weyl. Heisenberg commutation relations translated into the form of Weyl relations are the most evident basis for any quantization. On the other hand, Weyl relations and representations of the Heisenberg group are closely connected with complex tori and θ-functions [8], [5]. It seems understandable because "everywhere completely integrable" system splits into collection of infinite or finite dimensional tori. Evidently, in order to construct a solution of factorization equations one should note that in all the nontrivial cases these solutions are expressed in terms of Abelian functions or their natural degenerations, as we proved in [4], [26] (cf. [10]).

Consequently we can try to find an explicit expression for the elements of factorized S-matrix using Abelian functions. It is known that one can substitute the existence of the law of addition by the conditions of periodicity or quasiperiodicity of Picard type [1], [7], [8]. However these conditions of quasiperiodicity are nothing but the relations of Weyl type defining θ-functions. In other words it is quite natural to reformulate factorization equations in a different way using the standard realization of Weyl relations over 2g dimensional vector space V and the lattice L in V. Roughly speaking, we consider an S-matrix which is nontrivial over complex torus V/L;

this condition of nontriviality is replaced by the condition of triviality over V plus consistency conditions connected with translations by elements of L. These consistency conditions are, naturally, of Weyl type. Using this approach we formulate the factorization conditions for completely X-symmetric S-matrices. For our purposes we use physically oriented exposition of P. Cartier [1] of Weyl relations, Heisenberg group representations and θ-functions, that had been designed for treatment of the models of field theory.

The language we use in this paper is the classical one but it should be noted that the underlying structure is, in fact, much more rich. First of all the Picard relations should be naturally replaced by line bundles, while the triviality over V and nontriviality over V/L lead us at once to Picard stacks and category theory approach. We leave this fascinating subject to other publications of L. Breen and the authors.

In the beginning of the paper (§1) we briefly remind the reader of the definition of the factorization and unitarity equations for quantum S-matrix. We refer readers to [3], [4], [9], [16] for more detailed exposition of factorized S-matrices. We discuss the Picard quasiperiodicity relations and θ-function definition in §2 together with quasiperiodicity relations for S-matrices. θ-functions appear again in §3 through the representation theory approach for Heisenberg group, as proposed by Weil, see [8], [20].

This approach to θ-functions, though rather straight-forward, is the most efficient one, since it defines in a unified way the dimensions of spaces of graded θ-functions; projective imbeddings of Abelian varieties and law of addition on them. We follow the approach of Cartier [1], but expositions in [5], [7], [8], [19], [20] are of big use in detailed studies.

§4 is devoted to description of concrete realizations of induced representations of Heisenberg groups in terms of $e \times e$ matrices; and in §5 in terms of these representations we express the completely X-symmetric S-matrices. One important class of completely X-symmetric S-matrices is canonically associated with Abelian varieties of a given dimension g and induced representations of Heisenberg groups corresponding to torsion subgroups of an Abelian variety. The expression of the completely X-symmetric S-matrices is given in §5. In an important case when as Abelian variety is an elliptic curve ($g = 1$) we can establish that the factorization equations are satisfied for S-matrices from Conjecture 5.15. For $g > 1$ the factorization equations seem not to be satisfied by the corresponding canonical S-matrices (unlike [10]), though these matrices still give rise to completely integrable systems.

In the second chapter of the paper we are mainly concerned with applications of factorized S-matrices to quantum and classical completely integrable systems. We define the notion of S-matrices associated with quantum isospectral deformation equations. For classical two-dimensional isospectral deformation equations instead of quantum factorized S-matrix one uses so called semiclassical factorized s-matrix [26], [18].

We show that with the same quantum or classical factorized S-matrix is associated a large family of different two dimensional completely integrable systems possessing this S-matrix. We present a list of examples of two dimensional completely integrable system generated by the simplest S-matrices. At the same time there is one, canonical spectral problem (both continuous or lattice one), attached to a given factorized S-matrix. We describe this system and particular attention is devoted to the case when a factorized S-matrix comes from the representation of §5 for the case of an elliptic curve. We examine the structure of these interesting equations called in the continuous case elliptic higher Korteveg de Vries (KdV) equations. These new classes of equations offer a lot for future investigation. In the lattice case we obtain an immediate generalization of Baxter eight-vertex model (or XYZ-model) for $\mathbb{Z}/\mathbb{Z}N$-spin models on two dimensional lattices. In connection with these lattice models we again discuss factorization equations from the point of view of Baxter's star-triangle relations [2] and algebraic identities connecting θ-functions. This part of the paper serves as an introduction to the future publications on the geometry of generalizations of factorization equations and obstructions to factorization of S-matrices.

SECTION 1. FACTORIZED S-MATRICES SATISFYING QUASIPERIODICITY RELATIONS. θ-FUNCTIONS AND REPRESENTATIONS OF HEISENBERG GROUP

This section is dealing with a straightforward algebraic approach to θ-functions and quasiperiodicity relations defining them. This framework is used for the sake of simplicity in order to describe factorized S-matrices expressed in terms of θ-functions. The definition of factorization equations is given in §1 and studied in §5, while θ-functions are dealt with in §§2, 3. We use elementary representation theory approach in §§3, 4.

§1.

First of all, we remind the geometric definition of the factorization equations on S-matrix [3], [4], [10]. We start with a category $\mathcal{C} = \{V(\theta)\}$ of vector spaces of dimension e and an S-matrix $R(\theta_1, \theta_2)$ defines an isomorphism of tensor product of spaces $V(\theta)$

$$R(\theta_1, \theta_2): V(\theta_1) \otimes V(\theta_2) \to V(\theta_2) \otimes V(\theta_1) \tag{1.1}$$

We consider $R(\theta_1, \theta_2)$ as an $e^2 \times e^2$ matrix with elements $R(\theta_1, \theta_2)_{ab,cd}$ or $S_{ab}^{cd}(\theta_1, \theta_2)$ following the notation of [3]. The unitarity and factorization equations are conditions on $R(\theta_1, \theta_2)$

to generate a structure of monoid in the category $\mathcal{C}$. The unitarity equations are the following

$$R(\theta_1,\theta_2)R(\theta_2,\theta_1) = \mathbb{I}_{n^2} \tag{1.2}$$

for an identity $n^2 \times n^2$ matrix $\mathbb{I}_{n^2}$. The factorization equations, necessary for the associativity arise from the following diagram:

$$\begin{array}{ccc} & V_{\theta_1} \otimes V_{\theta_2} \otimes V_{\theta_3} & \\ \swarrow & & \searrow \\ V_{\theta_2} \otimes V_{\theta_1} \otimes V_{\theta_3} & & V_{\theta_1} \otimes V_{\theta_3} \otimes V_{\theta_2} \\ \downarrow & & \downarrow \\ V_{\theta_2} \otimes V_{\theta_3} \otimes V_{\theta_1} & & V_{\theta_3} \otimes V_{\theta_1} \otimes V_{\theta_2} \\ \searrow & & \swarrow \\ & V_{\theta_3} \otimes V_{\theta_2} \otimes V_{\theta_1} & \end{array}$$

We need certain notations that simplify formalism. We consider an S-matrix $R(\theta_i,\theta_j)$ as a linear nonsingular map

$$R(\theta_i,\theta_j)\colon V_{\theta_i} \otimes V_{\theta_j} \to V_{\theta_j} \otimes V_{\theta_i}$$

Then for three different i, j, k from $\{1,2,3\}$ we denote by $I \otimes R(\theta_i,\theta_j)$ or $R(\theta_i,\theta_j) \otimes I$ the map of

$$V_{\theta_k} \otimes V_{\theta_i} \otimes V_{\theta_j} \to V_{\theta_k} \otimes V_{\theta_j} \otimes V_{\theta_i}$$

or of

$$V_{\theta_i} \otimes V_{\theta_j} \otimes V_{\theta_k} \to V_{\theta_j} \otimes V_{\theta_i} \otimes V_{\theta_k}$$

identical in V_{θ_k}.

Next, by $I^{\tau}(\theta_i,\theta_j)$ we denote an "identical" S-matrix, being a trivial map

$$I^{\tau}(\theta_i,\theta_j): V_{\theta_i} \otimes V_{\theta_j} \to V_{\theta_j} \otimes V_{\theta_j}$$

given by a map of generators $e_a \otimes e_b$ of $V_{\theta_i} \otimes V_{\theta_j}$ into $e_b \otimes e_a$ of $V_{\theta_j} \otimes V_{\theta_i}$.

The classical factorization equations for the S-matrix R, as the diagram is showing, can be written as [3], [43]:

$$\begin{aligned} &(R(\theta_2,\theta_3) \otimes I)\,(I \otimes R(\theta_1,\theta_3)) \\ &\quad \times (R(\theta_1,\theta_2) \otimes I) \\ &\quad = (I \otimes R(\theta_1,\theta_2))(R(\theta_1,\theta_3) \otimes I) \\ &\quad \times (I \otimes R(\theta_2,\theta_3)) \end{aligned} \tag{1.3}$$

where I is $e \times e$ identity matrix. By a factorized S-matrix $R(\theta_1,\theta_2)$ one usually understands S-matrix satisfying (1.2) and (1.3) and depending only on $\theta_1 - \theta_2$ where θ is considered as complex (scalar or vector) variable.

§2.

We have shown in [4] that, since factorization equations have the form of law of addition, any "nontrivial" factorized S-matrix is expressed in terms of Abelian functions (for precise statement see [4]). It had been emphasized in [4] that this general result is helpful in explicit determination of factorized S-matrices. We pursue this goal below. First of all, it is very convenient to replace Abelian functions by θ-functions.

As usual by an Abelian function in $\mathbb{C}^g$ we understand a meromorphic function in $\mathbb{C}^g$ which is 2g periodic with a lattice of periods L (of rank 2g in $\mathbb{C}^g$). There is the famous relation between θ-functions and bilinear forms. We can formulate relations between Abelian functions, theta functions and Hermitean forms on $\mathbb{C}^g$ in a single statement. For this, in order to unify our notations, we identify $\mathbb{C}^g$ with a vector space V of dimension 2g over $\mathbb{R}$. In this case one has a natural anti-involution J on V (i.e. multiplication by i). An auxiliary result we are using is taken from Siegel [7] and has the following form

PROPOSITION 2.1. Let L be a lattice in V (= $\mathbb{C}^g$). Let us assume that there is an Abelian function $\mathfrak{A}$ in $\mathbb{C}^g$ with a lattice of periods containing L. Then there are two entire functions Θ_1 and Θ_2 in $\mathbb{C}^g$ that are relatively prime and satisfy functional equations of Picard type:

$$\Theta_1(\vec{z} + \lambda) = \Theta_1(\vec{z}) \exp (w_\lambda(\vec{z}))$$
$$\Theta_2(\vec{z} + \lambda) = \Theta_2(\vec{z}) \exp (w_\lambda(\vec{z})) \tag{2.2}$$

where $w_\lambda(\vec{z})$ is a linear function in $\vec{z}$, for all $\vec{z} \in \mathbb{C}^g$ and $\lambda \in L$. The function $\mathfrak{A}(\vec{z})$ is the ratio of $\Theta_1(\vec{z})$, $\Theta_2(\vec{z})$:

$$\mathfrak{A}(\vec{z}) = \frac{\Theta_1(\vec{z})}{\Theta_2(\vec{z})}$$

The structure of linear functions $w_\lambda(\vec{z})$ is governed by a simplectic structure on V in the following way. There exists alternating $\mathbb{R}$-valued, $\mathbb{R}$-bilinear form $B(x,y)$ on $V \times V$ such that $B(x,y)$ takes integer values on $L \times L$. This bilinear form $B(x,y)$ has several representations. First of all

$$B(x,y) = k(x,y) - k(y,x) \tag{2.3}$$

for all $x,y \in V$, where $k: V \times V \to \mathbb{C}$ is a $\mathbb{R}$-bilinear form, which is $\mathbb{C}$-linear in the second variable. Linear function $w_\lambda(\vec{z})$ can be represented in the following form

$$w_\lambda(\vec{z}) = 2\pi\sqrt{-1}(k(\lambda,\vec{z}) + f(\lambda)) \tag{2.4}$$

for $\vec{z} \in \mathbb{C}^g$ (= V) and $\lambda \in L$, where $f(\lambda)$ satisfy the following compatibility conditions

$$f(\lambda_1 + \lambda_2) - f(\lambda_1) - f(\lambda_2) \equiv k(\lambda_2,\lambda_1) \pmod{\mathbb{Z}} \tag{2.5}$$

for all $\lambda_1,\lambda_2 \in L$.

The bilinear form $B(x,y)$ is also generated by a $\mathbb{C}$-valued Hermitean form $h(x,y)$ on $V \times V$, such that $h(\lambda_1,\lambda_2) - h(\lambda_2,\lambda_1) \in 2\sqrt{-1}\,\mathbb{Z}$ for all $\lambda_1,\lambda_2 \in L$. Namely,

$$h(x,y) = B(x,Jy) + JB(x,y)$$

or (2.6)

$$B(x,y) = \mathrm{Im}(h(x,y))$$

DEFINITION 2.7. An entire function $\Theta(\vec{z})$ in $\mathbb{C}^g$ is called a θ-function if it satisfies functional relation (2.2)

$$\Theta(\vec{z} + \lambda) = \Theta(\vec{z})\exp(w_\lambda(\vec{z}))$$

for a linear function $w_\lambda(\vec{z})$ defined in (2.4) and satisfying (2.3) and (2.5).

Equations (2.2) defining θ-functions can be represented in a form more convenient for expression in terms of line bundles. For this one defines a multiplicator $\rho(\lambda)$ satisfying the following relations

$$|\rho(\lambda)| = 1$$

$$\rho(\lambda_1)\rho(\lambda_2) = \rho(\lambda_1 + \lambda_2)\exp\{\pi\sqrt{-1}\ B(\lambda_1,\lambda_2)\} = \pm\ \rho(\lambda_1 + \lambda_2)$$

for all $\lambda_1, \lambda_2 \in L$.

Then, after multiplication by a natural exponential factor, $\Theta(\vec{z})$ is transformed into canonical θ-function $\Theta_0(\vec{z})$ satisfying the functional equation corresponding to Hermitean form $h(x,y)$ and multiplicator $\rho(\lambda)$:

$$\Theta_0(\vec{z} + \lambda) = \rho(\lambda)\exp\{\pi h(\lambda,\vec{z}) + \frac{\pi}{2}h(\lambda,\lambda)\}\Theta_0(\vec{z}) \tag{2.8}$$

for all $\vec{z} \in \mathbb{C}^g$ and $\lambda \in L$.

The relation between Hermitean form $h(x,y)$ and bilinear form $B(x,y)$ is given by (2.6). The multiplicator $\rho(\lambda)$ is again expressed in terms of previously introduced quantities

$$\rho(\lambda) = \exp\{2\pi\sqrt{-1}(c_\lambda - d(\lambda))\} \tag{2.9}$$

where

$$c_\lambda = f(\lambda) - \frac{1}{2}\ k(\lambda,\lambda) \tag{2.10}$$

and $d(\lambda) = g(i\lambda) + ig(\lambda)$, where $g(\lambda)$ is a $\mathbb{R}$-linear form on $V\ (= \mathbb{C}^g)$ and $g(\lambda) = \operatorname{Im} c_\lambda$ for $\lambda \in L$.

This way of introduction of Abelian functions using proposition 2.1 is very convenient since it enables us to substitute the verification of Abelian function properties by a simple check of quasiperiodicity equations (2.2) or (2.8). We use these approach in order to find S-matrices that are expressed in terms of Abelian functions. Instead of Abelian functions we speak about elements of S-matrix as being entire θ-functions. The structure of S-matrix and θ-functions that are its elements can be found, if one writes equations of Picard type on S-matrices.

Picard type of equations on S-matrices $R(\theta_1 - \theta_2)$ can be found using simple conditions of invariance of S-matrices under translations by elements of the lattice L.

For this we briefly explain which elementary transformations do not change factorization and unitarity properties [4] of S-matrices.

We use the geometric interpretation of the S-matrix $R(\theta_1 - \theta_2)$ from §1 as an $e^2 \times e^2$ matrix generating an isomorphism of tensor product of vector spaces.

$$R(\theta_1 - \theta_2): V(\theta_1) \otimes V(\theta_2) \to V(\theta_2) \otimes V(\theta_1) \tag{2.11}$$

where the vector space $V(\theta)$ of the dimension e is generated by $\vec{A}(\theta) = (A_1(\theta),\dots,A_e(\theta))$.

E.g. making the simultaneous change $\vec{A}^t(\theta) = B.\vec{A}^t(\theta)$ of the bases of all vector spaces $V(\theta)$, we obtain very simple

LEMMA 2.12. If S-matrix $R(\theta_1 - \theta_2)$ satisfies unitarity and factorization equations, then for any constant, invertible $e \times e$ matrix B and any scalar linear function $v(\theta_1 - \theta_2)$ in $\theta_1 - \theta_2$, the S-matrix

$$e^{v(\theta_1-\theta_2)} \cdot (B \otimes B)\cdot R(\theta_1 - \theta_2)\cdot (B \otimes B)^{-1} \tag{2.13}$$

again satisfies factorization and unitarity equations.

Here θ_1, θ_2 are now considered as vector variables from $\mathbb{C}^g$; naturally $v(\theta_1 - \theta_2)$ is considered as a linear functional of $\theta_1 - \theta_2$ on $\mathbb{C}^g$ and, moreover, vanishing when $\theta_1 = \theta_2$.

Lemma 2.12 is sufficient in order to construct S-matrices satisfying quasiperiodicity conditions of Picard type (i.e. built from θ-functions). We take the lattice L in $V = \mathbb{C}^g$ and consider those S-matrices $R(\theta)$ $(= R(\theta_1 - \theta_2))$ for $\theta \in V$ that satisfy the following

ASSUMPTION 2.14. For any $\lambda \in L$ one has quasiperiodicity relations

$$\begin{aligned} R(\theta + \lambda) &= e^{v_\lambda(\theta)} (B_\lambda \otimes \mathrm{I\!I}) R(\theta) (B_\lambda \otimes \mathrm{I\!I})^{-1} \\ &= e^{v_\lambda(\theta)} (\mathrm{I\!I} \otimes B_\lambda)^{-1} R(\theta) (\mathrm{I\!I} \otimes B_\lambda) \end{aligned} \tag{2.15}$$

for $\theta \in V(= \mathbb{C}^g)$.

The geometric interpretation of assumption 2.14 is very clear and was outlined in the introduction. It means that translations by elements of L does not change the nature of S-matrix $R(\theta)$. Indeed transformation (2.13) together with interpretation (2.11) mean that the linear map (2.11) is unchanged; only the basis, in which $R(\theta)$ is written, changes. In this sense (2.15) means that the structure of $R(\theta)$ is invariant under translations by all elements of L.

The consistency relations between representations (2.15) at once lead to the Proposition 2.1.

Indeed, one finds in the notations (2.15) that

$$v_\lambda(\vec{\theta}) = 2\pi\sqrt{-1}(k(\lambda, \vec{\theta}) + f(\lambda)) \tag{2.16}$$

and for all $\lambda, \lambda' \in L$:

$$B_{\lambda+\lambda'} = B_\lambda \cdot B_{\lambda'} \cdot \exp\{2\pi\sqrt{-1}\ k(\lambda, \lambda')\} \cdot \gamma_{\lambda,\lambda'} \tag{2.17}$$

for some constants $\gamma_{\lambda,\lambda'}$. Simple homological considerations show that in (2.17) constants $\gamma_{\lambda,\lambda'}$ can be taken as units. One sees that (2.17) gives rise to representation (2.4) with functional equation (2.15). These matrices (2.17) are constructed explicitly below in §§3-4 using Cartier exposition of representation theory approach to Picard relations [1].

§3.

Here and in §4 we again introduce θ-functions by means of quasi-periodicity relations. This time it is done using the properties of representations of Heisenberg group G, cf. [1] or [8]. Using simple definition of an induced representation of G, we arrive to our main objects that are: lattice L, the one complementary to it L', sequence of integers $e_1,\dots,e_g$ giving a polarization of $\mathbb{C}^g/L$ and the family of operators A_λ, commuting with $\pi(G)$ and satisfying Weyl commutation relations. These objects are used in §5 to express completely X-symmetric S-matrices canonically associated with Abelian varieties.

We start with an even-dimensional over $\mathbb{R}$, vector space V of the dimension $2g$ and a nondegenerate alternating bilinear form $B(x,y)$ on $V \times V$. E.g. we can consider a symplectic basis for V with respect to $B(x,y)$:

$$\{P_1,\dots,P_g,Q_1,\dots,Q_g\}$$

such that

$$B(P_i,Q_j) = \delta_{ij}; \quad B(P_i,P_j) = B(Q_i,Q_j) = 0$$

for all $i,j = 1,\dots,g$.

The main object is a group G, which is a set of pairs $(t,v) \in \mathbb{R} \times V$ with the multiplication

$$(t,v)\cdot(t',v') = (t + t' + \frac{1}{2}B(v,v'),v + v')$$

The Lie algebra of G is denoted by $\tilde{g}$. We can naturally imbed V into $\tilde{g}$ by identifying v with the element (o,v). We denote (o,v) by e^v following the relation between the Lie group and the Lie algebra. Naturally the element (t,o) is denoted by $i(t)$ or e^{tz} with $z = (1,0)$. We have the main commutation relations in $\tilde{g}$:

$$[v,v'] = B(v,v')\cdot z, \quad [z,v] = 0$$

for $v, v' \in V$. In the symplectic basis $\{P_1,\ldots,P_g,Q_1,\ldots,Q_g\}$, we obtain a basis $\{z,P_1,\ldots,P_g,Q_1,\ldots,Q_g\}$ of $\tilde{g}$ with the only nonzero relations between basic elements:

$$[P_j,Q_j] = z: \quad j = 1,\ldots,g$$

We remark that the center and, simultaneously, the commutator subgroup of G is Z being an image of $\mathbb{R}$ under the homomorphism i. The characters of Z are given by the formula

$$\chi_\lambda(i(t)) = \exp(2\pi i\ \lambda t)$$

for $\lambda \in \mathbb{R}$.

Let L be a lattice in V such that $B(\lambda,\mu)$ takes integral values on $L \times L$. By the complementary lattice we understand

$$L' = \{v \in V: B(v,\lambda) \in \mathbb{Z} \text{ for all } \lambda \in L\}$$

Naturally $L' \supset L$. In order to describe index $[L':L]$ we can always choose a symplectic basis of $\tilde{g}$: $\{P_1,\ldots,P_g,Q_1,\ldots,Q_g,z\}$ and positive integers $e_1,\ldots,e_g$ such that

$$L = \{\Sigma_{i=1}^g n_iP_i + \Sigma_{i=1}^g m_iQ_i: n_i \in \mathbb{Z} \quad m_i \equiv 0 (\bmod e_i) \quad i = 1,\ldots,g\} \tag{3.1}$$

and

$$L' = \{\Sigma_{i=1}^{g} n_i P_i + \Sigma_{i=1}^{g} m_i Q_i : m_i \in \mathbb{Z} \qquad (3.2)$$
$$e_i n_i \equiv 0 \pmod 1$$
$$i = 1,\ldots,g\}$$

Of course, $[L':L] = (e_1 \ldots e_g)^2$.

Following standard procedure [19] one takes the symplectic basis $\{P_1,\ldots,P_g,Q_1,\ldots,Q_g\}$ in such a way that e_i divides e_{i+1}: $i = 1,\ldots,g-1$. This is the only restriction we impose so far on e_i.

We need a real valued function $F(\lambda)$ defined on L and satisfying the following congruence

$$F(\lambda+\mu) \equiv F(\lambda) + F(\mu) + mB(\lambda,\mu) \pmod 2 \qquad (3.3)$$

for $m \geq 1$ and $\lambda,\mu \in L$.

One of the solutions of this functional equation has the form

$$F(\lambda) = \pm m(n_1 m_1 + \ldots + n_g m_g) + a_1 n_1 + \ldots + a_g n_g \qquad (3.4)$$
$$+ e_1^{-1} b_1 m_1 + \ldots + e_g^{-1} b_g m_g$$

for $a_i, b_i \in \mathbb{R} \pmod 1$.

The main vector space we are looking at is the following one denoted by Cartier as $H_{L,m,F}$. This is the space of functions Φ on V that are Borel measurable on V; have the finite norm $\int_P |\Phi(v)|^2 dv$ for a fundamental domain P of L in V and satisfying the functional equations

$$\Phi(v + \lambda) = \exp(2\pi i(\tfrac{1}{2} F(\lambda) + \tfrac{m}{2} B(v,\lambda)) \cdot \Phi(v)$$

for $v \in V$, $\lambda \in L$.

We define an action of G in $H_{L,m,F}$ as follows

$$(U_{v_1}\Phi)(v) = \Phi(v + v_1)\cdot\exp(2\pi i \frac{m}{2} B(v,v'))$$

where $U_{v_1} = \pi(e^{v_1})$ is an operator on $H_{L,m,F}$ corresponding to an element e^{v_1} of G.

A very interesting set of operators is given by operators $A_{\lambda'}$ for $\lambda' \in L'$. These are operators commuting with all operators from $\pi(G)$. For simplicity's sake we put here and everywhere below $m = 1$. The operators $A_{\lambda'}$ are defined as translation operators using the formula

$$(A_{\lambda'}f)(g) = f(e^{\lambda'}\cdot g)$$

for $\lambda' \in L'$. In other words, $A_{\lambda'}$ are defined on $H_{L,1,F}$ as

$$(A_{\lambda'}\Phi)(v) = \exp(\pi i\ B(\lambda',v))\Phi(v + \lambda')$$

for $\Phi \in H_{L,1,F}$. By the definition of $H_{L,1,F}$ we have

$$A_{\lambda} = \exp(\pi i\ F(\lambda))\cdot I \quad \text{for} \quad \lambda \in L$$

LEMMA 3.5. Let S be any set of representatives of the cosets $L' : L$, i.e. $|S| = e^2$ for $e = e_1 \ldots e_g$. Then operators $A_s : s \in S$ form a basis of the algebra of all operators in $H_{L,1,F}$ commuting with $\pi(G)$. The operators $A_{\lambda'}$ satisfy the following commutation relations,

$$A_{\lambda'}\cdot A_{\mu'} = \exp(\pi i\ B(\lambda',\mu'))A_{\lambda'+\mu'} \tag{3.6}$$

for $\lambda',\mu' \in L'$.

REMARK 3.7. The quantity $\exp(\pi i(B(\lambda',\mu')))$ is a root of unity of degree not more than $2e$.

The main result about the representations of G (or, at least, one of the main results) is given by the following fine statement [1]. First, let us make a few comments about notations. Let $(\pi,\mathcal{H})$ be any representation of G such that the function $\Phi_{a,b}(g) = (a,\pi(g)b)$ is continuous on G, and we denote by $\mathcal{H}_\infty$ the subspace of those $a \in \mathcal{H}$ that $\varphi_{a,b}$ is of the C^∞-class for any $b \in \mathcal{H}$. By $\mathcal{H}_{-\infty}$ we denote the set of all continuous antilinear forms on $\mathcal{H}_\infty$ together with the natural imbedding of $\mathcal{H}$ into $\mathcal{H}_\infty$.

THEOREM 3.8. Let L be any lattice in V such that B takes integral values on $L \times L$, and let F be any solution of the functional equation (3.3) with $m = 1$. Let L' be the lattice complementary to L, and put $[L':L] = e^2$. Finally, let $(\widetilde{w},\mathcal{H})$ be any irreducible representation of G such that

$$\widetilde{w}(i(t)) = \exp(2\pi i\ t)\cdot I$$

for all real t. Then we have an induced representation $D_{L,F}$ of G which is isomorphic to e copies of $(\widetilde{w},\mathcal{H})$.

The set of solutions of the equations

$$\widetilde{w}(e^\lambda)\cdot t = \exp(\pi i\ F(\lambda))\cdot t$$

for $\lambda \in L$, form an e-dimensional subspace of $\mathcal{H}_{-\infty}$.

We explain in detail what an induced representation $D_{L,F}$ of G means. In general, if H is a closed subgroup of G, and χ is a character on H, we define a Hilbert space $\mathcal{H}_\chi$ consisting of all functions f on G satisfying the following conditions:

i) f is Borel measurable on G;

ii) $f(hg) = \chi(h)\cdot f(g)$ for g in G and $h \in H$;

iii) $\int |f(g)|^2 dg$ is finite.

With every g in G there is associated a unitary operator $\pi_\chi(g)$ on $\mathcal{H}_\chi$ by

$$(\pi_\chi(g)\cdot f)(g') = f(g'g)$$

The pair $(\pi_\chi, \mathcal{H}_\chi)$ is a representation on G, called the representation induced by the character χ on H. The representation $D_{L,F}$ defined above is a particular case of this construction. Namely, the role of H is played by a subgroup Γ_L of elements of the form $i(t)e^\lambda$ for $t \in \mathbf{R}$, $\lambda \in L$ and the character $\psi_{m,F}$ is of the natural form (an extension of the character from $\mathbb{Z}$):

$$\psi_{m,F}(i(t)\cdot e^\lambda) = \exp(2\pi i\ mt)\exp(\pi i\ F(\lambda))$$

It should be noted, and this is rather important, that the algebra of operators on $\mathcal{H}_{L,1,F}$ commuting with $\pi(\Gamma_L)$ is naturally isomorphic to the algebra of all $e \times e$ matrices, and the representation $D_{L,F}$ above splits into e components.

We are left only with the definition of some irreducible representation of G. First of all the classification of all representations of G, according to the theorem of von Neumann-Stone, leaves us with only two possibilities. There is only one (up to unitary equivalence), irreducible representation of G which is nontrivial on the center Z of G. There are also irreducible representations trivial on the center Z. These are one dimensional representations given by characters of G:

$$\tilde{w}_u(t,v) = \exp(2\pi i\ B(u,v))$$

for a fixed $u \in V$. In order to get a self contained description of the induced representation $D_{L,F}$ we need to describe an example of a single irreducible (Fock) representation of G nontrivial on the center Z[1].

In order to define the Fock representation it is necessary to introduce on V a complex structure.

We denote by V_J, the complex vector space having V as the underlying real space in which J is scalar multiplication by i. On V_J there is a unique hermitian form h having B as an imaginary part, i.e.

$$h(v,v') = B(v,Jv') + i\cdot B(v,v')$$

Now we consider the Hilbert space F_J consisting of functions holomorphic on V_J. Let us define the Frechet derivative by a formula

$$\theta_x f(v) = \lim_{t\to 0} \frac{1}{t}[f(v + tx) - f(v)]$$

Then F_J consists of all C^∞ functions Φ on V satisfying the properties

$$\theta_{Jx}\Phi = i\cdot\theta_x\Phi \quad \text{for every } x \text{ in } V$$

and

$$\int_V e^{-\pi\lambda h(v,v)} |\Phi(v)|^2 dv < \infty$$

The first of these conditions means that Φ is holomorphic on V_J. The scalar product is defined by the formula

$$(\Phi,\Phi') = \int_V e^{-\pi\lambda H(v,v)} \Phi(v)\overline{\Phi'(v)}dv$$

Now the representation $\tilde{w}_J$ is given by the formula

$$\tilde{w}_J(i(t)e^v) = \exp(2\pi i\ \lambda t)\cdot U_v$$

with

$$(U_v\Phi)(v_1) = e^{-\pi\lambda[h(v,v)/2 + h(v,v_1)]}\Phi(v + v_1)$$

The infinitesimal representation $\tilde{w}'_J$ associated with $\tilde{w}_J$ is given by

$$\tilde{w}'_J(x)\cdot\Phi = \theta_x\cdot\Phi - \pi\lambda\, h_x\Phi$$

The properties of the Fock representation defined here are summarized in the following [1]:

THEOREM 3.9. Let J be any operator defined above and $\lambda \neq 0$ be a real number, then

i) The Fock representation $(\tilde{w}_J, F_J)$ is irreducible,

ii) If $(\tilde{w}, \mathcal{H})$ is any irreducible representation of G which is nontrivial on the center Z of G, then the vectors in $\mathcal{H}_{-\infty}$, annihilated by $\tilde{w}'(V)$ form a one dimensional subspace of $\mathcal{H}_{\infty}$.

Here V is the real underlying subspace of V_J.

EXAMPLE 3.10. We can give an immediate interpretation of the Fock representation in terms of complex variables. Let us choose a complex basis $P_1,\ldots,P_g$, of V_J such that $h(P_i,P_j) = \delta_{ij}$, and let us put $Q_i = J\cdot P_i$. Then $\{z,P_1,\ldots,P_g,Q_1,\ldots,Q_g\}$ forms a basis of $\tilde{g}$. We can denote by $z_1,\ldots,z_g$ complex variable (charte) on V_J corresponding to $P_1,\ldots,P_g$ (i.e. V_J is identified with $\mathbb{C}^g$). Then monomials

$$M_\alpha = \lambda^{n/2}\prod_{j=1}^{n}\frac{(\pi\lambda)^{\alpha_j/2}}{(\alpha_j!)^{1/2}}\, z_j^{\alpha_j}, \quad \alpha = (\alpha_1,\ldots,\alpha_g)$$

form an orthonormal basis of F_J. Then the infinitesimal operator $\tilde{w}'_J(P_j - iQ_j)$ is twice the derivative with respect to the complex variable z_j, and $\tilde{w}'_J(P_j + iQ_j)$ is a multiplication by $-2\pi\lambda\, z_j$.

The formalism presented above gives, for example, all θ-functions at once. Let us take a scalar $\lambda \neq 0$ in the definition of the Fock representation to be equal to one. We take the Fock representation $(\tilde{w}_J, F_J)$. Then the direct sum of e copies of the Fock representation generates an induced representation $D_{L,F}$. We obtain henceforth an e-dimensional subspace Θ_L of $(F_J)_{-\infty}$ consisting of functions θ satisfying an equation

$$\tilde{w}_J(e^{\lambda})\cdot\theta = \exp(\pi i\ F(\lambda))\theta \quad \text{for all} \quad \lambda \in L$$

This equation has the usual form of an equation defining θ-functions:

$$\theta(v) \equiv \theta(v+\lambda)\exp\{-\pi[\tfrac{1}{2}\ h(\lambda,\lambda) + h(\lambda,v) + iF(\lambda)]\}$$

The set of such θ-functions is, thus, e-dimensional.

§4.

Our main object is the family of operators $A_{\lambda'}$: $\lambda' \in L'$ satisfying the commutation relations

$$A_{\lambda'}\cdot A_{\mu'} = \exp(\pi i\ B(\lambda',\mu'))A_{\lambda'+\mu'} \tag{4.1}$$

for $\lambda',\mu' \in L'$ and

$$A_{\lambda} = \exp(\pi i\ F(\lambda))\cdot I \quad \text{for} \quad \lambda \in L \tag{4.2}$$

First of all, we obtain from this family of operators another family of operators satisfying certain norming conditions. We define "normed" operators $F_{\lambda'}$ as

$$F_{\lambda'} = \exp(-\pi i\ F(\lambda'))\cdot A_{\lambda'} \tag{4.3}$$

for $\lambda' \in L'$. Here $F(\lambda')$ can be chosen as the canonical solution of the functional equation (3.3), e.g. of the form

$$F_0(\lambda') = -\Sigma_{i=1}^{g} n_i m_i \tag{4.4}$$

for a canonical symplectic basis $\{P_1,\ldots,P_g,Q_1,\ldots,Q_g\}$ corresponding to $B(x,y)$ and

$$\lambda' = \Sigma_{i=1}^{g}(n_i P_i + m_i Q_i) \in V$$

Here for $\lambda' \in L'$; $e_i n_i \equiv 0 \pmod 1$, $m_i \equiv 0 \pmod 1$, $i = 1,\ldots,g$. The operators $F_{\lambda'}$ satisfy functional equations that can be derived from (4.1). To describe them we use the bilinear form $k(x,y)$ from §2 (2.3). We put for $x,y \in V$:

$$k(x,y) = \frac{1}{2}(B(x,y) - F_0(x) - F_0(y) + F_0(x + y)) \tag{4.5}$$

In particular, for $x,y \in V$,

$$B(x,y) = k(x,y) - k(y,x) \tag{4.6}$$

and for $\lambda = \Sigma_{i=1}^{g}(n_i P_i + m_i Q_i)$, $\lambda' = \Sigma_{i=1}^{g}(n_i' P_i + m_i' Q_i)$ we have

$$k(\lambda,\lambda') = -\Sigma_{i=1}^{g} n_i' m_i \tag{4.7}$$

Consequently, the operators $F_{\lambda'}$: $\lambda' \in L'$ defined by (4.3), satisfy the fundamental functional relations:

$$F_{\lambda'}F_{\mu'} = \exp\{2\pi i\ k(\lambda',\mu')\}F_{\lambda'+\mu'} \tag{4.8}$$

for all $\lambda',\mu' \in L'$ and

$$F_\lambda = I \quad \text{for} \quad \lambda \in L$$

In particular, we have the following commutation relations of the Weyl type

$$F_{\lambda'} \cdot F_{\mu'} = \exp\{2\pi i\, B(\lambda',\mu')\} F_{\mu'} \cdot F_{\lambda'} \tag{4.9}$$

for $\lambda',\mu' \in L'$. It is known that the family of operators $F_{\lambda'}$ where λ' runs over all representatives of cosets $L':L$, is isomorphic to an algebra of $e \times e$ matrices. We present now another representation for this family of operators, following the suggestion of L. Auslander and R. Tolimieri, [5], for $g = 1$, as $e \times e$ matrices.

We consider a vector space of dimension e to be a tensor product of g vector spaces $\mathbb{C}^{e_i}$ of dimensions e_i: $i = 1,\dots,g$ and $e_1 \dots e_g = e$. We take standard basis $f_1^{(i)},\dots,f_{e_i}^{(i)}$ in $\mathbb{C}^{e_i}$ and define a basis in

$$\mathbb{C}^e \simeq \otimes_{i=1}^g \mathbb{C}^{e_i}$$

as follows. For $F \in \Pi_{i=1}^g \mathbb{Z}/\mathbf{Z}_{e_i}$ we put

$$f_F = f_{F(1)}^{(1)} \otimes \dots \otimes f_{F(g)}^{(g)} = \otimes_{i=1}^g f_{F(i)}^{(i)}$$

With this choice of basis in $\mathbb{C}^e$, we define $2g$ linear operators A_i, B_i: $i = 1,\dots,g$ in the following way:

$$A_i \cdot f_F = \exp(2\pi\sqrt{-1}\,\frac{F(i)}{e_i}) f_F$$

and

$$B_i \cdot f_F = f_{T_i F},$$

where $(T_i F)(j) = F(j)$ if $i \neq j$ and $(T_i F)(i) = F(i) + 1 \pmod{e_i}$.

The operators A_i, B_i satisfy the basic relations

$$A_i^{e_i} = B_i^{e_i} = 1$$

$$B_i A_i = \exp(-2\pi\sqrt{-1}/e_i) A_i B_i \tag{4.10}$$

for $i = 1,\ldots,g$, and operators A_i, B_j commute for $i \neq j$; and

$$[A_i, A_j] = 0, \quad [B_i, B_j] = 0$$

$i,j = 1,\ldots,g$.

Then the algebra generated by operators A_i, B_j is isomorphic to $M_e(\mathbb{C})$. At the same time the operators A_i and B_i generate the canonical operators $F_{\lambda'}$. Let $\lambda' = \Sigma_{i=1}^g n_i P_i + \Sigma_{i=1}^g m_i P_i \in L'$. Then $m_i \equiv e_i$, $n_i \equiv 0 \pmod 1$ for $i = 1,\ldots,g$. One can define henceforth:

$$F_{\lambda'} = \Pi_{i=1}^g A_i^{e_i n_i} \cdot \Pi_{i=1}^g B_i^{m_i} \tag{4.11}$$

We take as before:

$$F_0(\lambda') = -\Sigma_{i=1}^g n_i m_i \tag{4.12}$$

Then we have the validity of the fundamental relations (4.8) defining the operators $F_{\lambda'}$ in the representation (4.11) above. They have the following form

$$F_{\lambda'} \cdot F_{\mu'} = \exp\{-2\pi\sqrt{-1} \cdot \Sigma_{i=1}^g n_i' m_i\} F_{\lambda'+\mu'} \tag{4.13}$$

which is identical to

$$F_{\lambda'} \cdot F_{\mu'} = \exp\{2\pi i \; k(\lambda', \mu')\} F_{\lambda'+\mu'}$$

if

$$\lambda' = \Sigma_{i=1}^g (n_i P_i + m_i Q_i) \quad \text{and} \quad \mu' = \Sigma_{i=1}^g (n_i' P_i + m_i' Q_i)$$

belong to L'.

In the future we will need certain hermitian properties of the operators A_i, B_i. These properties can be established without any difficulty starting from the definition of operators A_i and B_i:

$$A_i^+ = A_i^{-1} = A_i^{n-1} \tag{4.14}$$

$$B_i^+ = B_i^{-1} = B_i^{n-1} \tag{4.15}$$

$i = 1,...,g$. Hence we have the following very useful relation:

$$F_{\lambda'}^+ = F_{-\lambda'} \cdot \exp\{2\pi i\ k(\lambda',\lambda')\} \tag{4.16}$$

for any $\lambda' \in L'$.

This property of $F_{\lambda'}$ will be very important in the future.

Let us present a canonical way of writing representatives of cosets L':L. We denote them by the same letter $\lambda' \in L'$ (mod L) and write elements of L'/L as

$$\lambda' = \Sigma_{i=1}^{g} (n_i P_i + m_i Q_i)$$

for

$$n_i = \frac{\overline{n_i}}{e_i} \in \{0,1,...,e_i - 1\}, \quad m_i \in \{0,1,...,e_i - 1\}$$

and $i = 1,...,g$. These elements we denote as elements of L'/L with addition (mod L).

§5.

According to our result [4] any "nontrivial" factorized S-matrix is expressed in terms of Abelian (or θ-functions) or their natural degenerations. We present below explicit formulae for S-matrices expressed in terms of θ-functions provided these S-matrices satisfy conditions of complete X-symmetry.

For this we consider an index set $\{1,\ldots,e\}$ as a multiplet X equipped with a group structure of finite abelian group [3], [4]. In this case X can be represented as a finite direct sum of cyclic groups

$$X \cong \mathbb{Z}/\mathbb{Z}e_1 \oplus \ldots \oplus \mathbb{Z}/\mathbb{Z}e_g \tag{5.1}$$

for a sequence of integers $e_1,\ldots,e_g$ and $|X| = e = e_1 \ldots e_g$. The S-matrix $R(\theta_1 - \theta_2)$ according to its geometric sense (1.1) is represented by $X^2 \times X^2$ matrix and can be written as a linear combination of tensor products of elementary $X \times X$ matrices.

We can identify X with $\oplus_{i=1}^{g} \mathbb{Z}/\mathbb{Z}e_i$ and take in $\mathbb{C}^X \cong \mathbb{C}^e$ the basis f_F: $F \in X$ introduced above. Then the S-matrices (quantum and semiclassical ones) are naturally represented as linear combinations of tensor products of elementary $X \times X$ matrices in the basis f_F: $F \in X$. Let us write the elements of the matrix $R(\theta_1,\theta_2)$ in the form

$$R(\theta_1,\theta_2)_{xy,zv} = S(\theta_1,\theta_2)_{xy}^{zv} \tag{5.2}$$

Then matrix $R(\theta_1,\theta_2)$ has the form

$$R(\theta_1,\theta_2) = \Sigma_{x,y,z,v\in X}\, S(\theta_1,\theta_2)_{xy}^{zv} E_{xz} \otimes E_{yv}$$

for elementary matrices $E_{x_1x_2}$ of sizes $X \times X$ with $(E_{x_1x_2})_{y_1y_2} = \delta_{x_1y_1} \cdot \delta_{x_2y_2}$ in the basis f_F: $F \in X$ above. It is quite natnatural to look now not on an arbitrary S-matrix $R(\theta_1,\theta_2)$, but only on S-matrices we call completely X-symmetric [4].

DEFINITION 5.3. An S-matrix $R(\theta_1,\theta_2)$ (not necessarily factorized) is called completely X-symmetric if

$$S_{xy}^{zv} = 0 \quad \text{for} \quad x + y \neq z + v$$
$$S_{xy}^{zv} \equiv S_{x+x_0,y+x_0}^{z+x_0,v+x_0} \quad \text{for any} \quad x_0 \in X \tag{5.4}$$

where the addition of indices is understood according to the operations in the group X.

The system of operators $F_{\lambda'}$ introduced above allows us to express in a very short form the condition of the complete X-symmetry of S-matrix $R(\theta_1,\theta_2)$.

THEOREM 5.5. In order for S-matrix $R(\theta_1,\theta_2)$ to be completely X-symmetric it is necessary and sufficient to have the following form

$$R(\theta_1,\theta_2) = \Sigma_{\lambda'\in L'/L}\, \tau_{\lambda'}(\theta_1,\theta_2)\cdot F_{\lambda'} \otimes F_{-\lambda'} \tag{5.6}$$

for some scalar functions $\tau_{\lambda'}(\theta_1,\theta_2)$: $\lambda' \in L'/L$.

REMARK 5.7. The usual S-matrix depends on $|X|^4$ functional variables. The condition of the complete X-symmetry means that the S-matrix depends only on $|X|^2$ functional variables. In the representation for $R(\theta_1,\theta_2)$ (5.6) one has L'/L functional variables $\tau_{\lambda'}$ and, naturally $|L'/L| = e^2 = |X|^2$. In other words, the number of functional parameters is the same.

Proof of theorem 5.5. According to the results above, the linear space generated by matrices $F_{\lambda'}$: $\lambda' \in L'/L$ is of dimension e^2. This means that any $X \times X$ matrix can be represented as a linear combination of matrices $F_{\lambda'}$. In other words, we have the following representation for an arbitrary S-matrix $R(\theta_1,\theta_2)$

$$R(\theta_1,\theta_2) = \Sigma_{\lambda',\mu'\in L'/L}\, \tau_{\lambda',\mu'}(\theta_1,\theta_2) F_{\lambda'} \otimes F_{\mu'} \tag{5.8}$$

for scalars $\tau_{\lambda',\mu'}(\theta_1,\theta_2):\lambda',\mu' \in L'/L$. If we use the definition of the operator $F_{\lambda'}$, then we obtain the following expression for the action of $F_{\lambda'}$ on basic vectors $f_F : F \in X$. One has

$$\text{for } \lambda' = \Sigma_{i=1}^{g} (n_i P_i + m_i P_i) \in L'/L,$$

$$F_{\lambda'} = \Pi_{i=1}^{g} A_i^{e_i n_i} \Pi_{i=1}^{g} B_i^{m_i}$$

and we put $n_i' = e_i n_i \in \mathbb{Z}: i = 1,\ldots,g$. Then

$$F_{\lambda'} \cdot f_F = (\Pi_{i=1}^{g} A_i^{e_i n_i} \Pi_{i=1}^{g} B_i^{m_i}) \cdot f_F$$

$$= \Pi_{i=1}^{g} \exp\{2\pi\sqrt{-1}(F(i) + m_i)n_i\} f_{\Pi_{i=1}^{g} T_i^{m_i} F}$$

in the notations introduced above. We simplify these notations. For this let us denote by 1_i an element of X corresponding to the generator 1 of $\mathbb{Z}/\mathbb{Z}e_i$ in the decomposition of X into $\mathbb{Z}/\mathbb{Z}e_i \oplus \ldots \oplus \mathbb{Z}/\mathbb{Z}e_g \cong X$. Then we can write:

$$F_{\lambda'} \cdot f_F = \exp\{2\pi\sqrt{-1} \cdot \Sigma_{i=1}^{g} (F(i) + m_i)n_i\} f_{F+\Sigma_{i=1}^{g} m_i 1_i}$$

We remark that $F = \Sigma_{i=1}^{g} F(i) \cdot 1_i$ in $X(1_i \in X, F(i) \in \mathbb{Z})$. Consequently,

$$R(\theta_1,\theta_2)_{xy,zv}$$

$$= \Sigma_{\lambda',\mu' \in L'/L} \tau_{\lambda',\mu'}(\theta_1,\theta_2) \cdot (F_{\lambda'})_{xz} \cdot (F_{\mu'})_{yv}$$

We have

$$(F_{\lambda'})_{F',F} = \delta_{F',F+\Sigma_{i=1}^{g} m_i 1_i} \cdot \exp\{2\pi\sqrt{-1}\ \Sigma_{i=1}^{g} (F(i) + m_i)n_i\}$$

for $\lambda' \in L'/L$ we define $\varphi: \lambda' \to ((\lambda')_1, (\lambda')_2)$, where $(\lambda')_1 = \Sigma_{i=1}^{g} m_i \cdot l_i \in X$ and $(\lambda')_2 = \Sigma_{i=1}^{g} e_i n_i \cdot l_i \in X$ (for $\lambda' = \Sigma_{i=1}^{g}(n_i P_i + m_i Q_i) \in L'/L$). Naturally φ is an isomorphism. For $a = \Sigma_{i=1}^{g} a_i \cdot l_i$, $b = \Sigma_{i=1}^{g} b_i \cdot l_i \in X$ we put $a*b = \Sigma_{i=1}^{g} \frac{a_i b_i}{e_i}$ (mod 1).

In these notations we have

$$(F_{\lambda'})_{a,b} = \delta_{a,b+(\lambda')_1} \cdot \exp\{2\pi\sqrt{-1}(b + (\lambda')_1)*(\lambda')_2\}$$

In other words,

$$\begin{aligned} R(\theta_1, \theta_2)_{xy,zv} \qquad\qquad\qquad (5.9) \\ = \Sigma_{\lambda',\mu' \in L'/L,\, (\lambda')_1 = x-z,\, (\mu')_1 = y-v}\, \tau_{\lambda',\mu'}(\theta_1, \theta_2) \\ \times \exp\{2\pi\sqrt{-1}((z + (\lambda')_1)*(\lambda')_2 + (v + (\mu')_1)*(\mu')_2\} \end{aligned}$$

Let us assume now that S-matrix $R(\theta_1, \theta_2)$ is completely X-symmetric. This means that for any $x,y,z,v \in X$ such that $x + y \neq z + v$ in X we have

$$R(\theta_1, \theta_2)_{xy,zv} = 0$$

One can rewrite this as a system of equations on $\tau_{\lambda',\mu'}(\theta_1, \theta_2)$:

$$\begin{aligned} \Sigma_{\lambda',\mu' \in L'/L,\, (\lambda')_1 = a,\, (\mu')_1 = -b}\, \tau_{\lambda',\mu'}(\theta_1, \theta_2) \qquad\qquad (5.10) \\ \times \exp\{2\pi\sqrt{-1}((c + (\lambda')_1)*(\lambda')_2 + (d + (\mu')_1)*(\mu')_2\} \\ = 0 \end{aligned}$$

for all $a \neq b, c, d \in X$. We can add to this one more series of equations arising from the second of equations (5.4):

$$\Sigma_{\lambda',\mu'\in L'/L,(\lambda')_1=a,(\mu')_1=-b}\tau_{\lambda',\mu'}(\theta_1,\theta_2) \tag{5.11}$$

$$\times\ (\exp\{2\pi\sqrt{-1}((c+(\lambda')_1)*(\lambda')_2 + (d+(\mu')_1)*(\mu')_2)\}$$

$$-\ \exp\{2\pi\sqrt{-1}((c+a+(\lambda')_1)*(\lambda')_2 + (d+a+(\mu')_1)*(\mu')_2\})$$

$$= 0$$

for all $a,b,c,d,e \in X$.

One can show by direct computation, using determinants, that the system of equations (5.10) and (5.11) for $a \neq b$ has a nonzero determinant. In other words if $R(\theta_1,\theta_2)$ is completely X-symmetric, then

$$\tau_{\lambda',\mu'}(\theta_1,\theta_2) = 0$$

if $(\mu')_1 \neq -(\lambda')_1$. The second system of equations (5.11) shows at the same time that

$$\tau_{\lambda',\mu'}(\theta_1,\theta_2) = 0$$

if $(\mu')_2 \neq -(\lambda')_2$. As a result we have

$$\tau_{\lambda',\mu'}(\theta_1,\theta_2) = 0 \ :\ \lambda' + \mu' \neq 0 \tag{5.12}$$

In other words, we have for completely X-symmetric S-matrices the form predicted in the statement of the theorem 5.5. On the other hand, if an S-matrix has the form as in the statement of the theorem 5.5, then by (5.12) the S-matrix is completely X-symmetric.

It is easy to work with factorized S-matrices in the form proposed in theorem 5.5. In these notations the factorization equations can be written in particularly simply form and are presented below in connection with the star-triangle relations.

We emphasize that theorem 5.5 is valid for arbitrary, not necessarily factorized S-matrices. If one looks on "nontrivial" factorized S-matrices, then they are expressed in terms of Abelian functions or, in a unique way, in terms of θ-functions. We can combine in this case, the assumption 2.14 and the statement of this theorem 5.5. One sees then, that the set X is determined by a lattice L in 2g-dimensional space V and by bilinear form $B(x,y)$ on V. We have

$$[L':L] = (e_1 \dots e_g)^2 = e^2$$

and the set $X \times X$ is identified with $[L':L]$.

In assumption 2.14 we naturally take the matrices B_λ to coincide with the matrices F_λ since the functional relations (2.17) for $\gamma_{\lambda,\lambda'} = 1$ are the same.

More careful analysis actually shows that $\gamma_{\lambda,\lambda'}$ can be arbitrary constants satisfying natural consistency conditions. Assumption 2.14 together with representation (5.6) give us the set of functional relations on coefficients $\tau_{\lambda'}$. We assume $\tau_{\lambda'}(\theta_1,\theta_2)$ depend only on $\theta_1 - \theta_2 \in V$ $(= \mathbb{C}^g)$. Let us present the conditions of Assumption 2.14 in a more convenient form

$$R(\theta + \lambda) = e^{v_\lambda(\theta)}(B_\lambda \otimes \mathbb{I})R(\theta)(B_\lambda \otimes \mathbb{I})^{-1} \tag{5.13}$$

$$= e^{v_\lambda(\theta)}(\mathbb{I} \otimes B_\lambda)^{-1}R(\theta)(\mathbb{I} \otimes B_\lambda)$$

Let $v_\lambda(\theta) = 2\pi\sqrt{-1}k(\lambda,\theta)$ and $B_\lambda = F_\lambda$, where $\lambda \in L'$, $\theta \in V$. Then functional equations on $\tau_\lambda(\theta_1 - \theta_2) = \tau_\lambda(\theta)$: $\lambda \in L'/L$ have the following form

$$\tau_\lambda(\theta + \mu) = e^{2\pi\sqrt{-1}k(\mu,\theta)} \cdot e^{2\pi i B(\mu,\lambda)} \tau_\lambda(\theta) \qquad (5.14)$$

for $\lambda,\mu \in L'$ and $\theta \in V$ $(= \mathbb{C}^g)$.

These conditions on $\tau_{\lambda'}$ take the classical form of relations of the Picard type on θ-functions associated with lattice L' in V.

Assumption 2.14 on quasiperiodicity of $R(\vec{\theta})$ immediately gives us the representation of scalars $\tau_{\lambda'}(\vec{\theta})$ in terms of θ-functions associated with L' and V. This representation is unambiguous, if on introduces norming assumption on $R(\vec{\theta})$ at $\vec{\theta} = \vec{0}$. We chose $R(\vec{0})$ to be a unit S-matrix $\mathbb{I}^\tau$ [4] defined as

$$(\mathbb{I}^\tau)_{ij,k\ell} = \delta_{i\ell}\delta_{jk}$$

We are led to the following general statement.

CONJECTURE 5.15. If $X^2 \times X^2$ S-matrix $R(\vec{\theta}_1,\vec{\theta}_2)$ which is non-trivial, depends only on $\vec{\theta}_1 - \vec{\theta}_2 \in \mathbb{C}^g$ satisfies the norming condition $R(\vec{0}) = \mathbb{I}^\tau$, and is a completely X-symmetric factorized S-matrix, then it has the form

$$R(\vec{\theta}_1 - \vec{\theta}_2) = \Sigma_{\lambda' \in L'/L}\, \tau_{\lambda'}(\vec{\theta}_1 - \vec{\theta}_2) \cdot F_{\lambda'} \otimes F_{-\lambda'} \qquad (5.16)$$

where $\vec{\theta} \in \mathbb{C}^g$; L is a lattice in $V = \mathbb{C}^g$; $B(x,y)$ is an antisymmetric bilinear form on V, corresponding to L, and L' is the lattice complementary to L. The coefficients $\tau_{\lambda'}(\vec{\theta})$ are ratios of θ-functions corresponding to lattice L'

$$\tau_{\lambda'}(\vec{\theta}) = \frac{\wp_{\lambda'}(\vec{\theta} + \vec{\eta})}{\wp_{\lambda'}(\vec{\eta})} \exp(2\pi i k(\lambda',\lambda')) \qquad (5.17)$$

for a certain fixed vector $\vec{\eta} \in \mathbb{C}^g$.

We can give an explicit expression for the elements $S_{xy}^{zv}(\vec{\theta})$ of the S-matrix $R(\vec{\theta})$ from (5.16)-(5.17). For this we take the

Riemann matrix Ω, corresponding to $B(x,y)$, L' and V, in the reduced form, when $2g$ periods are

$$T_i = (0,\ldots,\underbrace{1}_{i\text{-th}},\ldots,0),\quad T_i^* = (\tau_{1,i},\ldots,\tau_{g,i})$$

$i = 1,\ldots,g$. The generalized θ-functions corresponding to this basis can be written as

$$\theta\left[\begin{matrix}\vec{A}_1\\ \vec{A}_2\end{matrix}\right](\vec{x}) = \Sigma_{\vec{n}\in\mathbb{Z}^g} \exp\{\pi i(\vec{n} + \vec{A}_2)^t B(\vec{n} + \vec{A}_2) + 2(\vec{n} + \vec{A}_2)(\vec{x} + \vec{A}_1)\}$$

for $\vec{A}_1, \vec{A}_2 \in \mathbb{C}^g$, and $g \times g$ matrix B corresponding to T_i^*: $i = 1,\ldots,g$. For $X = \oplus_{i=1}^{g} \mathbb{Z}/\mathbb{Z}e_i$, $|X| = e = e_1 \ldots e_g$ one has the elements $S_{xy}^{zv}(\vec{\lambda})$ of the S-matrix $R(\vec{\lambda})$ (5.16) in the following form:

$$S_{xy}^{zv}(\vec{\lambda}) = \Sigma_{\vec{A}\in\oplus_{i=1}^{g}\mathbb{Z}/\mathbb{Z}e_i} \frac{\theta\left[\frac{\vec{A}}{\overline{(x-y)}}\right](\vec{\lambda} + \vec{\eta})}{\theta\left[\frac{\vec{A}}{\overline{(x-y)}}\right](\vec{\eta})} \tag{5.18}$$

$$\times \exp\{2\pi\sqrt{-1}\cdot\Sigma_{i=1}^{g} \frac{(z_i - y_i)A_i}{e_i}\} \times \delta_{x+y,z+v}$$

with

$$\overline{(x-y)} = \left(\frac{x_1 - y_1}{e_1},\ldots,\frac{x_g - y_g}{e_g}\right)$$

The description of the whole class of completely X-symmetric S-matrices presented in the Conjecture 5.15, is certainly a complete description of those completely X-symmetric S-matrices that satisfy quasiperiodic properties of the Picard type.

It is by no means obvious why the S-matrix described in the Conjecture 5.15 is a factorized one. Only for $g = 1$ and

$e_1 = 2$ is this factorized S-matrix equivalent to the Baxter factorized S-matrix [2].

Our general results in this direction can be summarized in the following:

PROPOSITION 5.19. The completely X-symmetric S-matrix (5.16)-(5.18) is a factorized and unitary S-matrix when $g = 1$ and $e = e_1$ is an arbitrary integer ≥ 2.

In its original form 5.15 conjectured that a completely X-symmetric S-matrix (5.16) is always factorized. For $g \geq 2$ and $e_1 = \ldots = e_g = 2$ this was asserted in [10]. Attempts to prove it were made by A. Bovier (Bonn University Preprint 1981) and other authors; but computer experiments by authors and others suggest that the factorization equations are not satisfied for general Abelian varieties.

SECTION 2. ISOSPECTRAL DEFORMATION EQUATIONS ASSOCIATED WITH FACTORIZED S-MATRICES

In this part of the paper we study factorized S-matrices from the point of view of two dimensional completely integrable systems. There are many relations between factorized S-matrices and completely integrable systems in two dimensional space-time. First of all, factorized S-matrices arise as a result of computations of S-matrices for completely integrable systems [3], [9], [16]. On the other hand, S-matrices, satisfying factorization equations, themselves give rise to two dimensional completely integral systems in the continuous and lattice cases [2], [3], [4]. Factorized S-matrices determine Hamiltonian structure, conservation laws and quantization of completely integrable systems [3], [4], [6]. In §6 we present the definition of isospectral deformation equations associated with a given S-matrix. Examples of quantized two dimensional isospectral deformation equations and the associated S-matrices are given. We see that there are several entirely different classes of isospectral deformation equations associated with the same S-matrix. There is, however, one canonical class of systems, both continuous and lattice attached to a given S-matrix. We show how to construct these canonical classes of systems that are the best candidates for an explicit Bethe Ansatz (like the eight-vertex model of [2]). For classical systems instead of quantum S-matrices one can use a more simple classical object called factorized semiclassical s-matrix. We discuss semiclassical s-matrices and their relation with classical isospectral deformation equations. Results of Section 1 concerning the explicit construction of factorized S-matrices in terms of θ-functions are used for construction of semiclassical s-matrices expressed in terms of θ-functions. Entirely new class of two dimensional classical completely integrable systems is attached to these semiclassical s-matrices. For an elliptic curve these new systems include elliptic generalizations of principal chiral fields, σ-models as well as elliptic generalization of the Heisenberg spin chain (that, e.g. includes XYZ-model). At the end of the paper lattice models associated with factorized S-matrices are considered.

We propose a new class of models generalizing eight-vertex models for $\mathbb{Z}/\mathbb{Z}e_1 \oplus \ldots \oplus \mathbb{Z}/\mathbb{Z}e_g$ spin systems, where Boltzmann weights are expressed in terms of θ-functions.

§6.

We present below classes of families of isospectral deformation equations with which we can explicitly associate the corresponding S-matrices. We are going to explain exactly this association. First of all let us consider a single spectral problem

$$\frac{d\Phi_\lambda}{dx} = U(x;\lambda)\Phi_\lambda \tag{6.1}$$

for $n \times n$ matrices $U(x;\lambda)$ and Φ_λ having operator entries. These operator entries are considered to be elements of an algebra B of field operators, depending on x. We assume that the dependence of $U(x;\lambda)$ on λ is meromorphic on a Riemann surface Γ. For example Γ can be an open Riemann surface $\mathbb{C}$. The isospectral deformation flows associated with (6.1) can be defined using the monodromy matrix of (6.1). The monodromy matrix is defined through the fundamental solution of (6.1) denoted by $\Phi(x,x_0,\lambda)$ that depend on initial point x_0 and satisfy the norming condition

$$\Phi(x_0,x_0,\lambda) = \mathbb{I} \tag{6.2}$$

This fundamental solution satisfies the equation (6.1)

$$\frac{d}{dx}\Phi(x,x_0,\lambda) = U(x;\lambda)\Phi(x,x_0,\lambda) \tag{6.3}$$

Then the monodromy (scattering) matrix $\mathcal{T}(\lambda)$ can be defined as $\lim_{\substack{x_1\to+\infty \\ x_0\to-\infty}}\Phi(x_1,x_0,\lambda)$, if this expression has sense in view of equation (6.1) or in a similar form $\lim_{\substack{x_1\to+\infty \\ x_0\to-\infty}} A(x_1)\Phi(x_1,x_0,\lambda)A(X_0)^{-1}$ for an appropriate $A(x)$. We consider $\mathcal{T}(\lambda)$ as a generating function for an infinite sequence of commuting Hamiltonians describing the whole class of isospectral deformation equations. Naturally by Complete Integrability Property (C.I. Property) we understand the following commutation relations

$$[\mathrm{Tr}\ \mathcal{T}(\lambda), \mathrm{Tr}\ \mathcal{T}(\mu)] = 0 \tag{6.4}$$

for all λ and μ. Here the Tr is taken over the algebra B, i.e. $\mathrm{Tr}\ \mathcal{T}(\lambda)$ is an element of B. Following Baxter [2], [11] we demand a much stronger property than (6.4). We require the existence of $n^2 \times n^2$ matrix $R(\lambda,\mu)$ with <u>scalar</u> coefficients, nonsingular for generic λ, μ and such that

$$R(\lambda,\mu)(\mathcal{T}(\lambda) \otimes \mathcal{T}(\mu)) = (\mathcal{T}(\mu) \otimes \mathcal{T}(\lambda))R(\lambda,\mu) \tag{6.5}$$

Here the tensor product is again taken over the algebra B. We consider the property (6.5) as the major property and we call the matrix $R(\lambda,\mu)$ the S-matrix of linear problem (6.1).

One can clearly see that the linear problem (6.1) by itself does not define uniquely the notion of S-matrix. What one needs is the definition of an algebra of commutation relations between elements of $U(x;\lambda)$ and $U(y;\mu)$ that generates the structure of B.

We prefer to formulate the Baxter property (6.5) for the fundamental solution $\Phi(x,x_0,\lambda)$ since the S-matrix in (6.5) may be changed by some linear transformation in the infinite volume limit $x,x_0 \to \infty$.

DEFINITION 6.6. Let us consider linear problem (6.1) together with some consistent algebra B_0 of commutation relations between operator entries of $U(x;\lambda)$ and $U(y;\mu)$. We call a scalar $n^2 \times n^2$ matrix $R(\lambda,\mu)$ which is nonsingular for generic (λ,μ), the S-matrix for the system (6.1) and the data B_0 if

$$R(\lambda,\mu)(\Phi(x,x_0,\lambda) \otimes \Phi(x,x_0,\mu)) \tag{6.7}$$

$$= (\Phi(x,x_0,\mu) \otimes \Phi(x,x_0,\lambda))R(\lambda,\mu)$$

The property (6.7) will be referred everywhere as the Baxter property associated with (6.1) and data B.

The definition of the data B_0 and the choice of the S-matrix $R(\lambda,\mu)$ are not independent. More precisely, one determines another.

Neither is $R(\lambda,\mu)$ an arbitrary matrix, nor is B_0 an arbitrary set of relations. Indeed, B_0 should generate a nontrivial Lie algebra. This means that the Jacobi identities should be satisfied. For $R(\lambda,\mu)$ this restriction on B_0 means that $R(\lambda,\mu)$

satisfies the set of equations being equivalent to factorization equations of §1 for S-matrices [3], [4], [10], [16]. However, even the knowledge of B_0 does not amount at once to the existence of Baxter relations (6.7). The reason for nontriviality of (6.7) is the fact that $\Phi(x,x_0,\lambda)$ is, in general, a nonlocal functional of $U(x;\lambda)$. It is not easy (if possible at all) to deduce at once from (6.1) the Baxter relations. In the simplest situations when (6.1) is the Dirac equation, it is possible to use the quantum Gelfand-Levitan equation to write the Neumann series for $\Phi(x.x_0,\lambda)$ and then use the Wick theorem to obtain (6.7). This approach going back to Honerkamp [22] and [23], gives (6.7) after extensive combinatorial considerations. Another approach [11], [18] replaces (6.1) by an approximate lattice equation and then makes use of the original Baxter approach on the lattice [3] in order to obtain (6.7) as an approximate lattice equation. This latter approach, though absolutely nonrigorous, seems to be rather to the point. The famous divergence problems and singular kernels, like δ-functions, immediately disappear when the problem is put on the lattice

We developed however, a completely rigorous mechanism [6],[25], [50],[51] that replaces any continuous spectral problem (6.1) by an exactly equivalent (and approximating it) lattice problem. This canonical correspondence between the continuous and lattice problem is given by Bäcklund-Darboux transformations [6],[25] of the linear problem (6.1) and may be formulated in terms of a simple Riemann boundary value problem [12],[24] or an appropriate infinite-dimensional Lie algebra [50]-[53].

Having replaced the linear problem (6.1) by a lattice one we can easily obtain the Baxter relation and, moreover, the algebra B_0 is generated by the corresponding lattice commutation relations, which certainly do not contain any δ-function.

One can compare this approach developed by us in [6],[25],[50]-[53 with the exact continuous problem (6.1) and the semiclassical analysis can be used. For this one should replace the algebra

B_0 by the system of Poisson brackets relations between the elements of $U(x;\lambda)$ and $U(y;\mu)$. In this case quantum relation (6.7) is replaced by a semiclassical relation [18], [26] and the S-matrix $R(\lambda,\mu)$ by the corresponding semiclassical s-matrix $r(\lambda,\mu)$. We develop this semiclassical approximation in detail below.

§ 7. Some Examples of Simple Factorized S-matrices and Corresponding Classical and Quantum Isospectral Deformation Equations

The simplest S-matrix with rational function elements already provides with a large family of important multicomponent two-dimensional isospectral deformation equations.

This is the S-matrix with "pair-two pairs" interaction

$$S^{zv}_{xy} \neq 0 \text{ if } (x,y) = (z,v)$$
$$\text{or } (x,y) = (v,z)$$

All S-matrices with these properties, satisfying the factorization equations are completely characterized [4].

In the completely X-symmetric case they have the form

$$S^{aa}_{aa}(\lambda) = S(\lambda)$$

$$S^{ab}_{ab}(\lambda) = e^{\tau(b-a)} \frac{sh(\mu\lambda)}{sh(\mu\lambda+\eta)} S(\lambda)$$

$$S^{ba}_{ab}(\lambda) = e^{\{\gamma(a-b)\pm\mu\}\lambda} \cdot \frac{sh\,\eta}{sh(\mu\lambda+\eta)} S(\lambda)$$

for constants μ, η, γ and odd $\tau(x)$. This class of S-matrices contains many interesting examples of primary importance.

EXAMPLE 7.1. Let $X = \mathbb{Z}/\mathbb{Z}2$, then S-matrix $R_1(\lambda)$ is completely X-symmetric and is of the form

$$R_1(\lambda)_{aa}^{bb} = \delta_{ab}$$

$$R_1(\lambda)_{ab}^{ab} = \frac{\operatorname{sh}\lambda}{\operatorname{sh}(\lambda+\eta)}$$

$$R_1(\lambda)_{ab}^{ba} = \frac{\operatorname{sh}\eta}{\operatorname{sh}(\lambda+\eta)}$$

This S-matrix is S-matrix of the following quantum systems:

7.1a. Sin-Gordon equation [11]

$$\varphi_{xt} = \sin\varphi$$

7.1b. Six vertex model or XXZ-model [28].

$$H = -\frac{1}{2}\Sigma_{n=1}^{N}(J\sigma_n^1\sigma_{n+1}^1 + J\sigma_n^2\sigma_{n+1}^2 + J'\sigma_n^3\sigma_{n+1}^3)$$

Moreover, in the case 7.1b. the local transfer matrix $\mathcal{L}_n(\lambda)$ for the problem

$$\psi_{n+1} = \mathcal{L}_n(\lambda)\psi_n$$

repeats the structure of $R_1(\lambda)$ i.e. the factorized S-matrix defines XXZ-model.

As degeneration of this matrix we obtain the following one:

EXAMPLE 7.2. Let $X = \mathbb{Z}/\mathbb{Z}2$. Then the simplest completely X-symmetric S-matrix $R_2(\lambda)$ has the form

$$R_2(\lambda)_{aa}^{bb} = \delta_{ab}$$

$$R_2(\lambda)_{ab}^{ab} = \frac{\lambda}{\lambda+\eta}$$

$$R_2(\lambda)_{ab}^{ba} = \frac{\eta}{\lambda+\eta}$$

This S-matrix is S-matrix of the following models:

7.2a. Nonlinear Schrödinger equation [22], [23], [43]

$$i\varphi_t = \varphi_{xx} + \varphi\varphi^*\varphi$$

7.2b. Heisenberg ferromagnet chain [11], [31]:

$$\frac{\partial \vec{S}}{\partial t} = \vec{S} \times \frac{\partial^2 \vec{S}}{\partial x^2} \quad \text{for} \quad \vec{S} = (S_1, S_2, S_3)$$

and $\|\vec{S}\| = 1$;

7.2c. Toda lattice:

$$\ddot{x}_i = e^{x_{n+1}-x_n} - e^{x_n - x_{n-1}}: \quad n = 1,\dots,N$$

and

$$x_{N+i} \equiv x_i$$

7.2d. XXX-model: the Hamiltonian H from 7.1 b with $J = J'$.

All models 7.2a-7.2c possess multicomponent generalizations having a lot of internal symmetries. These models again have factorized S-matrices, the structure of which repeats the structure of $R_2(\lambda)$.

EXAMPLE 7.3. Let $N \geq 2$ be arbitrary and $X = \mathbb{Z}/\mathbb{Z}N$. Then we have the following completely X-symmetric factorized S-matrix $R_{2,N}(\lambda)$:

$$R_{2,N}(\lambda)_{aa}^{bb} = \delta_{ab}$$

$$R_{2,N}(\lambda)_{ab}^{ab} = \frac{\lambda}{\lambda + \eta}$$

$$R_{2,N}(\lambda)_{ab}^{ba} = \frac{\eta}{\lambda + \eta}: \quad a,b \in X$$

All other elements of $R_{2,N}(\lambda)$ are zeros. Then the S-matrix

$R_{2,2M}(\lambda)$ is an S-matrix for the following multicomponent isospectral deformation (quantum) systems.

7.3a. Matrix nonlinear Schrödinger equation:

$$i\psi_t = \psi_{xx} + \psi\psi^+\psi$$

where ψ is $M \times M$ matrix.

7.3b. For coupled matrix nonlinear Schrödinger equation

$$i\Phi_t = \Phi_{xx} + \Phi\psi\Phi$$

$$-i\psi_t = \psi_{xx} + \psi\Phi\psi$$

where Φ is of the size $N \times M$ and ψ of the size $M \times N$, the S-matrix is

$$R_{2,M+N}(\lambda)$$

7.3c. Generalized Heisenberg chain [12]:

$$S_t = \frac{1}{2i}[S, S_{xx}], S^2 = I, \quad S^+ = S, \quad \text{tr } S = 0$$

7.3d. Matrix generalization of the Toda lattice:

$$\dot{B}_i + A_i B_i - B_i A_{i+1} = 0$$

$$\dot{A}_i = B_i - B_{i-1} : i = 1, \ldots, N$$

where $A_{N+i} \equiv A_i$, $B_{i+N} \equiv B_i$ (Polyakov model of the gauge theory in 1 + 1 dimensions on the lattice [29]).

If matrices A_i, B_i belong to $GL(N,\mathbb{C})$, then the S-matrix of the corresponding model is $R_{2,2N}(\lambda)$.

Naturally, all quantum systems whose Hamiltonian commutes with the one in 7.3a - 7.3d have the same S-matrix. E.g. generalized matrix modified KdV equation

$$\varphi_t = \varphi_{xxx} + 3(\varphi_x \psi \varphi + \varphi \psi \varphi_x)$$

$$\psi_t = \psi_{xxx} + 3(\psi_x \varphi \psi + \psi \varphi \psi_x)$$

possesses the same S-matrix as the nonlinear Schrödinger equation. The same concerns different flows commuting with the generalized Toda chain.

The simple structure of the S-matrix in the cases 7.3a-7.3c enables us to establish Bethe Ansatz for the models 7.3a, 7.3b, 7.3c. It is interesting to note that since the S-matrix $R_{2,N}(\lambda)$ is built from elementary S-matrices $R_2(\lambda)$, the Bethe Ansatz for multicomponent models is reduced to several applications of ordinary Bethe Ansatz in each of these models. For example, for the matrix nonlinear Schrödinger equation and matrices of sizes $M \times N$ we need to apply ordinary Bethe Ansatz $M \times N$ times. For $M = N$ it is enough to apply Bethe Ansatz only N times. Similar phenomenon was observed by Yang in [30], see [43].

For the model 7.3d we don't have Bethe Ansatz even in the scalar case. What S-matrix provides for 7.3d is the quantization and canonical relations (rather nontrivial!) and explicit form of quantum conservation laws. The Bethe eigenfunctions in this case are defined not in terms of elementary functions anymore and can be defined formally as functions in d variables satisfying d linear partial differential equations with rational functions coefficients. If Stokes multiplies are trivial we get the ordinary Bethe Ansatz and representation in terms of exponential functions. For Toda lattice the Stokes multipliers are already nontrivial.

A large class of multicomponent two dimensional isospectral deformation systems has the same factorized S-matrix with "pair-two pairs interaction". For example, principal chiral field equations in SO(N) case and O(N) nonlinear σ-model of Pohlmeyer [32], [33], are of this type. Gross-Neveu [34] model and multicomponent massive Thirring models [22] also have described above S-matrix with "pair-two pairs" interaction.

However the factorization property is not transferable from principal chiral field equations to their reductions or to $\mathbb{C}P^{N-1}$ σ-models. The reason is connected with the significant difference between the symplectic structure of the Hamiltonian system and its reduction. The S-matrix (quantum in the quantum case and semiclassical in the classical case) determines the symplectic structure [26]. However the quantization of the reduced classical system is different from reduction of quantization. This phenomenon is well known even on the classical level when reduced system may not be even completely integrable. Even if the reduced system is completely integrable it can be an entirely different Hamiltonian one. E.g. the KdV equation is a particular case of the coupled modified KdV equation ($\psi = 1$). However the Hamiltonian structure of the KdV equation with the Hamiltonian operator $\partial/\partial x$ is not induced by the Hamiltonian structure of the modified KdV with the Hamiltonian operator of the Darboux type

$$\begin{pmatrix} 0 & 1 \\ 1 & 0 \end{pmatrix}$$

Very interesting examples come from degeneration of Abelian S-matrices, when we put g periods of an Abelian variety $T_i^* \to \infty$: $i = 1,\dots,g$. These S-matrices are expressed in elementary functions and are rational functions of exponentials. Such S-matrices are suspected to be responsible for a very interesting class of two-dimensional completely integrable systems connected with root systems A_n, B_n, C_n etc., like two-dimensional Toda chain (A_n case) [35]

$$\frac{\partial^2}{\partial x \partial t} u_n = e^{u_{n+1}-u_n} - e^{u_n-u_{n-1}}$$

The corresponding S-matrices are by no means obvious and have a few non-zero elements. They are connected with cyclotomic polynomials. We present an example for $X = \mathbb{Z}/\mathbb{Z}3$, where Toda lattice indeed has the corresponding S-matrix as factorized one.

EXAMPLE 7.4 We have for $X = \mathbb{Z}/\mathbb{Z}3$ the following example of $\mathbb{Z}/\mathbb{Z}3$-symmetric but not completely $\mathbb{Z}/\mathbb{Z}3$-symmetric S-matrix $R_4(\lambda)$ which satisfy the factorization and unitarity conditions

$$R_4(\lambda)^{00}_{00} = \mathrm{sh}(3\lambda - 3\eta) - \mathrm{sh}\ 5\eta + \mathrm{sh}\ 3\eta + \mathrm{sh}\ \eta$$

$$R_4(\lambda)^{11}_{11} = R_4(\lambda)^{22}_{22} = \mathrm{sh}(3\lambda - 5\eta) + \mathrm{sh}\ \eta$$

$$R_4(\lambda)^{01}_{01} + R_4(\lambda)^{10}_{10} = R_4(\lambda)^{20}_{20} + R_4(\lambda)^{02}_{02}$$

$$= \mathrm{sh}(2\lambda - 5\eta) + \mathrm{sh}(\eta - 2\lambda) + \mathrm{sh}(\eta - \lambda) + \mathrm{sh}(\lambda - 5\eta)$$

$$R_4(\lambda)^{01}_{01} - R_4(\lambda)^{10}_{10} = R_4(\lambda)^{20}_{20} - R_4(\lambda)^{02}_{02}$$

$$= \mathrm{ch}(2\lambda - \eta) + \mathrm{ch}(\lambda - 5\eta) - \mathrm{ch}(\lambda - \eta) - \mathrm{ch}(2\lambda - 5\eta)$$

$$R_4(\lambda)^{10}_{01} = R_4(\lambda)^{01}_{10} = R_4(\lambda)^{20}_{02} = R_4(\lambda)^{02}_{20}$$

$$= \mathrm{sh}(3\lambda - 3\eta) + \mathrm{sh}\ 3\eta$$

$$R_4(\lambda)^{21}_{12} = R_4(\lambda)^{12}_{21} = \mathrm{sh}(3\lambda - \eta)\ \mathrm{sh}\ \eta$$

$$R_4(\lambda)^{12}_{00} + R_4(\lambda)^{00}_{21} = R_3(\lambda)^{00}_{12} + R_4(\lambda)^{21}_{00}$$

$$= \text{sh}(2\lambda - 4\eta) + \text{sh}(\lambda + 4\eta) - \text{sh}\ \eta - \text{sh}\ 2\eta$$

$$R_4(\lambda)^{12}_{00} - R_4(\lambda)^{00}_{21} = R_4(\lambda)^{00}_{12} - R_4(\lambda)^{21}_{00}$$

$$= \text{ch}(2\lambda - 4\eta) - \text{ch}(\lambda + 4\eta) + \text{ch}\ \lambda - \text{ch}\ 2\lambda$$

$$R_4(\lambda)^{12}_{12} + R_4(\lambda)^{21}_{21} = \text{sh}(2\lambda - 5\eta) - \text{sh}(2\lambda + 3\eta)$$

$$- \text{sh}(\lambda - \eta) + \text{sh}(\lambda + \eta) + \text{sh}(\lambda - 5\eta) - \text{sh}(\lambda - 3\eta)$$

$$R_4(\lambda)^{12}_{12} - R_4(\lambda)^{21}_{21} = \text{ch}(2\lambda + 3\eta) - \text{ch}(2\lambda - 5\eta)$$

$$+ \text{ch}(5\eta - \lambda) - \text{ch}(3\eta - \lambda) - \text{ch}(\lambda - \eta) + \text{ch}(\lambda + \eta)$$

We end this section with a nondegenerate example of an elliptic S-matrix that is probably the most famous one.

Baxter S-matrix [2], [3], corresponds to the case $g = 1$, $e_1 = 2$ (i.e. $X = \mathbf{Z}/\mathbf{Z}2$) of the §5. Then the factorized S-matrix $R_3(\lambda)$ has the form:

$$R_3(\lambda)^{11}_{11} = \text{sn}(\lambda + \eta;k)$$

$$R_3(\lambda)^{22}_{11} = k\ \text{sn}(\eta;k)\text{sn}(\lambda;k)\text{sn}(\lambda + \eta;k)$$

$$R_3(\lambda)^{21}_{12} = \text{sn}(\eta;k)$$

$$R_3(\lambda)^{12}_{12} = \text{sn}(\lambda;k)$$

and other relations that follow from the complete X-symmetry.

The most known example of the system having $R_3(\lambda)$ as its S-matrix is the eight vertex model of Baxter [2], [40], [43], [49]:

$$H = -\frac{1}{2}\Sigma_{n=1}^{N}(J_x\sigma_n^1\sigma_{n+1}^1 + J_y\sigma_n^2\sigma_{n+1}^2 + J_z\sigma_n^3\sigma_{n+1}^3)$$

The continuous version of this model is the generalization of the Heisenberg chain [18]:

$$\frac{\partial}{\partial t}\tilde{S} = \tilde{S}\times\frac{\partial^2\tilde{S}}{\partial x^2} + \tilde{S}\times J\tilde{S}$$

$$\tilde{S} = (S_1,S_2,S_3) \quad \text{with} \quad \|\tilde{S}\| = 1 \text{ and}$$

$$J\tilde{S} = (J_1S_1,J_2S_2,J_3S_3)$$

§8. Lattice models and Baxter's lemma.

We remind (see §6) that each new factorized S-matrix gives rise to a large family of different completely integrable Hamiltonian systems, see examples of §7. We mean family, since each of the systems (and they are all different) itself generates a large family of Hamiltonians commuting with it. Different systems having the same factorized S-matrix can be classified according to different representations of "Zamolodchikov algebra" [3], [4]. Since there are many nonequivalent representations, there are different systems, like nonlinear Schrödinger equation and Toda lattice associated with the same S-matrix $R_2(\lambda)$ from §7.

There is, however, a canonical, unique, system associated with a given factorized S-matrix. This is the isospectral deformation equation, where the structure of local transfer matrix repeats the structure of S-matrix $R(\lambda)$. These canonical systems associated with a given factorized S-matrix exist in the continuous case (following definitions of §6), but they exist also in the lattice case as well. Moreover in the lattice case the relationship between local transfer matrix and S-matrix is more transparent.

Let us explain, first of all, how discrete isospectral deformation equations are constructed and how factorized S-matrices are associated with them. Then we show how, having a factorized S-matrix, we construct the canonical lattice isospectral deformation system associated with this S-matrix [2], [3], [4].

Following the accepted point of view, we consider a linear spectral (possibly singular) problem with operator coefficients. We take a Hilbert space H and a ring of operators M(H) on it.

The spectral problem we are dealing with has the following local form

$$\psi_{k+1} = \mathcal{L}_k(\lambda)\cdot\psi_k \tag{8.1}$$

where ψ_k are $\ell \times \ell$ matrices from $M_\ell(M(H))$ (i.e. having elements from M(H)). We can take an initial condition for (8.1):

$$\psi = \underbrace{\begin{pmatrix} \mathbb{I} & & & \\ & \cdot & & \\ & & \cdot & \\ & & & \cdot \\ & & & & \mathbb{I} \end{pmatrix}}_{\ell} \Bigg\} \ell$$

where $\mathbb{I}$ is a unit in M(H). The local transfer matrices $\mathcal{L}_k(\lambda)$ are subject to several natural restrictions [12]:

a) elements of matrices $\mathcal{L}_k(\lambda)$ and $\mathcal{L}_{k'}(\lambda)$ are commuting (as elements of M(H)), when $k \neq k'$;

b) the elements of the matrix $\mathcal{L}_k(\lambda)$ are meromorphic functions of λ on a Riemann surface Γ of (genus g), with the coefficients from M(H). The poles on Γ of the coefficients of $\mathcal{L}_k(\lambda)$ (and their orders) are independent of k.

For (8.1) we can define a corresponding monodromy operator and for $N \leq \infty$, we put

$$\mathcal{T}_N(\lambda) \overset{\text{def}}{=} \psi_{N+1}$$

i.e. $\mathcal{T}_N(\lambda)$ is an ordered product (cf. condition a)) of local transfer matrices

$$\mathcal{T}_N(\lambda) = \Pi_{k=1}^{N} \mathcal{L}_k(\lambda) \tag{8.2}$$

Quantum isospectral deformation systems arise as commutation relations on the elements of $\mathcal{L}_k(\lambda)$ (or rather on the residues of coefficients of $\mathcal{L}_k(\lambda)$ at singularities on Γ). By "complete integrability" we understand the following.

C.I. PROPERTY 8.3. For any λ, μ on Γ we have

$$[\mathrm{Tr}(\mathcal{T}_N(\lambda)), \mathrm{Tr}(\mathcal{T}_N(\mu))] \equiv 0 \tag{8.4}$$

where $[\ldots]$ is a commutator in $M(H)$ and $\mathrm{Tr}(A)$ for $A \in M_\ell(M(H))$ is a trace over $M(H)$. This property we attribute in the explicit form both classically and quantumly to Kostant [13]. Similar identities have been studied by a number of people in recent times. The simplest criteria for C.I. Property 8.3 to be satisfied were given by Baxter in his studies of the eight-vertex model [2]:

BAXTER LEMMA 8.5. Let us assume that there is a matrix $R(\lambda,\mu) \in M_{\ell^2}(\mathbb{C})$ which is non-singular for a generic (λ,μ), such that the following identity is satisfied for the local transfer matrices $\mathcal{L}_k(\lambda)$:

$$R(\lambda,\mu)\cdot(\mathcal{L}_k(\lambda) \otimes \mathcal{L}_k(\mu)) = (\mathcal{L}_k(\mu) \otimes \mathcal{L}_k(\lambda))\cdot R(\lambda,\mu) \tag{8.6}$$

for all k. Then the similar functional identity is satisfied for the monodromy matrices

$$R(\lambda,\mu)\cdot(\mathcal{T}_N(\lambda)\otimes\mathcal{T}_N(\mu)) = (\mathcal{T}_N(\mu)\otimes\mathcal{T}_N(\lambda))\cdot R(\lambda,\mu) \qquad (8.7)$$

and in particular, C.I. Property 8.3 is satisfied.

Here $A \otimes B$ for $A, B \in M_\ell(M(H))$ is a Kronecker (tensor) product over the ring $M(H)$.

The equation (8.6) together with the conditions a), b) above gives us commutation relations on coefficients of $\mathcal{L}_k(\lambda)$ under which there is a family of commuting Hamiltonians in (8.4). It may look, however, as if the choice of $R(\lambda,\mu)$ is arbitrary. This is not true, and simple assumptions of nontriviality imply that $R(\lambda,\mu)$ is a factorized S-matrix. First of all it is quite natural to assume that $R(\lambda,\mu)$ depends only on $\lambda - \mu$ and $R(\lambda,\lambda) = \mathbb{I}$. In this case the equation (8.7) can be represented as

$$\mathcal{T}_N(\lambda)\otimes\mathcal{T}_N(\mu) = R(\lambda,\mu)^{-1}\cdot(\mathcal{T}_N(\mu)\otimes\mathcal{T}_N(\lambda))\cdot R(\lambda,\mu) \qquad (8.8)$$

However the condition (8.8) is nothing but commutation relations (1.1) written for ℓ^2 elements of $\mathcal{T}_N(\lambda)$ (i.e. we are dealing with an ℓ^2-plet of "particles"). The corresponding S-matrix is nothing but a product of two S-matrices $R(\lambda,\mu)^{-1}$ and $R(\lambda,\mu)$ in the sense of [4]. If monomials $(\mathcal{T}_N(\lambda_1))_{i_1,j_1},\ldots$ $\ldots,(\mathcal{T}_N(\lambda_s))_{i_s,j_s}$ are linearly independent for $s \leq 3$, then factorization equations are satisfied for a product of S-matrices $R(\lambda,\mu)^{-1}$ and $R(\lambda,\mu)$.

Naturally, there are many nonequivalent local transfer matrices $\mathcal{L}_k(\lambda)$ satisfying (8.6) that are rational in λ for the same fixed S-matrix. There is, however, a way to construct one single canonical completely integrable system associated with the factorized S-matrix $R(\theta)$ [3]. For this we must construct a local transfer matrix $\mathcal{L}_k(\lambda)$. We choose H as a Fock space $H = H_0 \otimes\ldots\otimes H_0 = \otimes_{n=1}^{N} H_0$. According to our conventions a) and b), elements of different $\mathcal{L}_k(\lambda)$ should commute. This and the universality of $\mathcal{L}_k(\lambda)$ in k leads to the following representation of $\mathcal{L}_k(\lambda)$ in terms of a fixed matrix $\mathcal{L}(\lambda)$ from $M_\ell(M(H_0))$

$$\mathcal{L}_k(\lambda) = \mathbb{I}_\ell \otimes \ldots \underbrace{\otimes\ \mathcal{L}(\lambda)\ \otimes}_{\text{k-th place}} \ldots \otimes \mathbb{I}_\ell \tag{8.9}$$

In other words, $\mathcal{L}_k(\lambda)$ acts nontrivially only on the k-th component of the Fock space and on this k-th component it acts as $\mathcal{L}(\lambda)$. Next we take H_0 as a Hilbert space of dimension ℓ and define its elements in terms of the S-matrix $R(\theta)$. In other words, $\mathcal{L}(\lambda)$ is a $\ell \times \ell$ matrix with entries being again $\ell \times \ell$ matrices. The assumption is to take $\mathcal{L}(\lambda)$ as an $\ell^2 \times \ell^2$ matrix, repeating the form of $R(\theta)$. Namely, we consider

$$\mathcal{L}(\lambda) = (L_{ij}(\lambda))_{i,j=1,\ldots,\ell} \tag{8.10}$$

where $L_{ij}(\lambda)$ are elements of $M(H_0)$ written in a fixed basis of H_0 as $\ell \times \ell$ matrices

$$(L_{ij}(\lambda))_{\alpha\beta} = S_{\alpha i}^{j\beta}(\theta) = R(\theta)_{\alpha i, j\beta} \tag{8.11}$$

Then for $\mathcal{L}_k(\lambda)$ defined through (8.9)-(8.11), the Baxter condition (8.6) is exactly equivalent to the factorization equations (1.3). In other words, the C.I. Property 8.3 will be satisfied for a statistical mechanics model defined by (8.1) with $\mathcal{L}_k(\lambda)$ in (8.9). In the continuous case it is equally interesting E.g. let us present the canonical isospectral deformation equation associated with completely X-symmetric S-matrix written in terms of θ-functions and induced representations of the group G from §§4,5.

The completely X-symmetric S-matrix from §5 corresponding to lattices L':L in V_{2g} is written in the form:

$$R(\lambda) = \Sigma_{\ell\in L'/L}\ \tau_\ell(\lambda) F_\ell \otimes F_{-\ell} \tag{8.12}$$

where F_ℓ satisfy Weyl relations and $\tau_\ell(\lambda)$ are ratios of θ-functions corresponding to L. Then the isospectral deformation equation is defined as follows:

$$\frac{d}{dx}\Phi_\lambda = U(\lambda)\Phi_\lambda \tag{8.13}$$

where

$$U(\lambda) = \Sigma_{\ell\in L'/L}\ \tau_\ell(\lambda) u_\ell(x) F_{-\ell}$$

Here $u_\ell(x)$ are scalars in classical case and in quantum case-field operators. The canonical commutation relations between $u_\ell(x)$ and $u_{\ell'}(y)$ are of Weyl type and exactly repeat commutation relations between F_ℓ and $F_{\ell'}$, e.g.

$$[u_\ell(x), u_{\ell'}(y)] = \{e^{\pi ik(\ell,\ell')} - e^{\pi ik(\ell',\ell)}\} \times u_{\ell+\ell'}(x)\delta(x - y) \tag{8.14}$$

for $\ell, \ell' \in L'/L$. Then in the case of factorized S-matrix $R(\lambda)$, the monodromy matrix $\Phi(x,y,\lambda)$ satisfying the norming condition $\Phi(y,y,\lambda) = \mathbb{I}$ we have the Baxter completely integrable property (see §6)

$$R(\lambda - \mu)(\Phi(x,y,\lambda) \otimes \Phi(x,y,\mu)) = (\Phi(x,y,\mu) \otimes \Phi(x,y,\lambda))R(\lambda - \mu) \tag{8.15}$$

In particular,

$$[\mathrm{Tr}\ \Phi(x,y,\lambda), \mathrm{Tr}\ \Phi(x,y,\mu)] = 0$$

In classical case we replace quantum S-matrix $R(\lambda)$ by its semiclassical approximation.

$$R(\lambda) = \mathbb{I} + \hbar \mathbb{I}^{T} \cdot r(\lambda) + O(\hbar^2) \tag{8.16}$$

We replace [.,.] by Poisson brackets {.,.} defined between $u_\ell(x)$, $u_{\ell'}(y)$ in the same way as above. Then the Baxter relation is replaced by the semiclassical one

$$\{\Phi(x,y,\lambda) \otimes \Phi(x,y,\mu)\} \tag{8.17}$$

$$= -[r(\lambda - \mu), \Phi(x,y,\lambda) \otimes \Phi(x,y,\mu)]$$

and we have

$$\{\mathrm{tr}\, \Phi(x,y,\mu), \mathrm{tr}\, \Phi(x,y,\lambda)\} = 0 \tag{8.18}$$

The semiclassical approximation (8.16) and associated classical systems are studied in more detail in the next section.

§9. Classical Systems Associated with Factorized s-matrices Expressed in Elliptic Functions

Here we use expressions of completely X-symmetric factorized S-matrices of §5 in the case of genus g = 1 (elliptic curve). We construct several classes of classical two dimensional completely integrable systems associated with these S-matrices. In the classical case we use semiclassical approximation $r(\theta)$ to factorized S-matrix $R(\theta)$, that is called factorized s-matrix. We briefly explain how factorized s-matrices induce symplectic structure on two dimensional completely integrable systems associated with spectral problem on a Riemann surface. In particular, semiclassical factorization equations turn out to be equivalent to the Jacobi identity for the corresponding Poisson brackets [26]. Examples of equations associated with elliptic factorized s-matrices include: elliptic principal chiral field equations, elliptic KdV equation, elliptic generalization of Heisenberg spin chain, etc.

In §6 it is shown that factorized S-matrices correspond via Baxter relation to quantum two dimensional completely integrable systems. Similarly, semiclassical approximations to S-matrices, as defined in (8.16), correspond to classical two-dimensional isospectral deformation equations. These relations are presented below.

In the classical case the most general form of singular spectral problem on a Riemann surface Γ has the form [12]

$$\frac{d\Phi_\lambda}{dx} = U(x;\lambda)\Phi_\lambda \tag{9.1}$$

where $U(x;\lambda)$ and $\Phi_\lambda(x)$ are $n \times n$ matrices and $U(x;\lambda)$ is a rational function of λ on Γ with singularities at fixed points $\lambda_1^0,\ldots,\lambda_m^0$. The spectral problem (9.1) gives rise to isospectral deformation flows[1)] that are usually determined by the fundamental solution $\Phi(x,x_0,\lambda)$ of (9.1) (called the monodromy matrix of (9.1)). We suppose the fundamental solution $\Phi(x,x_0.\lambda)$ to satisfy the initial condition

$$\Phi(x_0,x_0,\lambda) = \mathbb{I} \tag{9.2}$$

Since we don't impose any boundary conditions on (9.1), the quantities generating isospectral deformation flows are given by $\mathrm{tr}\,\Phi(x,x_0,\lambda)$. For nonsingular spectral problem over $\mathbb{C}$ and asymptotically vanishing potentials one takes for Hamiltonian densities the traces of scattering matrices $\lim_{x_1\to+\infty,x_0\to-\infty} A(x_1)\Phi(x_1,x_0,\lambda)A(x_0)^{-1}$ for an appropriate $A(x)$.

[1)] The words "isospectral..." have here very general sense, since for some singular spectral problems the notion of spectrum needs specific clarifications in each particular case

The main property of commutation of Hamiltonian flows generated by the spectral problem (9.1) is formulated in the generally accepted form (cf. (8.18)):

$$\{\mathrm{tr}\ \Phi(x,x_0,\lambda),\mathrm{tr}\ \Phi(x,x_0,\mu)\} = 0 \tag{9.3}$$

for any λ, μ and fixed x_0 [13]. However the Poisson brackets $\{.,.\}$ are not defined yet. It is their definition that provides Hamiltonian structure for an isospectral deformation equation and establishes the existence of infinitely many commuting Hamiltonian flows via the Complete Integrability (C.I.) Property (9.3). In order to define Poisson brackets between arbitrary functionals of $U(x;\lambda)$ it is enough to define Poisson brackets between different elements of $U(x;\lambda)$ and $U(y;\mu)$. Let us assume that functional variables in $U(x;\lambda)$ are all its elements $U(x;\lambda)_{ab}$: $a,b = 1,\ldots,n$. Then the general definition of Poisson brackets between functionals $F(\lambda)$ and $G(\mu)$ can be given as follows:

$$\{F(\lambda),G(\mu)\}_1 = \int dx\int dy\ \Sigma^n_{a,b,c,d=1} \frac{\delta F}{\delta U(x;\lambda)_{ab}} \{U(x;\lambda)_{ab},U(y;\mu)_{cd}\}_1 \frac{\delta G}{\delta U(y;\mu)_{cd}} \tag{9.4}$$

One can naturally rewrite these Poisson brackets $\{.,.\}_1$ in terms of relations between residues of the potential $U(x;\lambda)$ at $\lambda = \lambda_i^0$: $i = 1,\ldots,m$, i.e. to exclude the dependence on λ in (9.4). However, as we see later, the definition (9.4) is perfectly consistent in concrete cases.

It remains to give the definition of fundamental Poisson brackets $\{U(x;\lambda)_{ab},U(y;\mu)_{cd}\}_1$. In doing this we use for comparison the quantum situation (§6 and §8). Namely, we consider Poisson brackets relations as the limiting case of commutation relations that are defined by factorized S-matrices $R(\lambda,\mu)$. Hence we need a semiclassical limit of $R(\lambda,\mu)$. We remark that in all interesting cases $R(\lambda,\mu)$ has the form $R(\lambda - \mu)$ and depends on a certain free parameter η. Let us consider such a limit value of parameter η, where S-matrix $R(\lambda - \mu)$ turns to be a trivial one [4]. By a trivial S-matrix $R(\lambda - \mu)$ we understand such an $n^2 \times n^2$ matrix that

$$R(\lambda-\mu)ab,cd = \delta_{ad}\cdot\delta_{bc} \tag{9.5}$$

In what follows the trivial S-matrix defined by (9.5) is denoted by $\mathbb{I}^{\tau}$. Then the semi-classical s-matrix $r(\lambda,\mu)$ is the leading linear term in η-expansion of $R(\lambda,\mu)$. In order to stress the semiclassical approach we take η to be proportional to $\hbar$ and then write

$$R(\lambda,\mu) = \mathbb{I}^{\tau} + \hbar\, \mathbb{I}^{\tau} r(\lambda,\mu) + O(\hbar^2) \tag{9.6}$$

as $\hbar \to 0$. In this case commutator $[.,.]$ is substituted by $\hbar\{.,.\}$. Then the main Baxter property (see §6,8) (8.15) is substituted by the corresponding relation involving $r(\lambda,\mu)$ and Poisson brackets between elements of $\Phi(x,x_0,\mu)$ and $\Phi(x,x_0,\mu)$ proposed by Sklanin [18]

$$\{\Phi(x,x_0,\lambda) \otimes \Phi(x,x_0,\mu)\} = \tag{9.7}$$

$$= [\Phi(x,x_0,\lambda) \otimes \Phi(x,x_0,\mu), r(\lambda,\mu)]$$

instead of (8.15). Here the $n^2 \times n^2$ matrix $\{\Phi \otimes \Phi'\}$ contains all Poisson brackets between elements of Φ and Φ'

$$\{\Phi \otimes \Phi'\}_{ab,cd} = \{\Phi_{ac}, \Phi'_{bd}\} \tag{9.8}$$

The $n^2 \times n^2$ matrix $\{\Phi \otimes \Phi'\}$ is the semiclassical limit of

$$(\Phi \otimes \Phi') - \amalg^T (\Phi' \otimes \Phi) \amalg^T$$

We consider relations (9.7) in the continuous case as an approximation to the Baxter relations (8.15). They are the most important for us relations and they generalize the famous Poisson brackets relations between elements of scattering matrix for nonsingular spectral problems.

The local formula (8.6) gives us a condition under which relations (9.7) can be established if one specifies Poisson bracket relations between elements if $U(x;\lambda)$ and $U(y;\mu)$. In the notations (9.8) we have

LEMMA 9.9. If the Poisson brackets $\{...\}_1$ between elements of $U(x;\lambda)$ and $U(y;\mu)$ satisfy the identity

$$\{U(x;\lambda) \otimes U(y;\mu)\}_1$$
$$= \delta(x-y)[r(\lambda,\mu), U(x;\lambda) \otimes \mathbb{I} + \mathbb{I} \otimes U(x;\mu)] \qquad (9.10)$$

then the property (9.7) for the fundamental solution $\Phi(x,x_0,\lambda)$ of (9.1)-(9.2) is satisfied. In particular, the C.I. property (9.3):

$$\{\mathrm{tr}\ \Phi(x,x_0,\lambda), \mathrm{tr}\ \Phi(x,x_0,\mu)\}_1 = 0$$

is valid.

REMARK 9.11. It is clear from (9.4) that definition (9.10) gives the definition of Poisson brackets $\{.,.\}_1$ in terms of potentials $U(x;\lambda)_{ab}$ only. Definition (9.10) is the definition of "local" Poisson brackets [11] in the sense that it does not contain derivatives of delta function.

Lemma 9.9 is valid without any restrictions on $n^2 \times n^2$ scalar matrix $r(\lambda,\mu)$. However the definition (9.10) may not generate a Lie algebra, if $r(\lambda,\mu)$ is arbitrary scalar matrix. We show that necessary and sufficient conditions for Poisson brackets (9.4), (9.10) to generate a Lie algebra are the conditions of factorization and unitarity for semiclassical s-matrix. These conditions are semiclassical approximations to the ordinary unitarity and factorization conditions on quantum S-matrix of §1.

We formulate these conditions in matrix notations keeping in mind the geometric sense of S-matrix $R(\lambda,\mu)$ of §1.

The semiclassical s-matrix is the approximation to the quantum S-matrix up to the first order. Namely, if $R(\theta_1,\theta_2)$ depends on a parameter $\hbar$ which is usually denoted (as in §5) by η, then we can assume that as $\hbar \to 0, R(\theta_1,\theta_2)$ turns into the identity S-matrix $\mathbb{I}^T$ in the expansion (9.6): $R(\lambda,\mu) = \mathbb{I}^T + \hbar\, \mathbb{I}^T r(\lambda,\mu) + O(\hbar^2)$ as $\hbar \to 0$. Then the unitarity conditions on semiclassical s-matrix $r(\lambda,\mu)$ have the form

$$\mathbb{I}^T r(\lambda,\mu) + r(\mu,\lambda)\mathbb{I}^T = 0 \tag{9.12}$$

where $\mathbb{I}^T$ is defined in (9.5). The semiclassical factorizations equations one obtains from (1.3) by substituting (9.6) and keeping the leading order $O(\hbar^2)$. Here are the semiclassical factorization equations on elements of $r(\lambda,\mu)$

$$\begin{aligned}
&(\mathbb{I}^T \otimes I)(I \otimes \mathbb{I}^T r(\theta_{13}))(\mathbb{I}^T r(\theta_{12}) \otimes I) \\
&- (I \otimes \mathbb{I}^T r(\theta_{12}))(\mathbb{I}^T r(\theta_{13}) \otimes I)(I \otimes \mathbb{I}^T) \\
&+ (\mathbb{I}^T r(\theta_{23}) \otimes I)(I \otimes \mathbb{I}^T)(\mathbb{I}^T r(\theta_{12}) \otimes I) \\
&- (I \otimes \mathbb{I}^T r(\theta_{12}))(\mathbb{I}^T \otimes I)(I \otimes \mathbb{I}^T r(\theta_{23})) \\
&+ (\mathbb{I}^T r(\theta_{23}) \otimes I)(I \otimes \mathbb{I}^T r(\theta_{13}))(\mathbb{I}^T \otimes I) \\
&- (I \otimes \mathbb{I}^T)(\mathbb{I}^T r(\theta_{13}) \otimes I)(I \otimes \mathbb{I}^T r(\theta_{23})) = 0
\end{aligned} \tag{9.13}$$

identical in θ_1, θ_2, θ_3. Equations (9.13) can be considered as a sum of three commutators written in a proper form. Moreover the structure of (9.13) resembles slightly the structure of (9.10) which is also a sum of three commutators. In [26] we proved

PROPOSITION 9.14. In order the Poisson brackets $\{.,.\}_1$ defined in (9.4) and (9.10) give rise to a Lie algebra it is necessary and sufficient that the $n^2 \times n^2$ matrix $r(\lambda,\mu)$ from (9.10) satisfies the semiclassical unitarity equations (9.12) and semiclassical factorization equations (9.13).

The proof of this proposition is performed by direct computation of Poisson brackets of $\{U(x;\lambda)_{ab}, U(y;\mu)_{cd}\}_1$, and $U(z;\eta)_{ef}$ according to the definition of Jacobi identity. In fact, (9.13) (with (9.12) in mind) is another representation of Jacobi identity for Poisson brackets defined by (9.10).

For simplicity we call $n^2 \times n^2$ matrix $r(\lambda,\mu)$ a factorized s-matrix if $r(\lambda,\mu)$ satisfies the semiclassical unitarity and factorization equations (9.12)-(9.13).

It is clear from 9.9 and 9.14 that having any factorized semiclassical s-matrix $r(\lambda,\mu)$ one defines a canonical Hamiltonian structure (9.10) associated with isospectral deformation equation (9.1), where an infinite system of commuting Hamiltonians exists in view of (9.3). One can think about each of the members of this family as giving rise to two dimensional isospectral deformation equations.

The direct analysis of the equation (9.12)-(9.13) can be performed. In particular, we see (cf. [26]) that a nontrivial factorized s-matrix $r(\theta_1,\theta_2)$ that depends only on the difference $\theta_1 - \theta_2$ is expressed in terms of either rational, exponential or elliptic functions. In each of these cases $r(\theta_1-\theta_2)$ is a meromorphic function of $\theta_1 - \theta_2$ in $\mathbb{C}$. The equation (9.13) can be rewritten in terms of pole expansions of $r(\theta)$. For example, in the particular case, when we do know that $r(\theta)$ is an elliptic function, one can claim that the poles of this function form a torsion subgroup of E.

The easiest way to obtain semiclassical factorized s-matrices is to deduce them in the limit (9.6) of quantum factorized S-matrix $R(\lambda,\mu)$. We use here the expression of S-matrix (5.16) in the elliptic curve case $g = 1$, when we know that the S-matrix is factorized. In other words we are working with an elliptic curve E and its torsion subgroup of points of order $n \overset{\text{def}}{=} e = e_1$.

The expression of s-matrix $r(\lambda,\mu)$ corresponding to this case can be obtained from the construction of §5. Elements of $r(\lambda,\mu)$ are rational functions on E. We present below a large class of spectral problems (9.1) for $\Gamma = E$, where the Poisson brackets (9.4), (9.10) are defined using $r(\lambda,\mu)$ and satisfy C.I. property (9.3). In other words, the corresponding systems generated by tr $\Phi(x,x_0,\lambda)$ are completely integrable two dimensional systems associated with semiclassical $\mathbb{Z}/\mathbb{Z}n$-symmetric factorized s-matrix, arising from the elliptic curve E and its n-torsion subgroup. We take an elliptic curve E over $\mathbb{C}$ in its Weierstrass representation given in affine form by a cubic $Y^2 = 4X^3 - g_2X - g_3$, satisfied by $(\wp'(u),\wp(u))$. The elliptic curve E is represented as $\mathbb{C}/L$ where the lattice L is $\mathbb{Z}\omega_1 \oplus \mathbb{Z}\omega_2$ with Im $\omega_2/\omega_1 > 0$. The n-torsion subgroup E_n of E is the subgroup of n^2 points

$$(\wp'(\frac{i\omega_1}{n} + \frac{j\omega_2}{n}),\wp(\frac{i\omega_1}{n} + \frac{j\omega_2}{n})): 0 \leq i,j \leq n-1$$

The corresponding larger lattice $L_n = \mathbb{Z}\omega_1/n \oplus \mathbb{Z}\omega_2/n$ gives rise to cosets L_n/L playing, as in §5, the role of $X \times X = \mathbb{Z}/\mathbb{Z}n \times \mathbb{Z}/\mathbb{Z}n$. Corresponding to L_n/L there is a set of "normalized" operators F_λ: $\lambda \in L_n/L$, arising from an induced representation of the Lie algebra $\tilde{g}$ for elliptic curve, $g = 1$. These operators obey the rule, $\lambda,\mu \in L_n$

$$F_\lambda \cdot F_\mu = \exp\{2\pi i k(\lambda,\mu)\} F_{\lambda+\mu} \tag{9.15}$$

where for $\lambda = \lambda_1\omega_1 + \lambda_2\omega_2$, $\mu = \mu_1\omega_1 + \mu_2\omega_2 \in L_n$,

$$k(\lambda,\mu) = -n\lambda_2\mu_1;\ B(\lambda,\mu) = n(\lambda_1\mu_2 - \lambda_2\mu_1) \tag{9.16}$$

These operators F_λ: $\lambda \in L_n$ are realized as $n \times n$ matrices following §5:

$$F_\lambda = A^{\lambda_1 \cdot n} \cdot B^{\lambda_2 \cdot n} \quad \text{for} \quad \lambda = \lambda_1\omega_1 + \lambda_2\omega_2 \in L_n \tag{9.17}$$

with

$$(A)_{ij} = \delta_{ij}\exp\left(\frac{2\pi\sqrt{-1}i}{n}\right);\ (B_{ij}) = \delta_{i,j+1} (\mathrm{mod}\ n) \tag{9.18}$$

We are going to present the spectral problem that is induced by the semiclassical s-matrix generated by L and L_n. This spectral problem is defined for the values of spectral parameter on E with poles of the first order at points of E_n. For the description of rational functions on E we use Weierstrass ζ-function $\zeta(x)$, $\zeta'(x) = -\wp(x)$, and derivatives $\wp^{(j)}(x)$ of $\wp(x)$. Any function $f(\theta)$ on E having its poles only at points of E_n and all of the order k, can be written as

$$f(\theta) = \Sigma_{\lambda\in L_n/L}\, c_\lambda^{(0)} \zeta(\theta + \lambda) + \Sigma_{j=1}^{k} \Sigma_{\lambda\in L_n/L}\, c_\lambda^{(j)} \wp^{(j-1)}(\theta + \lambda) \tag{9.19}$$

with

$$\Sigma_{\lambda\in L_n/L} c_\lambda^{(0)} = 0$$

This decomposition (9.19) is used in order to define a spectral problem that has on E poles of arbitrary order k. Following the structure of semiclassical s-matrix associated with

L_n and L we present the initial spectral problem with the poles of the first order only on E in the form

$$\left(\frac{d}{dx} - U_0(x,\theta)\right)\Phi(x,\theta) = 0 \tag{9.20}$$

where $n \times n$ matrix $U_0(x,\theta)$ is expressed in terms of the matrices F_λ (9.17), (9.18) as

$$U(x,\theta) = \Sigma_{\mu \in L_n/L, \mu \neq 0} F_\mu u_\mu(x) \tag{9.21}$$

$$\cdot \Sigma_{\omega \in L_n/L} \zeta(\theta+\omega)\exp\{2\pi\sqrt{-1}B(\omega,\mu)\}$$

In other words the linear problem (9.20) depends on $n^2 - 1$ scalar functions $u_\mu(x)$. Naturally, in the quantum case $u_\mu(x)$ are field operators. In order to stress our generic relationship with the Baxter model [2] we denote elliptic functions of the form involved in (9.21) by w's. In general, for $j = 0,1,2,\ldots$ and $\mu \in L_n/L$, $\mu \neq 0$

$$w_\mu^{(j)}(\theta) = \Sigma_{\omega \in L_n/L} \wp^{(j-1)}(\theta+\omega)\exp\{2\pi\sqrt{-1}B(\omega,\mu)\} \tag{9.22}$$

where, formally, $\wp^{(-1)}(u) = \zeta(u)$. From the definition (9.16) it follows that the functions $w_\mu^{(0)}(\theta)$ are elliptic for $\mu \neq 0$, $\mu \in L_n/L$, since the sum of its residues in the fundamental domain of E is zero. The functions $w_\mu^{(j)}(\theta)$ are by the definition elliptic, for $j \geq 1$.

The semiclassical s-matrix determines at the same time Poisson brackets between functions $u_\mu(x), u_\lambda(y)$ for $\mu,\lambda \in L_n/L$. Namely, in the notations of (9.16) we put

$$\{u_\mu(x), u_\lambda(y)\} =$$

$$= \delta(x-y)\left(e^{2\pi i k(\mu,\lambda)} - e^{2\pi i k(\lambda,\mu)}\right)u_{\mu+\lambda}(x) \tag{9.23}$$

for $\mu, \lambda \in L_n/L$ (i.e. for μ, λ being elements of L_n (mod L)) and $\mu \neq 0$, $\lambda \neq 0$.

The initial spectral problem (9.20) gives rise to a large class of two dimensional completely integrable systems with $n^2 - 1$ variables $u_\mu(x)$. The Hamiltonians of these systems belong to a family of commuting Hamiltonians generated by the monodromy matrix $\Phi(x,y,\theta)$ of (9.20)

$$\operatorname{tr} \Phi(x,y,\theta) \tag{9.24}$$

Indeed, according to the result 9.9 all Hamiltonians (9.24) commute for fixed x, y and different θ's, if Poisson brackets are defined as in (9.23). We call this family of two-dimensional systems an elliptic Korteweg-de-Vries equation or simply elliptic KdV. It can be also called an elliptic sin-Gordon equation. The reasons for this is the following. When E degenerates into a rational, unicursal curve, i.e. the Abelian variety E is substituted by $\mathbb{A}^1$ (additive group), then the spectral problem (9.20) turns out to be a matrix linear differential operator of the first order with the single pole in $\mathbb{P}\mathbb{C}^1$. If this pole is at ∞ we come to the situation familiar from KdV or nonlinear Schrödinger equations [12]. However if pole is at zero or at any other finite point, then this gives rise to sin-Gordon equation ($n = 2$) or different nonlinear σ-models.

The most interesting class of elliptic KdV or sin-Gordon equations arises when this equation can be written as a commutativity condition of two linear problems, one of which is the problem (9.20) itself

$$\begin{aligned} \frac{d\Phi(x,\theta)}{dx} &= U(x,\theta)\Phi(x,\theta) \\ \frac{d\Phi(x,\theta)}{dt} &= V(x,\theta)\Phi(x,\theta) \end{aligned} \tag{9.25}$$

Then the two-dimensional equations have the form

$$\frac{d}{dt}U(x,\theta) - \frac{d}{dx}V(x,\theta) + [U(x,\theta),V(x,\theta)] = 0 \tag{9.26}$$

The class of equations (9.26) is indeed a rich one, if one takes $V(x,\theta)$ in (9.25) as a rational function of θ on E with poles of the order k. It is most natural to take the set of poles to be a translation of E_n. E.g. we can present an equation we call an elliptic principal chiral field. This equation is obtained from (9.25), (9.26), when $V(x,\theta)$ has the same form as $U(x,\theta + \alpha)$ for $\alpha \in E$, $\alpha \neq 0$ with different scalar coefficients.

In other words we take $V(x,\theta)$ in the following form

$$V(x,\theta) = \Sigma_{\mu\in L_n/L,\mu\neq 0}\, F_\mu v_\mu(x) \cdot \Sigma_{\omega\in L_n/L}\, \zeta(\theta+\omega+\alpha)\exp\{2\pi\sqrt{-1}B(\omega,\mu)\} \tag{9.27}$$

for $\alpha \notin L_n$. If the potential $U(x,\theta)$ is defined as in (9.21) and $V(x,\theta)$ as in (9.27), then equations (9.26) which we call equations of elliptic principal chiral field can be written as $2n^2 - 2$ equations on $2n^2 - 2$ variables $u_\mu(x,t)$, $v_\mu(x,t)$

$$\frac{\partial}{\partial t}u_\nu(x,t) + \Sigma_{\eta\neq 0,\nu}\, K(\alpha,\nu-\eta)\{e^{2\pi ik(\eta,\nu-\eta)} - e^{2\pi ik(\nu-\eta,\eta)}\} \times u_\eta v_{\nu-\eta} = 0$$

$$\frac{\partial}{\partial x}v_\nu(x,t) + \Sigma_{\eta\neq 0,\nu}\, K(-\alpha,\nu-\eta) \times \{e^{2\pi ik(\eta,\nu-\eta)} - e^{2\pi ik(\nu-\eta,\eta)}\}v_\eta u_{\nu-\eta} = 0 \tag{9.28}$$

Here we denote

$$K(\alpha,\lambda) = \Sigma_{\omega\in L_n/L}\, \zeta(\alpha+\omega)e^{-2\pi iB(\omega,\lambda)} \tag{9.29}$$

The system (9.28)-(9.29) is called the elliptic principal chiral field and one should, perhaps, indicate briefly why such name is given. The first reason for this is a deep geometric one connected with Kählerian manifolds. However there is also an immediate formal explanation for this name. One can consider degeneration of an elliptic curve, e.g. when the module k of E tends to zero. In this case the functions $w_{\mu}^{(0)}(\theta)$ all tend to $1/\theta$ as the parameter on $\mathbb{C}$. Then the linear problems (9.25) for $U(x,\theta)$ and $V(x,\theta)$ defined as in (9.21) and (9.27) turns into a linear spectral problem on $\mathbb{C}$:

$$\begin{aligned} \frac{d\Phi_\theta}{dx} &= \frac{U}{\theta}\,\Phi_\theta \\ \frac{d\Phi_\theta}{dt} &= \frac{V}{\theta+\alpha}\,\Phi_\theta \end{aligned} \tag{9.30}$$

where U and V are $n \times n$ matrices that are linear combinations of F_{μ}, $\mu \neq 0$, $\mu \in L_n/L$ with scalar function coefficients. In other words, U and V are arbitrary traceless matrices. It is known that the consistency condition for linear problem (9.30) is called principal chiral field equation [24] (principal chiral field equation for an algebra g), if U and V are arbitrary traceless matrices (from an algebra g). In other words equations (9.28) are natural generalizations of the principal chiral field equations (9.30), if one considers the corresponding spectral problem (9.25) over E instead of (9.30) over $\mathbb{C}$. In particular, one can propose invariant restrictions on u_{μ}, v_{μ} in (9.28) in order to generate different σ-models corresponding to Grassmanian manifolds over elliptic curves.

The most natural object for elliptic generalizations is matrix Heisenberg spin system. It should be remarked that the 8-vertex model is an elliptic generalization of the Heisenberg ferromagnetic [18]. In the same way we now generalize an arbitrary matrix Heisenberg system. The general matrix

Heisenberg chain had been introduced in our paper [12], example 1.2. This system arises as the consistency condition of two linear problems of the following sort

$$\frac{d}{dx}\Phi_\lambda = \frac{iS}{\lambda}\Phi_\lambda$$
$$\frac{d}{dt}\Phi_\lambda = \left(\frac{T}{\lambda} + \frac{2iS}{\lambda^2}\right)\Phi_\lambda \qquad (9.31)$$

The two dimensional equations corresponding to (9.31) have the form

$$-2S_x + [S,T] = 0, \quad iS_t = T_x \qquad (9.32)$$

One obtains from (9.32) a matrix Heisenberg spin system if we impose an invariant restriction on $S : S^2 = \mathbb{I}$. In this case equations (9.32) take the familiar form

$$S_t = \frac{1}{2i}[S, S_{xx}], \quad S^2 = \mathbb{I} \qquad (9.33)$$

We had noted in [12] that the system (9.33) is gauge equivalent to the matrix nonlinear Schrödinger equation. In order to obtain natural generalization of (9.33) in the same way as the Baxter model is a generalization of Heisenberg ferromagnet it is necessary to consider linear problem (9.25) with $U(x,\theta)$ as in (9.21) and $V(x,\theta)$ with poles of the second order at E_n. This way we imitate the structure of linear problem (9.31). Consequently, $V_1(x,\theta)$ has the following form

$$V_1(x,\theta) = \Sigma_{\mu\in L_n/L,\mu\neq 0}\, v_\mu^0(x) F_\mu w_\mu^{(0)}(\theta) + \Sigma_{\mu\in L_n/L,\mu\neq 0}\, v_\mu^1(x) F_\mu w_\mu^{(1)}(\theta) \qquad (9.34)$$

The system (9.26) has the following form

$$\frac{d}{dt}U(x,t,\theta) - \frac{d}{dx}V_1(x,t,\theta) \tag{9.35}$$

$$+ [U(x,t,\theta),V_1(x,t,\theta)] = 0$$

The structure of v^1_μ is resembling that of u_μ

$$v^1_\mu = a\cdot u_\mu : \mu \in L_n/L, \mu \neq 0 \tag{9.36}$$

for some scalar $a \neq 0$. The system (9.35)-(9.36) is the exact elliptic generalization of nonreduced system (9.32). Here is e.g. the equation determining $v^0_\mu(x,t)$

$$\Sigma_{\eta\neq 0,\nu} \{e^{2\pi ik(\eta,\nu-\eta)} - e^{2\pi ik(\nu-\eta,\eta)}\}u_\eta u_{\nu-\eta} \tag{9.37}$$

$$\times (\Sigma_{\omega\in L_n/L,\omega\neq 0} \zeta(\omega)e^{2\pi iB(\omega,\eta)})$$

$$+ a^{-1}\Sigma_{\eta\neq 0,\nu}\{e^{2\pi ik(\eta,\nu-\eta)} - e^{2\pi ik(\nu-\eta,\eta)}\}u_\eta v^0_{\nu-\eta}$$

$$= \frac{\partial}{\partial x}u_\nu$$

for any $\nu \in L_n/L$, $\nu \neq 0$. Imposing on the system (9.35)-(9.36) the same kind of restrictions as on (9.32) one obtains an elliptic generalization of Heisenberg spin system. E.g. for $n = 2$ these restrictions are the following

$$u^2_{(0,1)} + u^2_{(1,0)} - u^2_{(1,1)} \equiv 1$$

where $(i,j) \in \mathbb{Z}/\mathbb{Z}2 \oplus \mathbb{Z}/\mathbb{Z}2$ is identified with $i\frac{\omega_1}{2} + j\frac{\omega_2}{2}$ from L_2/L. For general n the number of these restrictions is much larger and they may be taken in the following form (cf. also (9.28))

$$\Sigma_{\mu,\eta\in L_n/L;\mu+\eta=\nu,\mu\neq 0,\eta\neq 0} e^{2\pi ik(\mu,\eta)} u_\mu u_\eta = 0 \qquad (9.38)$$

for $\nu \in L_n/L$, $\nu \neq 0$ and

$$\Sigma_{\mu\neq 0} e^{-2\pi ik(\mu,\mu)} u_\mu u_{-\mu} = 1 \qquad (9.39)$$

Moreover further restrictions can be added to (9.38)-(9.39), corresponding to the restrictions on (9.33) of the form $S = \mathbb{I} - 2P$, where P is one-dimensional projector. This way one gets various elliptic generalizations of the nonlinear Schrödinger equation.

§10. Generalized Eight-vertex and XYZ-models Associated with Factorized S-matrices and Their Geometric Interpretation

In this section we investigate a general class of two-dimensional completely integrable models of statistical mechanics of the Ising model type, generated by a factorized S-matrix $R(\theta)$. Factorization equations imply a "star-triangle" type of relations, like in eight-vertex model, and are the basis of the introduction of completely integrable models. These models, as those in §8, are uniquely associated with a given factorized S-matrix. Since in models we are considering spin variables take values f om the set X (the initial multiplet considered in §§ 1,5), we describe n-spin valued Potts models which correspond to the case $X = \mathbb{Z}/\mathbb{Z}n$ or to elliptic curve E and its n-torsion subgroup from §5 and §9. Together with the generalized eight-vertex model we study associated generalized XYZ-models of the type $\mathcal{H} = \Sigma_k\{\Sigma_{\lambda\in L'/L} J_\lambda F_{\lambda,k}\cdot F_{-\lambda,k+1}\}$. For these Hamiltonians we present Lax representation

$$\frac{d}{dt}\mathcal{L}_k(\theta) = \mathcal{G}_{k+1}(\theta)\mathcal{L}_k(\theta) - \mathcal{L}_k(\theta)\mathcal{G}_k(\theta)$$

which we deduced in two different ways, using factorization

equations. Also we show that generalized XYZ-models are, in fact, equivalent to generalized eight-vertex model, in the sense that $\mathcal{H}$ appears as $\partial/\partial\theta \log \operatorname{Tr} \mathcal{T}_N(\theta)$ at $\theta = 0$. These considerations are quite general and apply to every factorized S-matrix, following [40].

The "star-triangle" relation, that is equivalent to the factorization equations, itself implies a clear geometric picture. In the elliptic case $g = 1$, $e = 2$, it implies the existence of a certain foliation of $\mathbf{P}^3$ into elliptic curves with a distinguished point on them. Similarly, for $g = 1$, $e \geq 2$ factorization equations imply the existence of an imbedding of a module space of elliptic curves with a level structure into a higher-dimensional projective space.

A new class of models of statistical mechanics arising from a quantum S-matrix associated with n-torsion subgroup E_n of an elliptic curve E offers a wide family of different generalization of eight-vertex and XYZ-models. These models are natural generalization of XYZ-model in two directions. First of all, Pauli matrices σ^i are replaced by normed matrices $F_{\lambda'}$: $\lambda' \in E/E_n$, $\lambda' \neq 0$, defined by their commutation relations. What is the most important, the ordinary Ising two-valued spin models are replaced by n-spin configurations with the cyclic $\mathbb{Z}/\mathbb{Z}n$-symmetry. This $\mathbb{Z}/\mathbb{Z}n$-symmetry is naturally to be expected in view of complete $\mathbb{Z}/\mathbb{Z}n$-symmetry of the corresponding quantum S-matrix (see §5).

The generality of exposition requires to follow Baxter proposal [41] and to consider an arbitrary, nonregular planar lattice. This is the lattice which is generated by an intersection of N straight lines on the Euclidean plane. In order to follow natural geometric pattern we assume that our N straight lines start and end at the boundary of a large circle C in the plane; their intersection points lie inside C and no three lines are allowed to intersect at a common point.

The intersections of N lines form sides of a graph called a lattice $\mathcal{L}$. The line segments between sites form edges of $\mathcal{L}$. Each site is the end point of four edges (i.e. locally lattice is a regular one). We have in total $N(N-1)/2$ sides and $N(N-1)$ angles associated with intersections of lines. According to the general point of view, borrowed from the theory of S-matrices, we consider the set X of allowed values of spin variables, where X is equipped with a structure of a topological group. The models of statistical mechanics associated with the lattice $\mathcal{L}$, we are considering, are formed by assignment to every face an element of the set X. In order to form a statistical sum we need also Boltzmann weights assigned to every site as a function of four faces surrounding the site. Instead of assigning the spin variables from X to faces we can equally assign them to edges forming a given site. The canonical model of statistical mechanics corresponding to $\mathcal{L}$ and a given factorized S-matrix $R(\theta)$ arises when the Boltzmann weight of the site with edges $x,y,z,v \in X$ and the angle of the intersection $\theta_{12} = \theta_1 - \theta_2$ is chosen as

$$S^{zv}_{xy}(\theta_{12}) = R(\theta_{12})_{xy,zv},$$

according to the following picture

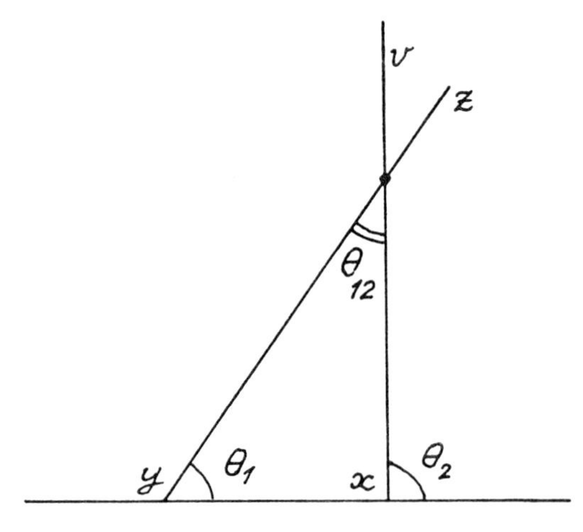

FIGURE 1

Now the statistical sum Z is defined as a sum over all possible configurations of spins assigned to all faces of $\mathcal{L}$, where every configuration is counted with the weight equal to the product of Boltzmann weights corresponding to all sites of a given configuration. In other words the statistical sum Z can be written following this definition in terms of the elements of S-matrix

$$
\begin{aligned}
&Z(\{\theta_j : j = 1,\ldots,N\}) \qquad (10.1)\\
&\quad = \Sigma_{x_i \in X}\, S_{x_1 x_2}^{x_3 x_4}(\theta_{12}) S_{x_3 x_4}^{x_5 x_6}(\theta_{13}) S_{x_5 x_6}^{x_7 x_8}(\theta_{14}) \ldots \\
&\qquad \ldots S_{x_{2N-3} x_{2N-2}}^{x_{2N-1} x_{2N}}(\theta_{1N}) \ldots
\end{aligned}
$$

Here in (10.1) $\theta_1,\ldots,\theta_N$ are the angles formed by N lines defining the lattice $\mathcal{L}$ as presented in Figure 2.

The interpretation of this statistical sum Z (10.1) in the best way is given using geometric sense of S-matrix as pre-

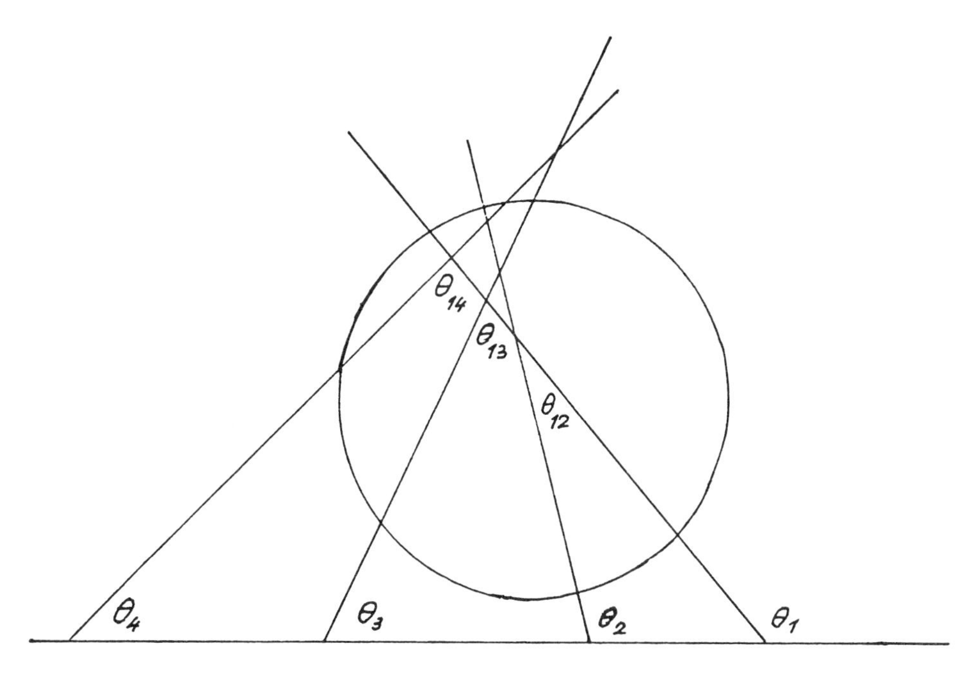

FIGURE 2

sented in §1. Namely, the statistical sum Z in (10.1) can be naturally considered as a trace of a matrix defining a linear map between vector spaces

$$V(\theta_1) \otimes \ldots \otimes V(\theta_N) \to V(\theta_N) \otimes \ldots \otimes V(\theta_1)$$

More precisely, let us assume that an S-matrix $R(\theta_1,\theta_2)$ depends only on the difference $\theta_1 - \theta_2$. Then $R(\theta_i - \theta_j)$ is a matrix defining an isomorphism of spaces $V(\theta_i) \otimes V(\theta_j) \to V(\theta_j) \otimes V(\theta_i)$. Using elementary matrices $R(\theta_1 - \theta_2)$ one can form a large $|X|^N \times |X|^N$ matrix

$$R(\theta_1,\ldots,\theta_N)$$

defining a desired isomorphism:

$$R(\theta_1,\ldots,\theta_N)\colon V(\theta_1) \otimes \ldots \otimes V(\theta_N) \to \to V(\theta_N) \otimes \ldots \otimes V(\theta_1) \tag{10.2}$$

At this point we need factorization equations because it is by no means obvious why in (10.2) an isomorphism $R(\theta_1,\ldots,\theta_N)$ does not depend on the order in which the elementary isomorphisms $R(\theta_1-\theta_j)$ are applied. In order to require independence of $R(\theta_1,\ldots,\theta_N)$ of the order of elementary isomorphisms it is necessary and sufficient to assume the validity of unitarity equations (1.2) and factorization equations (1.3). Certainly, the validity of unitarity and factorization equations is necessary. The fact that the unitarity and factorization equations are sufficient to imply the uniqueness of $R(\theta_1,\ldots,\theta_N)$, is a statement from monoidal categories. This statement is quite analogical to the ordinary algebra statement that the associativity axioms allows to write multiplication without any brackets. This categorial framework is developed elsewhere [54].

Now if Boltzmann weights $S_{xy}^{zv}(\theta_{12})$ are elements of a factorized S-matrix, then the statistical sum Z (10.1) is the trace of $|X|^N \times |X|^N$ matrix $R(\theta_1,\dots,\theta_N)$

$$Z(\{\theta_j: j = 1,\dots,N\}) = \mathrm{tr}\, R(\theta_1,\dots,\theta_N) \tag{10.3}$$

which is the sum of $|X|^N$ elements.

The main result of the Baxter paper [41] is the statement that the statistical sum Z in (10.3) for a factorized S-matrix $R(\theta)$ does not depend on the positions of the lines defining $\mathcal{L}$, i.e. parallel shifts of lines in $\mathcal{L}$ does not change the statistical sum. This is a natural consequence of the uniqueness of the matrix $R(\theta_1,\dots,\theta_N)$, which is itself the consequence of the unitarity and factorization equations. Moreover Baxter in the eight-vertex model case proves, much more strong statement [41]: the main statistical mechanics properties of models associated with the same factorized S-matrix $R(\theta)$ are <u>independent</u> of the way the lattice $\mathcal{L}$ is built (i.e. of the angles and positions of lines). Here by the main statistical properties we understand free energy $-\beta f = \lim_{N\to\infty} \frac{1}{|\mathcal{L}|} \ln Z_{\mathcal{L}}$ and all local n-point correlation functions as $N\to\infty$. This is obviously an important theorem, which shows that models of statistical mechanics are determined by spin variables and Boltzmann weights but not by the metric shape of the lattice. The statement of Baxter for the eight-vertex model where $|X| = 2$ is easily generalized for an arbitrary lattice $\mathcal{L}$. If the initial S-matrix $R(\theta)$ is factorized, then the main statistic mechanics properties are the same for all lattices $\mathcal{L}$ as $N \to \infty$, with the same Boltzmann weights

$S^{zv}_{xy}(\theta_{12})$. This property is called by Baxter Z-invariance property.

In particular we can start now with a regular square (or rectangular) lattice with N sites. In the case of $|X| = 2$ we can represent statistical sum in the famous 2-spin form $\sigma = \pm 1$:

$$Z = \Sigma_{\sigma_1=\pm 1} \Sigma_{\sigma_2=\pm 1} \cdots \Sigma_{\sigma_N=\pm 1} \Pi w(\sigma_i, \sigma_j, \sigma_k, \sigma_\ell) \tag{10.4}$$

where the product is taken now over all faces of the lattice. For each face, i, j, k, ℓ are the sites around it like on the figure 3.

The function $w(\sigma_i, \sigma_j, \sigma_k, \sigma_\ell)$ is a Boltzmann weight associated with a given face, unlike in (10.1). Examples of w in (10.4) which correspond to factorized S-matrices can be deduced from Baxter [41], [42], [43]. Actually in view of Baxter property of $\mathbb{Z}$-invariance there are known only two different classes of Boltzmann weights corresponding to factorized S-matrices and integrable models of statistical mechanics. These are the examples of eight-vertex model (§7, example 7.4), including Heisenberg XYZ-model and Ising model; and a new example of hard hexagon model [42].

Let us write explicitly the corresponding Boltzmann weights since it is very easy to do.

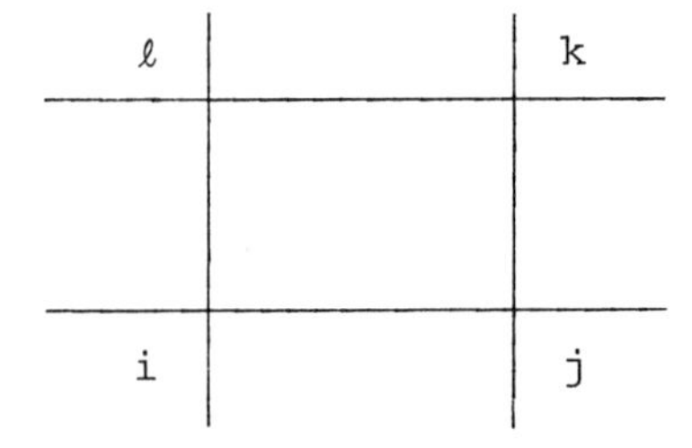

FIGURE 3

EXAMPLE 10.5. For eight-vertex model one can put

$$w(\sigma,\sigma',\sigma'',\sigma''') = a^{\sigma\sigma''}\beta^{\sigma'\sigma'''}\gamma^{\sigma\sigma'\sigma''\sigma'''} \tag{10.6}$$

for $\sigma,\sigma',\sigma'',\sigma''' = \pm 1$ and constants α, β, γ.

EXAMPLE 10.7. For hard hexagon model [42] the Boltzmann weights in (10.4) are defined as follows

$$w(\sigma,\sigma',\sigma'',\sigma''') = \zeta^{4-\sigma-\sigma'-\sigma''-\sigma'''} \cdot g_{\sigma\sigma'} g_{\sigma'\sigma''} g_{\sigma''\sigma''} g_{\sigma'''\sigma} \cdot g_{\sigma'\sigma'''},$$

for $\sigma,\sigma',\sigma'',\sigma''' = \pm 1$, an arbitrary constant ζ and

$$g_{1,1} = g_{1,-1} = g_{-1,1} = 1, \quad g_{-1,-1} = 0$$

For $|X| > 2$ the possibilities of choice of Boltzmann weights leading to completely integrable systems are more broad since the class of solutions of star-triangle relations is larger.

Exact solution of a model of statistical mechanics associated with a lattice $\mathscr{L}$ and a factorized S-matrix $R(\theta)$ is facilitated by the introduction of an auxiliary spectral problem and a transfer matrix associated with it. General definition of the local transfer matrix and corresponding lattice and continuous problems was already given in §8. Here we represent local transfer matrices from slightly different point of view directly relating them to Boltzmann weights and statistical sums.

We consider a Fock space $H_N = \otimes_{i=1}^{N} H_i$, $H_i = \mathbb{C}^n$, $|X| = n$. In this case, for example, the matrix $R(\theta_1,\ldots,\theta_N)$ can be considered as a linear operator on H_N. We consider this matrix with the prescribed numeration of its elements: with row numeration by possible vertical configuration of the lattice and column numeration by possible horizontal elements of the lattice. If one takes a canonical basis $e_1, \dots, e_n$ in $H_i = \mathbb{C}^n$, then in H_N we have a basis labeled by multindices $(\alpha) \in X^N$ as follows

$$e_{(\alpha)} = e_{\alpha_1} \otimes \dots \otimes e_{\alpha_N}$$

In this basis one introduces a transfer matrix T as $|X|^N \times |X|^N$ matrix acting on H_N with elements defined as follows

$$T_{(\alpha),(\beta)} = \Sigma_{\gamma_1 \in X} \dots \Sigma_{\gamma_N \in X} \Pi_{j=1}^{N} S_{\alpha_j \gamma_j}^{\beta_j \gamma_{j+1}} \tag{10.8}$$

where, in the periodic case, $\gamma_{N+1} = \gamma_1$.

Using notations (10.8) it is possible to rewrite the statistical sum Z from (10.1) in the following form

$$\begin{aligned} Z &= \Sigma_{(\alpha^1)} \dots \Sigma_{(a^N)} T_{(\alpha^1),(\alpha^2)} \times \dots \times T_{(\alpha^N),(\alpha^1)} \\ &= \mathrm{Tr}(T^N) \end{aligned} \tag{10.9}$$

In other words in order to compute statistical sum Z it is enough to compute eigenvalues of the transfer matrix T as the matrix of size $|X|^N \times |X|^N$. The computation of free energy is even a simpler task, since it requires only to find asymptotics as $N \to \infty$ of the largest by absolute value eigenvalue of T.

Certainly the transfer matrix T, as it is defined in (10.8), depends on the choice of Boltzmann weights S_{xy}^{zv}. This is actually a way of introduction of a spectral parameter θ, since S_{xy}^{zv} depends on θ. However as Examples 10.5 and 10.7 show Boltzmann weights may be independent of θ still providing completely integrable systems. The previous discussion provides us with a geometric way of introducing a spectral parameter θ into any such completely integrable system. For this we need only to use Z-invariance principle of Baxter. Namely one visualizes the dependence of Boltzmann weights $S_{xy}^{zv}(\theta)$ from θ if instead of a square lattice one considers a parallelogram lattice with the same angle θ, $0 < \theta < \pi/2$ at the base of the fundamental cell. The Z-invariance property insures that this lattice is actually equivalent to a square one. For the parallelogram lattice and transfer matrix T depends on the angle θ and the main C.I. Property of Baxter is satisfied for this family of transfer matrices

$$T(\theta)T(\theta') = T(\theta')T(\theta) \qquad (10.10)$$

i.e. transfer matrices are commuting for different angles (another form of the Z-invariance property). The transfer matrices $T(\theta)$ in the notations of §8 are nothing but traces of the monodromy matrices $\mathcal{T}_N(\theta)$. From this point of view relation (10.10) is itself a consequence of a more general Baxter property (8.7), which on its own is merely an equivalent of factorization equation. In other words equation (10.10) shows that the quantities $T(\theta)$ generate an infinite system of commuting Hamiltonians. In particular $T(\theta)$ generate an infinite system of completely integrable models of statistical mechanics of the Ising model type. These Hamiltonians have similarity with XYZ-model and as in XYZ-model case they appear through the expansion of $\partial/\partial\theta \log T(\theta)$ in powers of θ around the point $\theta = 0$.

In the case of $n = 2$ when spin variables are $\sigma = \pm 1$ or in the case, when X is a product of 2-cyclic groups, one still has the same form of the Hamiltonian as in XYZ-model case. For $n \geq 2$ we turn, however, to quantum Potts models. We assume X to be a cyclic group $\mathbb{Z}/\mathbb{Z}n$. In this case it is more natural to consider the values of spin variables to be roots of unity of the form $\exp(2\pi im/n)$. We relate the corresponding Potts model with operators F_λ : $\lambda \in L_n$ defined in §9 by formulae (9.17), (9.18) and satisfying fundamental commutation relations (9.15). The usual Potts model is defined as a one dimensional lattice ferromagnet model with spin values $\zeta_n^m : m = 0,1,\ldots,n-1$ for $\zeta_n = \exp(2\pi i/n)$. Denoting spin variable on the k-th site by σ_k, the general form of the potential energy, that is assumed to be local, is

$$V = \Sigma_k \, V(\sigma_k \sigma_{k+1}^*)$$

At the same time the kinetic energy is responsible for the transition between the states with different configurations $\{\sigma_k\}$. It is easier to represent the Hamiltonian using the corresponding Fock space. So instead of working with spin variables we can consider a diagonal unitary operators $\sigma_{\tilde{k}}$ acting on $|\{\sigma_j\}\rangle$ as follows [47]

$$\sigma_{\tilde{k}} |\{\sigma_k\}\rangle = \sigma_k |\{\sigma_k\}\rangle$$

Similarly the transition process between the states with different configurations $\{\sigma_k\}$ is defined through the shift operators $s_{\tilde{k}}$

$$s_{\tilde{k}} |\{\ldots \sigma_{k-1}, \sigma_k, \sigma_{k+1} \ldots\rangle = |\ldots \sigma_{k-1}, \zeta_n \sigma_k, \sigma_{k+1} \ldots\rangle$$

Hence, all $\sigma_{\tilde{k}}$ and all $s_{\tilde{k}}$ commute on different sites, while nontrivial commutation relations are [47]

$$(\sigma_{\tilde{k}})^n = (s_{\tilde{k}})^n = 1$$

$$\sigma_{\tilde{k}} \cdot s_{\tilde{k}} = \zeta_n s_{\tilde{k}} \cdot \sigma_{\tilde{k}}; \quad \sigma_{\tilde{k}} (s_{\tilde{k}})^+ = \zeta_n^{-1} \cdot (s_{\tilde{k}})^+ \sigma_{\tilde{k}}$$

The Hamiltonians, as the sum of the potential and kinetic energy is written as

$$\mathcal{H} = \Sigma_k\{T(s_{\tilde{k}}) + V(\sigma_{\tilde{k}}(\sigma_{k\tilde{+}1})^+)\}$$

This class of Hamiltonians is known as Potts models [47]. Potts models are natural generalizations of Ising type models with $\sigma = \pm 1$, $n = 2$. Potts models are realized as Hamiltonians in a Fock space H_N in a way analogical to Ising model. However instead of Pauli matrices σ^i for Ising model we consider the operator F_λ defined in (9.17)-(9.18). In particular, the role of operator $S^\sim$ is played by the matrix B and the role of $\sigma^\sim$ is played by matrix A; both A and B are defined by (9.18). In the Fock representation, operator variables are the following

$$A_k = I \otimes \ldots \otimes \underbrace{A}_{k} \otimes \ldots \otimes I$$

$$B_k = I \otimes \ldots \otimes \underbrace{B}_{k} \otimes \ldots \otimes I$$

where $\sigma_{\tilde{k}}$ is substituted for A_k, $s_{\tilde{k}}$ by B_k. The Potts-type Hamiltonians, that are completely integrable, have the following special form [47]

$$\mathcal{H}_0 = \Sigma_k\{\Sigma_{i=0}^{n-1} B_k^i + \alpha\cdot\Sigma_{i=0}^{n-1} A_k^i(A_{k+1}^+)^i\}$$

using F_λ notations, we can represent these Hamiltonians in a general form as follows

$$\mathcal{H}_1 = \Sigma_k\{\beta\cdot\Sigma_{i(\mathrm{mod}\ n)} \zeta_n^{k_i} F_{i/\omega_2,k} + \alpha\,\Sigma_{i(\mathrm{mod}\ n)} \zeta_n^{k_i} F_{i/\omega_1,k} F_{-i/\omega_1,k+1}\}$$

This class of Hamiltonians can be considered as a natural generalization of X-Hamiltonians. Similarly, there can be constructed generalizations of XY-Hamiltonians (as in the Ising case). The general form of XYZ-Hamiltonians for n-spin configurations is presented below. We should comment on one special feature of Hamiltonians under consideration. For two-spin models the general term of potential part of Hamiltonian is $\sigma_k^i \sigma_{k+1}^i$. For Potts type systems the generic term in the potential part is, however, $F_{\lambda,k} F^+_{\lambda,k+1}$ or (equivalently) $F_{\lambda,k} \cdot F_{-\lambda,k+1}$. This difference appears because only for $n = 2$, all elements of X or of L_n/L are of the second order.

Completely integrable models of statistical mechanics arising from a factorized S-matrix $R(\theta)$ (as Boltzmann weights) not only belong to an infinite family of commuting Hamiltonians generated by $T(\theta)$ but also possess Lax type of representation for each of them. Namely, original local spectral problem

$$\psi_{k+1} = \mathcal{L}_k(\theta) \cdot \psi_k \tag{10.11}$$

is amended by an additional time dependent spectral problem

$$\frac{d}{dt} \psi_k = \mathcal{G}_k(\theta) \cdot \psi_k \tag{10.12}$$

The consistence condition of (10.11)-(10.12) is a system of nonlinear equations for elements of $\mathcal{L}_k(\theta)$, $\mathcal{G}_k(\theta)$

$$\frac{d}{dt} \mathcal{L}_k(\theta) = \mathcal{G}_{k+1}(\theta)\mathcal{L}_k(\theta) - \mathcal{L}_k(\theta)\mathcal{G}_k(\theta) \tag{10.13}$$

We call system (10.13) the Lax representation for the model of statistical mechanics with Hamiltonian $\mathcal{H}$, if system (10.13) is equivalent to the evolution according to the Hamiltonian $\mathcal{H}$. Following the Liouville theorem, the evolution according to $\mathcal{H}$ of any operator on H_N is defined as

$$\frac{df}{dt} = [f,\mathcal{H}] \tag{10.14}$$

It turns out that for Hamiltonians $\mathcal{H}$ arising from an expansion of $\log T(\theta)$ at $\theta = 0$, Lax representation (10.13) exists and the quantities $\mathcal{L}_k(\theta)$, $\mathcal{A}_k(\theta)$ are local in variables of the system. Certainly the initial spectral problem (10.11) is the same as in §8 with $\mathcal{L}_k(\theta)$ repeating the structure of $R(\theta)$. The auxiliary problem (10.12) is constructed separately in a way to get (10.13) equivalent to (10.14). We present below two different methods of finding Lax representation (10.13) for the first nontrivial member of the family of Hamiltonians generated by $\log T(\theta)$, namely, for X-spin generalizations of XYZ-model.

The generalized XYZ-models, we are considering, are Hamiltonians in H_N with X-spin variables substituted by the corresponding operator variables

$$f_{\lambda,k} = I \otimes \ldots \otimes \underbrace{F_\lambda}_{k} \otimes \ldots \otimes I$$

where $F_\lambda : \lambda \in L'/L$ are normed operators introduced in (4.11) for lattice L in V $(= \mathbb{C}^g)$ and complimentary lattice L', $[L':L] = (e_1 \ldots e_g)^2 = n^2$. For $g = 1$ see (9.18). In these variables the form of the Hamiltonian we consider is the following one

$$\mathcal{H} = \Sigma_{k=1}^{N}\{\Sigma_{\substack{\lambda \in L'/L \\ \lambda \neq 0}} J_\lambda f_{\lambda,k} \cdot f_{-\lambda,k+1}\} \tag{10.15}$$

or

$$\mathcal{H} = \Sigma_{k=1}^{N}\{\Sigma_{\substack{\lambda \in L'/L \\ \lambda \neq 0}} J'_\lambda \cdot f_{\lambda,k} \cdot f^{+}_{\lambda,k+1}\} \tag{10.15'}$$

for $J_\lambda = J'_\lambda \exp\{2\pi i k(\lambda,\lambda)\}$ and $f_{\lambda,N+1} = f_{\lambda,1}$: $\lambda \in L'/L$, $\lambda \neq 0$.

The evolution according to the Hamiltonian $\mathcal{H}$ (10.15) following (10.14) has such a form

$$\frac{d}{dt} f_{\mu,k} = \Sigma_{\rho\in L'/L,\rho\neq 0,\mu}\{e^{2\pi ik(\mu,\rho-\mu)} - e^{2\pi ik(\rho-\mu,\mu)}\} \times \times f_{\rho,k}(f_{\mu-\rho,k-1}\cdot J_{\mu-\rho} + f_{\mu-\rho,k+1}J_{\rho-\mu}) \quad (10.16)$$

$\mu \in L'/L$, $\mu \neq 0$. For simplicity we denote

$$\chi(\mu,\rho-\mu) = e^{2\pi ik(\mu,\rho-\mu)} - e^{2\pi ik(\rho-\mu,\mu)} \quad (10.17)$$

For the XYZ-model, $f_{\lambda,k}:\lambda \in L'/L$, are substituted by Pauli matrices σ_k^i and, since all elements of L'/L are of the second order, $J_\nu = J_{-\nu}$.

The constants $J_\lambda:\lambda \in L'/L$, $\lambda \neq 0$, are not independent. Only for $n = 2$, $g = 1$, all constants are arbitrary. The constants J_λ: $\lambda \in L'/L$, $\lambda \neq 0$ depend on parameters of Abelian varieties $\tau \in \mathfrak{S}_g$ and $\eta \in \mathbb{C}^g$.

We propose an alternative way to find relations between the constants J_λ that lead to systems $\mathcal{H}$(10.15) with first integrals. These relations are algebraic. Namely, let us consider the case, when the Hamiltonian $\mathcal{H}$ admits a Lax representation (10.13); thus obtaining the relations between J_λ.

A simple combinatorial analysis of (10.13) and equations (10.16) shows that the auxiliary spectral problem (10.12) can be taken with $\mathfrak{a}_k(\theta)$ of the following general structure

$$\mathfrak{a}_k(\theta) = \sum_{\substack{\lambda_1,\lambda_2\in L'/L \\ \lambda_1\neq 0,\lambda_2\neq 0 \\ \lambda_1+\lambda_2\neq 0}} \pi(\lambda_1,\lambda_2) f_{\lambda_1,k-1} f_{-\lambda_1-\lambda_2,k} f_{\lambda_2,k+1} + \sum_{\substack{\lambda\in L'/L \\ \lambda\neq 0}}\{\beta(\lambda) f_{-\lambda,k-1} f_{\lambda,k} + \beta'(\lambda) f_{\lambda,k}\, f_{-\lambda,k+1} + \gamma(\lambda) f_{-\lambda,k-1} f_{\lambda,k+1}\} \quad (10.18)$$

for constants $\pi(\lambda_1,\lambda_2)$, $\beta(\lambda)$, $\beta'(\lambda)$, $\gamma(\lambda)$ depending on a spectral parameter θ.

The spectral problem (10.11) is chosen according to the rule of §8, i.e. repeating the structure of $R(\theta)$

$$\mathcal{L}_k(\theta) = \Sigma_{\lambda\in L'/L} \tilde{w}_\lambda(\theta) f_{\lambda,k} f_{-\lambda,k+1}$$

Since parameters $\tilde{w}_\lambda(\theta)$ are homogeneous, we take $\mathcal{L}_k(\theta)$ in the form

$$\mathcal{L}_k(\theta) = I + \Sigma_{\lambda\in L'/L,\lambda\neq 0} \alpha(\lambda) f_{\lambda,k} f_{-\lambda,k+1} \tag{10.19}$$

Inserting (10.18), (10.19) into (10.13) we obtain a system of equations on $F_{\lambda,k}$ equivalent to (10.16) only if the following system of equations relating $\alpha(\lambda)$, $\beta(\lambda)$, $\beta'(\lambda)$, $\gamma(\lambda)$, $\pi(\lambda_1,\lambda_2)$ is satisfied

$$\gamma(\lambda) + \beta'(-\lambda)\alpha(-\lambda)e^{-2\pi ik(\lambda,\lambda)} + \Sigma_{\lambda_1\neq 0,-\lambda} \pi(-\lambda-\lambda_1,\lambda)\alpha(\lambda_1)e^{-2\pi ik(\lambda+2\lambda_1,\lambda_1)} = 0$$

$$\beta'(\lambda) + \gamma(-\lambda)\alpha(-\lambda)e^{2\pi ik(\lambda,\lambda)} + \Sigma_{\lambda_1\neq 0,\lambda} \pi(\lambda_1,-\lambda)\alpha(-\lambda_1)e^{-2\pi ik(2\lambda_1-\lambda,\lambda_1)} = 0$$

$$\gamma(-\lambda) + \Sigma_{\lambda_1\neq 0,-\lambda} \pi(\lambda,-\lambda-\lambda_1)\alpha(-\lambda_1)e^{-2\pi ik(\lambda_1,2\lambda_1+\lambda)} + \beta(-\lambda)\alpha(\lambda)e^{-2\pi ik(\lambda,\lambda)} = 0 \tag{10.20}$$

$$\Sigma_{\lambda\neq 0}(\beta(\lambda)-\beta'(\lambda))\alpha(-\lambda)e^{-4\pi ik(\lambda,\lambda)} = 0$$

$$\beta(\lambda) - \beta'(-\lambda) + \Sigma_{\mu\neq 0,-\lambda}\{\beta(\mu+\lambda)\alpha(\mu)e^{4\pi ik(\mu+\lambda,\mu)} - \beta'(-\mu-\lambda)\alpha(\mu)e^{-4\pi ik(\mu,\mu+\lambda)}\} = 0$$

$$\beta(\lambda) + \gamma(\lambda)\alpha(\lambda)e^{2\pi ik(\lambda,\lambda)} \tag{10.20}$$

$$+ \Sigma_{\lambda_1\neq 0,\lambda}\ \pi(-\lambda,\lambda_1)\alpha(\lambda_1)e^{2\pi ik(\lambda_1,\lambda-2\lambda_1)} = 0$$

$$\alpha(\lambda)J_{-\lambda_1}\chi(\lambda,\lambda_1) = \pi(\lambda,\lambda_1) + \beta'(-\lambda_1)\alpha(\lambda)e^{2\pi ik(\lambda_1,\lambda)}$$

$$+ \Sigma_{\mu\neq 0,\lambda,-\lambda-\lambda_1}\ \pi(\lambda-\mu,\lambda_1)\alpha(\mu)e^{2\pi ik(\lambda_1+2\lambda-2\mu,\mu)}$$

$$+ \gamma(\lambda_1)\alpha(\lambda+\lambda_1)e^{2\pi ik(\lambda_1,\lambda+\lambda_1)}$$

$$\alpha(-\lambda_1)J_\lambda\chi(\lambda_1,\lambda) = -\pi(\lambda,\lambda_1) - \beta(\lambda)\alpha(-\lambda_1)e^{2\pi ik(\lambda_1,\lambda)}$$

$$- \Sigma_{\mu\neq 0,-\lambda_1,-\lambda-\lambda_1}\pi(\lambda,\lambda_1+\mu)\alpha(\mu)e^{-2\pi ik(\mu,\lambda+2\lambda_1+2\mu)}$$

$$- \gamma(-\lambda)\alpha(-\lambda-\lambda_1)e^{-2\pi i(\lambda+\lambda_1,\lambda)}$$

for $\lambda,\lambda_1 \in L'/L$, $\lambda \neq 0$, $\lambda_1 \neq 0$, $\bar{\lambda} + \lambda_1 \neq 0$. This system of equations determines $J_\lambda : \lambda \neq 0$ as the conditions of consistency of (10.20) for some $\alpha(\lambda)$, $\beta(\lambda)$, $\gamma(\lambda)$, $\pi(\lambda,\lambda_1)$.

There is another form of the Lax representation of $\mathcal{H}$ (10.15), now with $\mathcal{L}_k(\theta)$, $\mathcal{G}_k(\theta)$ as $n \times n$ matrices with entries being operators (from $M(H_N)$) on H_N.

The local transfer matrices $\mathcal{L}_k(\theta)$ are now $n \times n$ matrices with elements being operators on H_N. Following the recipie of §8 (that $\mathcal{L}_k(\theta)$ should repeat the structure of $R(\theta)$) we take $\mathcal{L}_k(\theta)$ in the following form

$$\mathcal{L}_k(\theta) = \Sigma_{\lambda\in L'/L}\ w_\lambda(\theta)f_{-\lambda,k}\cdot F_\lambda \tag{10.21}$$

or

$$\mathcal{L}_k(\theta) = \Sigma_{\theta\in L'/L}\ v_\theta(\theta)\cdot f^+_{\lambda,k}\cdot F_\lambda \tag{10.21'}$$

where $f_{\lambda,k}$ are operator variables on H_N and F_λ are $n \times n$ matrices, defined in (4.11): $\lambda \in L'/L$ and

$$v_\lambda(\theta) = w_\lambda(\theta)\cdot e^{2\pi ik(\lambda,\lambda)} : \lambda \in L'/L \tag{10.22}$$

It should be remarked that, in fact, the expressions (10.21) and (10.19) are alike, since the algebraic relations satisfied by $f_{\lambda,k}$ and F_λ are, certainly, the same. We define now the system of equations on $w_\lambda(\theta)$, necessary in order that Baxter C.I. Property of §6 and §8 be satisfied. In order to satisfy C.I. Property 8.3, or, what is the same, commutativity of transfer-matrices $T(\theta)$ in (10.10), we need, according to 8.5, to satisfy (8.6):

$$R(\theta',\theta)(\mathcal{L}_k(\theta') \otimes \mathcal{L}_k(\theta)) \tag{10.23}$$

$$= (\mathcal{L}_k(\theta) \otimes \mathcal{L}_k(\theta'))R(\theta',\theta)$$

We choose $R(\theta,\theta')$ in the form repeating the structure of $\mathcal{L}_k(\theta)$. In order to drop θ's we denote

$$\mathcal{L}_k(\theta) = \Sigma_\lambda\ v_\lambda e^{2\pi ik(\lambda,\lambda)} f_{-\lambda,k}\cdot F_\lambda$$

$$\mathcal{L}_k(\theta') = \Sigma_\lambda\ v'_\lambda e^{2\pi ik(\lambda,\lambda)} f_{-\lambda,k}\cdot F_\lambda \tag{10.24}$$

$$R(\theta',\theta) = \Sigma_\lambda\ v''_\lambda e^{2\pi ik(\lambda,\lambda)}\cdot F_{-\lambda} \otimes F_\lambda$$

with $v_\lambda = v_\lambda(\theta)$, $v'_\lambda = v_\lambda(\theta')$, $v''_\lambda = v_\lambda(\theta - \theta')$ for the S-matrix $R(\theta - \theta')$ depending only on $\theta - \theta'$: $\lambda \in L'/L$. Substituting (10.24) into (10.23) we obtain a system of equations, which is an analog of "star-triangle" relation [2], [41] (because w_λ play the role of Boltzmann weights). The system of equations on v_λ, v'_λ, v''_λ implying (10.23) is the following one:

$$\Sigma_{\lambda \in L'/L} v''_\lambda \{ v_{\mu+\lambda} v'_{\eta-\lambda} e^{2\pi i B(\lambda,\mu)} - v'_{\mu+\lambda} v_{\eta-\lambda} e^{2\pi i B(\lambda,\eta)} \} = 0 \tag{10.25}$$

for all μ, $\eta \in L'/L$ with $B(x,y) = k(x,y) - k(y,x)$ and $B(x,y)$ integer-valued on $L \times L$ (cf. §4). Since algebraic relations satisfied by $f_{\lambda,k}$ and F_λ are the same, we have in addition to (10.23) one more form of the "star-triangle" relation

$$(\mathcal{L}_{k-1}(\theta)\cdot\mathcal{L}_k(\theta'))s_k = s_k(\mathcal{L}_k(\theta')\cdot\mathcal{L}_{k-1}(\theta))$$
$$s_k = \Sigma_{\lambda \in L'/L} v''_\lambda e^{2\pi i k(\lambda,\lambda)} \cdot f_{\lambda,k-1} \cdot f_{-\lambda,k} \tag{10.26}$$

Indeed, if one rewrites (10.26) as a system of equations on $n \times n$ matrices for a given s_k (which is an operator on H_N) we obtain the system of equations on v_λ, v'_λ, v''_λ ($\lambda \in L'/L$) equivalent to the system (10.25). One should note a certain difference between $R(\theta',\theta)$ in (10.24) and s_k in (10.26).

We consider the first of the equations (10.26) at $\theta' = \theta$ and take a derivative with respect to θ'.

$$\mathcal{L}_{k-1}(\theta)\mathcal{L}'_k(\theta)s_k(\theta,\theta) + \mathcal{L}_{k-1}(\theta)\mathcal{L}_k(\theta)s'_k(\theta,\theta) \tag{10.27}$$
$$= s_k(\theta,\theta)\mathcal{L}'_k(\theta)\mathcal{L}_{k-1}(\theta) + s'_k(\theta,\theta)\mathcal{L}_k(\theta)\mathcal{L}_{k-1}(\theta)$$

with $\mathcal{L}'_k = \partial/\partial\theta \mathcal{L}_k$ and $s'_k(\theta,\theta) = \partial/\partial\theta' \, s_k|_{\theta'=\theta}$. Now we use the norming of $R(\theta',\theta)$ proposed in §5

$$R(\theta,\theta) = \mathbb{I}^T$$

in other words for the coefficients v''_λ we have

$$v''_\lambda(\theta,\theta) = 1 : \lambda \in L'/L \tag{10.28}$$

Now we use the fact that under the restriction (10.28), the operators s_k from (10.26) play the role of transposition operators, changing $f_{\lambda,k-1}$ to $f_{\lambda,k}$ and vice versa.

Namely we consider an operator

$$P_k = \frac{1}{n} s_k(\theta,\theta) \tag{10.29}$$

cf. (10.26) and (10.28). This operator P_k on H_N changes $f_{\lambda,k-1}$ and $f_{\lambda,k}$

$$\begin{aligned} P_k f_{\lambda,k-1} P_k &= f_{\lambda,k} \\ P_k f_{\lambda,k} P_k &= f_{\lambda,k-1} \end{aligned} \tag{10.30}$$

while $P_k f_{\lambda,j} P_k = f_{\lambda,j}$ if $j \notin \{k-1,k\}$. The last condition is certainly true by (10.26). In order to prove (10.30) we use the fundamental relations (4.8)

$$f_{\lambda,j} \cdot f_{\mu,j} = \exp\{2\pi i k(\lambda,\mu)\} f_{\lambda+\mu,j}$$

the commutativity of $f_{\lambda,j}$ and $f_{\mu,i}$ for $i \neq j$ and one more useful formula. Namely, as a consequence of our definition of L'/L and $B(x,y)$ from §4 we get the following identity for roots of unity:

$$\Sigma_{\lambda \in L'/L} e^{2\pi i B(\omega,\lambda)} = \begin{cases} 0, & \text{if } \omega \not\equiv 0 \pmod{L} \\ n, & \text{if } \omega \equiv 0 \pmod{L} \end{cases} \tag{10.31}$$

for all $\omega \in L'$ and $n = e_1 \ldots e_g$. We obtain (10.30) for P_k in (10.29), when we rewrite them using fundamental relations (4.8) and the identity (10.31).

We can combine (10.27) and (10.30). In order to apply P_k to (10.27) we remark that the definition of P_k (either through

(10.26), (10.28), (10.29) or through (10.30)) shows that P_k is a projector: $P_j^2 = P_j$. The relation we obtain from (10.27) and (10.30) can be represented in a short form as (cf. [49])

$$[\frac{1}{n} P_k s_k'(\theta,\theta), \mathcal{L}_k(\theta)\mathcal{L}_{k-1}(\theta)] \tag{10.32}$$
$$= \mathcal{L}_k(\theta)\mathcal{L}_{k-1}'(\theta) - \mathcal{L}_k'(\theta)\mathcal{L}_{k-1}(\theta)$$

Now $\Sigma_k\ 1/n\ P_k \cdot s_k'(\theta,\theta)$ is, up to an additive constant, our Hamiltonian $\mathcal{H}$ of the form (10.15). Indeed, let us rewrite $\Sigma_k \cdot\ 1/n\ P_k \cdot s_k'(\theta,\theta)$ using the definition (10.26), (10.28), (10.29) in the explicit form. We get then

$$\Sigma_k \frac{1}{n} P_k s_k'(\theta,\theta) = \tag{10.33}$$
$$= \frac{1}{n^2} \Sigma_k\{\Sigma_{\lambda \in L'/L}(\Sigma_{\rho \in L'/L}\ \tilde{v}''_\rho e^{2\pi i B(\rho,\lambda)})$$
$$\times e^{2\pi i k(\lambda,\lambda)} f_{\lambda,k-1} \cdot f_{-\lambda,k}\}$$

where we denote

$$\tilde{v}'' = \partial/\partial\theta'\, v''_\rho(\theta',\theta)|_{\theta'=\theta}$$

In other words we obtain

$$\Sigma_k \frac{1}{n} P_k s_k'(\theta,\theta) \tag{10.34}$$
$$= \Sigma_k\{\Sigma_{\lambda \in L'/L, \lambda \neq 0}\ J_\lambda f_{\lambda,k-1} \cdot f_{-\lambda,k}\} + C \cdot I_N$$

with an explicit definition of constants J_λ, namely for $\lambda \in L'/L$, $\lambda \neq 0$

$$J_\lambda = \frac{1}{n^2} \Sigma_{\rho \in L'/L} \frac{\partial}{\partial\theta'} w''_\rho(\theta',\theta)|_{\theta'=\theta} \cdot e^{2\pi i k(\lambda+\rho,\lambda-\rho)} \tag{10.35}$$

with $w''_\rho : \rho \in L'/L$ from $R(\theta',\theta)$ in (10.24). In particular, for the S-matrix $R(\theta' - \theta)$ from (5.16)-(5.17) the expression for $J_\lambda : \lambda \in L'/L$, $\lambda \neq 0$ is entirely in terms of θ-function $\theta[\vec{x}]$ and of parameter $\vec{\eta} \in \mathbb{C}^g$. The formulae (10.35) together with (5.17) give us the transcendental representations for the coupling constants $J_\lambda : \lambda \in L'/L$, $\lambda \neq 0$, instead of the algebraic in (10.20).

We see that for the Hamiltonian $\mathcal{H}$ (10.15) with coupling constants J_λ from (10.35), if conditions (10.25) are satisfied then there is a Lax representation (10.13) for $n \times n$ matrices $\mathcal{L}_k(\theta)$, $\mathfrak{a}_k(\theta)$, having elements being operators on H_N. Indeed, $\mathcal{L}_k(\theta)$ is already presented above in (10.21) ((10.21')). Now we prove the existence of $\mathfrak{a}_k(\theta)$, local in operators $f_{\lambda,j}$ (as in [49]).

We claim that there exists an $n \times n$ matrix $\mathfrak{a}_k(\theta)$ with elements depending on $f_{\lambda,k-1}$, $f_{\mu,k}$ such that

$$\begin{aligned} \mathfrak{a}_k \mathcal{L}_{k-1} &= -[\tfrac{1}{n} P_k s'_k, \mathcal{L}_{k-1}] + \mathcal{L}'_{k-1} \\ \mathcal{L}_k \mathfrak{a}_k &= +[\tfrac{1}{n} P_k s'_k, \mathcal{L}_k] + \mathcal{L}'_k \end{aligned} \qquad (10.36)$$

Indeed, one sees that the consistency condition for the system of equations (10.36) is, in fact, equivalent to identity (10.32)

The definition shows that $\mathfrak{a}_k(\theta)$ can be expressed entirely in terms of $f_{\lambda,k-1}, f_{\mu,k}$.

We show now that the representation (10.13) with $\mathcal{L}_k(\theta)$ as in (10.21) and $\mathfrak{a}_k(\theta)$ as in (10.36) is equivalent to the evolution (10.16) according to the Hamiltonian $\mathcal{H}$ (10.15) with coupling constants J_λ from (10.35); or to the evolution according to the equivalent Hamiltonian (10.34). Indeed, we have by (10.36)

$$[\mathcal{L}_k, \mathcal{H}] = [\mathcal{L}_k, \frac{1}{n} P_{k+1} s'_{k+1} + \frac{1}{n} P_k s'_k] = \mathcal{C}_{k+1}\mathcal{L}_k - \mathcal{L}_k \mathcal{C}_k$$

while by (10.14)

$$\frac{d\mathcal{L}_k}{dt} = [\mathcal{L}_k, \mathcal{H}]$$

Hence the Lax representation (10.13) is established for a Hamiltonian $\mathcal{H}$ (10.15) with coupling constants J_λ from (10.35) or for the equivalent Hamiltonian (10.34), where

$$C = \frac{1}{n^2} \Sigma_{\rho \in L'/L} \frac{\partial}{\partial \theta'} w''_\rho(\theta', \theta)\Big|_{\theta'=\theta} e^{-2\pi i k(\rho,\rho)}$$

We conclude that for the generalized XYZ-Hamiltonians $\mathcal{H}$ (10.15) associated with the factorized S-matrix $R(\theta)$, the Lax representation (10.13) is established in two different ways. The coupling constants $J_\lambda : \lambda \in L'/L, \lambda \neq 0$ in (10.15) corresponding to completely integrable system are determined either algebraically (cf. system (10.20)) or transcendentally (10.35) in terms of S-matrix elements. The latter approach is considered again below in connection with a generalized eight-vertex models.

We showed in the beginning of this chapter (cf. also [3]), how a general factorized S-matrix $R(\theta)$ gives rise to a completely integrable two-dimensional model of the eight-vertex type, with the Boltzmann weight defined using S-matrix elements. Moreover, this model generates a family of commuting Hamiltonians given by a functional $T(\theta) = \mathrm{Tr}\, \mathcal{T}_N(\theta)$. For the usual eight-vertex model ($n = 2$, $g = 1$) the class of local Hamiltonians arising from the expansion of $\log T(\theta)$ was studied by Sutherland and Luscher [40]; the first non-trivial member of this class is the usual XYZ-Hamiltonian (§7). In fact, the similar situation takes place in general, for an arbitrary family of statistical mechanics models, where Boltzmann weights are defined using $R(\theta)$. One should distinguish, however, between nonlocal Hamiltonians, that arise from an expansion of $T(\theta)$ and local Hamiltonians that arise from an expansion of $\log T(\theta)$ at $\theta = 0$. It is possible to prove, that expansion of $\log T(\theta)$ at $\theta = 0$ always gives rise to local Hamiltonians. We show now that the first nontrivial member of this class of Hamiltonians is a generalized XYZ-model of the form (10.15).

For this we start with an arbitrary factorized S-matrix $R(\theta)$. As a norming condition on $R(\theta)$, we demand, as in §5

$$R(0) = \mathbb{I}^{\tau} \tag{10.37}$$

where $(\mathbb{I}^{\tau})_{xy,zv} = \delta_{xv}\delta_{yz}$. We consider now the transfer matrix $T(\theta)$ and obtain

$$T_{(\alpha),(\beta)}(0) = \delta_{\alpha_1\beta_2}\delta_{\alpha_2\beta_3} \cdots \delta_{\alpha_N\beta_1} \tag{10.38}$$

Consequently, $T(0)$ is an operator of a cyclic permutation in the space H_N (cf. with operators P_k above in (10.30)). Let us now differentiate $T(\theta)$ at $\theta = 0$. We obtain then

$$\begin{aligned} &\frac{d}{d\theta} T_{(\alpha),(\beta)}(\theta)\big|_{\theta=0} \\ &= \Sigma_{k=1}^{N} \delta_{\alpha_1\beta_2} \cdots \delta_{\alpha_{k-2}\beta_{k-1}} \frac{d}{d\theta} S_{\alpha_k\alpha_{k-1}}^{\beta_k\beta_{k+1}}(\theta)\big|_{\theta=0} \\ &\qquad \cdot \delta_{\alpha_{k+1}\beta_{k+2}} \cdots \delta_{\alpha_N\beta_1} \end{aligned} \tag{10.39}$$

Let us now use both (10.38) and (10.39); then for the matrix elements of logarithmic derivative $\frac{d}{d\theta} \log T(\theta)\big|_{\theta=0}$ we obtain the formula:

$$\begin{aligned} &\Sigma_{k=1}^{N} \delta_{\alpha_1\beta_1} \cdots \delta_{\alpha_{k-1}\beta_{k-1}} \frac{d}{d\theta} S_{\alpha_{k+1}\alpha_k}^{\beta_k\beta_{k+1}}(\theta)\big|_{\theta=0} \\ &\qquad \cdot \delta_{\alpha_{k+2}\beta_{k+2}} \cdots \delta_{\alpha_N\beta_N} \end{aligned} \tag{10.40}$$

Now we can take the general form of the factorized S-matrix $R(\theta)$, written in terms of tensor product of some $|X| \times |X|$ matrices B_λ as:

$$\mathbb{I}^T R(\theta) = \Sigma_{\lambda_1,\lambda_2 \in L'/L}\, r_{\lambda_1,\lambda_2}(\theta) B_{\lambda_1} \otimes B_{\lambda_2}$$

In particular, we can write $\mathbb{I}^T R(\theta)$ as in (5.2) in terms of tensor product of elementary matrices E_{xy}. Expressions (10.39), (10.40) show that the Hamiltonian $\frac{d}{d\theta} \log T(\theta)|_{\theta=0}$ can be represented as local Hamiltonian depending on nearest neighbor interaction of the form

$$\mathcal{H} = \Sigma_k \{\Sigma_{\lambda_1,\lambda_2 \in L'/L}\, \tilde{r}_{\lambda_1,\lambda_2} B_{\lambda_1,k} B_{\lambda_2,k+1}\}$$

only for $B_{\lambda,k} = \mathbb{I} \otimes \ldots \otimes \underbrace{B_\lambda}_{k} \otimes \ldots \otimes \mathbb{I}$. The coupling constants $\tilde{r}_{\lambda_1,\lambda_2} = \frac{d}{d\theta} r_{\lambda_1,\lambda_2}(\theta)|_{\theta=0}$. In particular, let us consider the completely X-symmetric case for $R(\theta)$ given as in (5.6)

$$R(\theta) = \Sigma_{\lambda \in L'/L}\, \tau_\lambda(\theta) F_\lambda \otimes F_{-\lambda}$$

Then we get an explicit expression of the generalized XYZ-Hamiltonian $\frac{d}{d\theta} \log T(\theta)|_{\theta=0}$, coinciding with that given in (10.34).

We consider the elliptic case of $g = 1$, where $L'/L = L_n/L$ (see §9). Then the constants of interactions $J_\lambda : \lambda \in L_n/L$, $\lambda \neq 0$, corresponding to an n-torsion subgroup E_n of an elliptic curve E, have the form (with $\bar{\eta} \in \mathbb{C}$ not to be confused with quasi-periods; $\eta_i = 2\zeta(\omega_i/2)$) with $\eta(\rho) = a\eta_1 + b\eta_2$ for $\rho = a\omega_1 + b\omega_2 \in L_n$

$$J_\lambda^{(1)} = \frac{1}{n^2} e^{2\pi i k(\lambda,\lambda)} \cdot \Sigma_{\rho \in L_n/L}\, E^{2\pi i B(\rho,\lambda)} \{\zeta(\bar{\eta}+\rho) - \eta(\rho)\} \tag{10.41}$$

for $\lambda \in L_n/L$, $\lambda \neq 0$, where $J_\lambda^{(1)}$ indicates that (10.41) corresponds to the genus one. The formula (10.41) has actually only two free (nonhomogeneous) parameters: the parameter $\bar{\eta}$ on the elliptic curve E and a parameter of the elliptic curve E ($j(E)$ or $k(E)$). The situation is the same as for the ordinary XYZ-model, where expressions of interaction constants J_X, J_Y, J_Z depend on $\bar{\eta}$ and k and are very special cases of (10.41) for $n = 2$ [2], [43]

$$J_X = \frac{1+k\ \mathrm{sn}^2(\eta;k)}{\mathrm{sn}(\eta;k)} \qquad J_Y = \frac{1-k\ \mathrm{sn}^2(\eta;k)}{\mathrm{sn}(\eta;k)} \qquad J_Z = \frac{\mathrm{ch}(\eta;k)\mathrm{dn}(\eta;k)}{\mathrm{sn}(\eta;k)}$$

In order to get an explicit relation between the generalized XYZ-model (10.15) and the usual XYZ-model (§7, [2], [43]) we present a relation with $E_2 \subset E$. Namely, four standard Pauli matrices σ^i : $i = 1,2,3,4$ can be identified with four matrices F_λ : $\lambda \in L_2/L$

$$\sigma^1 = F_{\omega_2/2}, \quad \sigma^2 = -iF_{(\omega_1+\omega_2)/2}, \quad \sigma^3 = F_{\omega_1/2}, \quad \sigma^4 = F_0$$

Similarly, for an arbitrary n-torsion subgroup, $E_n \subset E$ we get an elliptic generalization of XYZ-model of the (10.15') form

$$\mathcal{H} = \frac{1}{4}\ \Sigma_{k=1}^{n}(\Sigma_{\lambda\in L_n/L,\lambda\neq 0}\ J'_\lambda \cdot f_{\lambda,k} \cdot f^+_{\lambda,k+1}) \tag{10.42}$$

the expressions for constants J'_λ borrowed from (10.42) can be written as elliptic functions of parameter $\bar{\eta}$

$$\begin{aligned} &J'_{(i_1\omega_1+i_2\omega_2)/n} \\ &= \Sigma_{j_1,j_2=0}^{n-1}\ e^{\frac{2\pi i}{n}(j_1i_2-j_2i_1)}\{\zeta(\bar{\eta} + \frac{j_1\omega_1+j_2\omega_2}{n}) \\ &\qquad - \eta_1\frac{j_1}{n} - \eta_2\frac{j_2}{n}\} \end{aligned} \tag{10.43}$$

Finally one should notice that the factor $e^{2\pi iB(\lambda',\mu')}$ appearing everywhere for $\lambda', \mu' \in L'/L$ is nothing but a Weil pairing of torsion elements λ', μ'.

We would like to remark that induced representations of G introduced above and, especially, the fundamental set of operators $A_{\lambda'}, \lambda' \in L'$ of the family of operators $F_{\lambda'}$, $\lambda' \in L'/L$ are naturally connected with a class of two dimensional lattice models admitting Kramers-Wannier duality transformations, associated with solvable finite groups in papers of Drouffe, Itzykson and Zuber [44], Monastirsky and Zamolodchikov [45] and explicitly introduced in the paper of Casalbuoni, Rittenberg and Yankelowicz [46]. In the latter paper the main object is a solvable group $\mathfrak{D}(n)$, whose elements $g^{\gamma}_{\alpha,\beta}$ are characterized by a triplet (α,β,γ) of elements of $\mathbb{Z}/\mathbb{Z}n$ and the multiplication rules are

$$g^{\gamma}_{\alpha,\beta} g^{\gamma'}_{\alpha',\beta'} = g^{\gamma+\gamma'+\alpha\beta'}_{\alpha+\alpha',\beta+\beta'}$$

All addition of indices is performed (mod n). We claim that the group $\mathfrak{D}(n)$ is naturally isomorphic to the group of operators $A_{\lambda'}: \lambda' \in L'$, where lattices L and L' are two dimensional, $g = 1$, and $e = e_1 = n$. Indeed, in terms of the operators $F_{\lambda'}$ we can put

$$g^{\gamma}_{\alpha,\beta} = \exp \frac{2\pi i\gamma}{n} \cdot F_{\lambda'}$$

where $\lambda' = \frac{\alpha}{n} P_1 + \beta Q_1$. Then it is obvious from the commutation relations satisfied by $F_{\lambda'}$, that so defined $g^{\gamma}_{\alpha,\beta}$ fulfil multiplication formulae defining $\mathfrak{D}(n)$. The relationship between $\mathfrak{D}(n)$ and $F_{\lambda'}$ is very useful since self-duality conditions for models of statistical mechanics generated by $\mathfrak{D}(n)$ [44], [45] give rise to a class of Hamiltonians generalizing 8-vertex model, that arises from just $\mathfrak{D}(2)$.

One can naturally ask the question whether it is possible to have multispin generalizations of eight-vertex model with many (more than two) free parameters using the same methods. In order to obtain such generalizations it is natural to look for Abelian rather than elliptic functions. In this case the Fock space H_N has a similar form

$$H_N = \Pi_{k=1}^{N} \otimes H_k, \quad H_k = \mathbb{C}^e = \otimes_{i=1}^{g} \mathbb{C}^{e_i}$$

and the Hamiltonian $\mathcal{H}$ is of the similar structure

$$\mathcal{H}_N = -\frac{1}{2} \Sigma_{k=1}^{N} \Sigma_{\lambda\in L'/L,\lambda\neq 0} J'_\lambda \cdot f_{\lambda,k} f^+_{\lambda,k+1}$$

The family of operators $F_\lambda : \lambda \in L'/L$ arises from a lattice L in $\mathbb{C}^g$, antisymmetric bilinear form $B(x,y)$ and the lattice L' complimentary to L with respect to $B(x,y)$, $[L':L] = (e_1 \ldots e_g)^2 = e^2$. The coupling constants J_μ in this case are determined by the following parameters: an Abelian variety of the dimension g given by a point in the Siegel space $\mathfrak{S}_g$ [19] and a distinguished point on an Abelian variety.

In future publication we will study definitions and properties of models of statistical mechanics (and their continuous counterparts) associated with multidimensional Abelian varieties. These systems are associated with completely X-symmetric S-matrices from §5. The geometric objects they correspond to arise from foliations of projective algebraic manifolds defining moduli of Abelian varieties with a given polarization and level structure.

REFERENCES

[1] P. Cartier, Proc. Symp. Pure Math. 9, Amer. Math. Soc., Providence, 361-387 (1965).

[2] R. Baxter, Ann. Phys., 76, 1, 25, 48 (1973).

[3] A.B. Zamolodchikov, Comm. Math. Phys., 69, 165 (1979).

[4] D.V. Chudnovsky, G.V. Chudnovsky, Phys. Lett. 79A, (1980)); 36-38, Phys. Lett. B, 98B(1981), 83-88.

[5] L. Auslander, R. Tolimieri, Abelian harmonic analysis, theta functions and function algebras on a nilmanifold, Springer lecture notes in mathematics, 436 (1975).

[6] D.V. Chudnovsky, G.V. Chudnovsky, Lett. Math. Physics, 4(1980), 373-380; J. Math. Pure Appl. 61(1981),No 1,1-16.

[7] C.L. Siegel, Topics in complex function theory, v. 3 Wiley (1973).

[8] A. Weil, Acta Math., 111, 143 (1964).

[9] D. Iagolnitzer, Lecture Notes Physics, Springer, 126, 1 (1980).

[10] J. Cherednick, Dolk. Acad. Sci. USSR, 249, 1095 (1979).

[11] E. Sklanin, L.A. Takhtadjan, L.D. Faddeev, Theor. Math. Phys. 40, n° 3, (1979).

[12] D.V. Chudnovsky, G.V. Chudnovsky, Z. Phys. D.5, 55, (1980).

[13] B. Kostant, Lecture Notes Mathematics, Springer, 570, 177 (1977).

[14] D.V. Chudnovsky, Phys. Lett. 74A (1979), 185; 73A, 292 (1979).

[15] G.V. Chudnovsky, Cargese Lectures, June 1979, in Bifurcation Phenomena in Math. Physics, D. Reidel, Publ. Company, 1980, 447-510; Les Houches Lectures, August 1979, Lecture Notes Physics, Springer, 126, 136-169; (1980).

[16] M. Karowsky, Lecture Notes in Physics, Springer 126, 344, (1980).

[17] D.V. Chudnovsky, Cargese Lectures, June 1979 in Bifurcation Phenomena in Math. Physics, D. Reidel, Publ. Company, 1980, 385-447; Les Houches Lectures, August 1979, Lecture Notes Physics, Springer, 126, 352-416, (1980).

[18] E.V. Sklanin, LOMI-Preprint E-3-1979, Leningrad, (1979)

[19] Jun-ichi Igusa, Theta functions, Springer, 1972).

[20] H. Cartan, Seminaire: Fonctions automorphes. Paris E.N. S. Secretariat Math. (1957-1958).

[21] S. Parke, Phys. Lett. B, 88B, 287, (1980).

[22] J. Honerkamp, Lecture Notes Physics, 126, 417-428, Springer Verlag (1980).

[23] H.F. Thacker, Fermilab-Pub-80/38-Thy. April (1980). Phys. Rev. D19, 3660. (1979).

[24] V. Zakharov, A. Mikhailov, ZETP 74, 1953, (1978).

[25] D. Chudnovsky, G.V. Chudnovsky, Saclay Preprint Lecture Notes, Math., v. 925 pp. 134-146.

[26] D.V. Chudnovsky, G.V. Chudnovsky, Lett. Math. Phys. 4, 485 (1980).

[27] D.V. Chudnovsky, G.V. Chudnovsky, Lett. Math. Phys. 5, 43 (1981).

[28] E. Leib, W. Linger, Phys. Rev. 130, 1605 (1963); E. Leib, D. Mattis, Phys. Rev. 125, 164, (1962).

[29] A. Poliakov, Caltech Lectures, 1978; Phys. Lett. 82B, 247, (1979).

[30] C.N. Yang, Phys. Rev. Lett., 19, 1312 (1967).

[31] P.P. Kulish, E.K. Sklanin, Phys. Lett., 70A, 461 (1979).

[32] K. Pohlmeyer, Comm. Math. Phys. 46, 267 (1976).

[33] K. Eichenherr and I. Honerkamp, Freiburg prepring 79/12 (1979).

[34] D. Gross, A. Neveu, Phys. Rev. D10, 3235 (1974).

[35] A.P. Fordy, J. Gibbons, Dublin Prepring IA S-STP-80-09, (to appear) 1980. A.V. Mikhailov, Pisma ZhETP, 30, 443(1979).

[36] N. Andrei, J. Lowenstein, Phys. Lett. 90B, 106, (1980); Phys. Lett. 91B, 401 (1980).

[37] B. Berg, Lecture Notes in Physics, 126, Springer-Verlag, 316-344 (1980).

[38] D.V. Chudnovsky, G.V. Chudnovsky J. Math. Phys., 22, 2518 (1981).

[39] J.H. Lowenstein, E.R. Spear (to appear).

[40] M. Luscher, Nucl. Phys., B135, 1 (1978).

[41] R.J. Baxter, Trans. Royal Soc. London, A289, 315 (1978).

[42] R.J. Baxter, J. Physics A13, L61 (1980).

[43] M. Gaudin, Modeles exacts en mechanique statistique: la methode de Bethe et ses generalisations, Note CEN-N-1559 (1), 1559 (2), (1972-1973).

[44] J. Drouffe, C. Itzykson, J. Zuber, Nucl. Phys., B147, 132 (1979).

[45] M. Monastyrsky, A. Zamolodchikov, ZhETPh, 77, 325, (1979).

[46] R. Casalbuoni, V. Rittenberg, S. Yankelowicz, Nucl. Phys. B170, 139 (1980).

[47] Y. Bashilov, S. Pokrovsky, Comm. Math. Phys., 76, 129 (1980).

[48] I. Schafarevich, Basic Algebraic geometry, Springer-Verlag (1974).

[49] M. Jackel, J.-M. Maillard, Seminarie E.N.S. Ex n° 2 (1979), see J. Phys. A15, 1309 (1982).

[50] D.V. Chudnovsky, G.V. Chudnovsky, Phys. Lett. 82A, 271. (1981)

[51] D.V. Chudnovsky, G.V. Chudnovsky, Phys. Lett. 84A, 353 (1981)

[52] D.V. Chudnovsky, G.V. Chudnovsky, Phys. Lett. 87A, 325 (1982).

[53] D.V. Chudnovsky, G.V. Chudnovsky, Phys. Rev. Lett. 47, 1093 (1981).

[54] D.V. Chudnovsky, in Problems of Mathmatics and Physics, Festschrift in honor of F. Gursey, 1981 (to appear).

RECURRENCES, PADÉ APPROXIMATIONS AND THEIR APPLICATIONS

David V. Chudnovsky and Gregory V. Chudnovsky

Department of Mathematics
Columbia University
New York, New York

INTRODUCTION

Here we discuss approximations to systems of analytic functions in the form of Hermite-Padé rational approximations. These approximations are used to study diophantine approximations of values of functions at particular (rational) points. Our aim is to improve the knowledge of approximations by rational numbers of such classical constants of analysis as $\log 2$, $\pi/\sqrt{3}$, $\pi, \ldots$ etc.

The key method, that we use, is the computation of consecutive (contiguous) Padé approximants by means of linear recurrences connecting them. Linear recurrences, relating contiguous Padé approximants, are important in arithmetic applications, where they define the asymptotic behavior of sequences of rational approximations to a given number. Simultaneously, matrix recurrences re-

---

This work was supported in part by the U.S. Air Force and AFOSR-81-0190 and by the Office of Naval Research.

lating contiguous Padé approximants, introduced by Mahler [1], represent local transfer matrices in the context of models of statistical mechanics and quantum field theory, which are solvable by an algebraic version of the inverse scattering method, cf. [3], [5], [9], [18]. [21]. This way we are establishing an important relationship between rational approximations to numbers and various (discrete) dynamical systems defined by recurrences, cf. [21].

We briefly describe Mahler's approach [1] on the representation of Padé approximation scheme in terms of matrix recurrences. Then on the basis of contiguous relations and monodromy considerations we present improved measures of irrationality of values of logarithmic and inverse trigonometric functions at rational points.

In §1 we formulate the fundamental properties of Padé approximations to systems of functions. The relationship between (inverse) spectral problems and matrix linear recurrences of Mahler [1] is considered. In §2 we illustrate the applications of Padé approximations to the study of arithmetic nature of values of functions. The classical approach, described in §2, uses Gauss contiguous relations to study values of logarithmic functions, see [2], [21]. In order to improve measures of irrationality of logarithms we present new Padé-type approximations with better convergence away from Padé approximation points. Matrix recurrences relating these new Padé-type approximations to logarithmic functions (and to the number $\log 2$) are presented in §2. In §3 we study new Padé-type approximations to $\frac{8}{\sqrt{z}} \operatorname{arctg} \frac{1}{\sqrt{z}}$ at $z = \infty$. This gives us a new "dense" system of rational approximations to $\frac{\pi}{\sqrt{3}}$, satisfying interesting explicit three term linear recurrences. We present the corresponding improvements for measures of irrationality of logarithms.

§1.

Mahler [1] was the first to define and study (generalized) Padé approximations (Hermite-Padé approximations) to systems of functions and to describe the matrix recurrence relations connecting consecutive (contiguous), Padé approximants. We describe these matrix recurrence relations and show their relations with the algebraic inverse scattering method, associated with completely integrable systems.

Let $f_1(z),\ldots,f_m(z)$ be m functions with definite formal power series expansions at $z = 0$ (e.g. $f_1(z),\ldots,f_m(z)$ may be analytic functions regular in the neighborhood of $z = 0$). Let us take m non-negative integers $n_1,\ldots,n_m$. Then there are m polynomials in z: $P_1(z),\ldots,P_m(z)$ of degrees $n_1,\ldots,n_m$, respectively, such that the following condition is satisfied. The function

$$R(z) = \Sigma_{i=1}^{m} P_i(z).f_i(z)$$

has a zero at $z = 0$ of order of at least

$$\Sigma_{i=1}^{m}\{n_i+1\} - 1$$

The existence of polynomials $P_1(z),\ldots,P_m(z)$, not all identically zero, follows from the trivial observation that the number of unknown coefficients of polynomials $P_i(z)$ is $\Sigma_{i=1}^{m}\{n_i+1\}$, and the number of linear equations defining these polynomials is $\Sigma_{i=1}^{m}\{n_i+1\} - 1$.

DEFINITION 1.1. Polynomials $P_1(z),\ldots,P_m(z)$ that satisfy the conditions above are called Padé approximants to functions $f_1(z),\ldots,f_m(z)$ with weights $n_1,\ldots,n_m$. The function $R(z) = \Sigma_{i=1}^{m} P_i(z)f_i(z)$ is called the remainder function in the

Padé approximation problem to functions $f_1(z),\ldots,f_m(z)$. The system of Padé approximants is called perfect if the order of zero of $R(z)$ at $z = 0$ is exactly $\Sigma_{i=1}^{m}\{n_i+1\} - 1$.

The assumption of the perfectness of a system of Padé approximations for all weights $n_1,\ldots,n_m$ can be formulated as the assumption that the degrees of polynomials $P_1(z),\ldots,P_m(z)$ are exactly $n_1,\ldots,n_m$, respectively. Similar definitions of Padé (rational) approximations can be generalized to approximations taken simultaneously at several points.

For applications of Padé approximations to number theory (going back to Hermite [2])one has to consider simultaneously a system of Padé approximants with weights $n_1,\ldots,n_m$ and systems of Padé approximants with weights $n_1,\ldots,n_i + 1,\ldots,n_m$ for $i = 1,\ldots,m$. We incorporate all the corresponding polynomial Padé approximants into a single $m \times m$ matrix. For this we denote Pade approximants $P_i(z)$ in the Padé approximation problem with weights $n_1,\ldots,n_m$ by $P_i(z|n_1,\ldots,n_m)$ for $i = 1,\ldots,m$. For fixed $N_1,\ldots,N_m$ we denote by $\Phi(z|N_1,\ldots,N_m)$ the following $m \times m$ matrix:

$$\Phi(z|N_1,\ldots,N_m) = (P_i(z|N_{j1},\ldots,N_{jm}))_{i,j=1}^{m},$$

where weights $N_{j1},\ldots,N_{jm}$ are

$$N_{jk} = \begin{cases} N_k & , \text{ if } j \neq k \\ N_k+1 & , \text{ if } j = k \end{cases}$$

for $j,k = 1,\ldots,m$. Since, in the Padé approximation problem, we have freedom of norming, we choose the norming in the way that

$$\det(\Phi(z|N_1,\ldots,N_m)) = z^{\sigma}$$

$$\sigma = \Sigma_{i=1}^{m}\{n_i+1\}$$

These notations were introduced by Mahler in 1935 in his fundamental paper on Hermite-Padé approximations (see [1]). Among other results, Mahler [1] noticed that for two systems of weights $N_1,\ldots,N_m$ and $N_1',\ldots,N_m'$ satisfying

$$\sigma = \Sigma_{i=1}^{m} N_i \leq \sigma' = \Sigma_{i=1}^{m} N_i'$$

the ratio of two matrices

$$P(z|\begin{matrix} N_1',\ldots,N_m' \\ N_1,\ldots,N_m \end{matrix}) = \Phi(z|N_1',\ldots,N_m')\cdot\Phi(z|N_1,\ldots,N_m)^{-1}$$

is an $m \times m$ matrix with polynomial entries which have degrees that are bounded by functions of $N_i' - N_i$: $i = 1,\ldots,m$. This matrix relation is an example of a general form of recurrence relations connecting different Padé approximants. Among these recurrence relations the most important are contiguous relations corresponding to the case when weights differ by one. These recurrences are characterized by the well known property of linearity in z. Namely, we consider two contiguous systems of weights (N_i) and $(N_i' = N_i + 1)$, $i = 1,\ldots,m$. In this case we find that the matrix $P(z|\begin{matrix} N_1+1,\ldots,N_m+1 \\ N_1,\ldots,N_m \end{matrix})$ is linear in z and

$$P(z|\begin{matrix} N_1+1,\ldots,N_m+1 \\ N_1,\ldots,N_m \end{matrix}) = z\cdot I + P(N_1,\ldots,N_m)$$

where I is a unit $m \times m$ matrix and $P(N_1,\ldots,N_m)$ is a constant matrix (note that the determinant of the transfer matrix $P(z|\begin{matrix} \cdots \\ \cdots \end{matrix})$ is z^m). For number-theoretic applications, we consider mainly the diagonal Padé approximations, when weights are equal $N_i = N$ for $i = 1,\ldots,m$. We define:

$$\Phi(z|N) = \Phi(z|N,\ldots,N)$$

Then the contiguous recurrence relations between consecutive Padé approximations take the form

$$\Phi(z|N + 1) = (z\cdot I + P_N)\Phi(z|N) \tag{1.1}$$

for a scalar $m \times m$ matrix P_N. Here as well as in similar circumstances, the transformation matrix $z\cdot I + P_N$ is called a local transfer matrix (cf. [3]). The sequence of local transfer matrices uniquely determines the sequence of polynomial Padé approximants. This is the most effective method of computation of Padé approximants (instead of solving systems of linear equations defining the remainder function).

Mahler [1] posed the following problem: when does a system of $m \times m$ scalar matrices P_N that defines local transfer matrices $z\cdot I + P_N$ give rise to a family of functions $f_1(z),\ldots,f_m(z)$, whose Padé approximants have contiguous relations (1.1)?

In order to answer this problem, one realizes that the equation (1.1) is a linear difference equation, whcih can be studied using the inverse scattering method. To find a system of functions $f_1(z),\ldots,f_m(z)$ for which the system (1.1) represents contiguous relations between consecutive Padé approximants, is to solve a direct spectral problem (i.e. to find the spectral measure corresponding to (1.1)). An important assumption which is not imposed explicitly is the analyticity of $f_1(z),\ldots,f_m(z)$ at $z = 0$. It is simple to show that for $\|P_N\|$ rapidly growing as $N \to \infty$, the corresponding functions $f_1(z),\ldots,f_m(z)$ are not necessarily analytic at $z = 0$. The restrictions on P_N to guarantee the analyticity of functions $f_1(z),\ldots,f_m(z)$ are of the following form:

1) The case of the rapidly decreasing "potential" $P(N) = P_N$. For example, one can assume that $\|P_N\|$ tends to zero as a geometric progression as $N \to \infty$.

2) The quasi-periodicity of the potential P(N) as a function of N. This can be formulated as a condition of existence of a basis of quasi-periods $T_1,\ldots,T_k$ such that for every $\varepsilon > 0$ there are infinitely many ε-quasi-periods T with

$$\|P(N + T) - P(N)\| < \varepsilon$$

In particular an important subclass of periodic potentials P(N) connected with algebraic functions $f_i(z)$ belongs to class 2).

For a solution of the direct (and inverse) spectral problems of (1.1) with potentials satisfying restrictions 1), one uses integral equations and the Riemann boundary value problem as a generalization of the inverse scattering method [5]. For a solution of the spectral problem with a potential satisfying restrictions 2) one uses instead an algebraic version of the inverse scattering method, which relies on the solution of the inverse Jacobi problem on imbedding of a curve into its Jacobian (in an algebraic or transcendental cases, cf. [4], [5], [6]).

The case m = 2 has its advantages since, only in this case, are the functions $f_1(z)$ and $f_2(z)$ (or, more precisely, only their ratio $f_2(z)/f_1(z)$) determined directly from the moment problem that arises from orthogonal polynomials defined by the three term recurrence relations derived from (1.1), cf. [17]. The analyticity of $f_2(z)/f_1(z)$ depends on the properties of the spectral measure.

§2.

Number-theoretic applications of Padé (rational approximations) are based on the specialization of the value of a function at $x = x_0$ to obtain a system of "good" rational approximations to a number. For example, if one knows the explicit recurrences determining the continued fraction expansion of $f(x)$, then, for a rational $x = x_0$, one can study the arithmetic nature of a number $f(x_0)$, provided that the coefficients of the recurrence are rational numbers. This method can determine completely the continued fraction expansion of the number $f(x_0)$ only in a few cases, but often it can be used to obtain some partial results. For example, expressions for elements of the continued fraction expansions of functions satisfying differential equations can be sometimes determined from the Euler procedure of the continued fraction expansions of solutions of Ricatti equations. One of the most famous is Euler's continued fraction expansions for the exponential functions. E.g. the expansion

$$\mathrm{th}(y^{-1}) = \cfrac{1}{y + \cfrac{1}{3y + \cfrac{1}{5y + \dots}}}$$

was used by Lambert to prove the irrationality of π.

Euler's construction became the first step in Hermite's explicit determination of simultaneous Padé approximants to arbitrary exponential functions $e^{\omega_k x}$ [2]. Hermite's formulas are still the most simple from all the known ones. We present here the description of diagonal Pade approximants to e^x at $x = 0$, following Siegel [8]. The polynomials $P_n(x)$, $Q_n(x)$ of degrees n such that the (remainder) function

$$R_n(x) = P_n(x)e^x + Q_n(x)$$

has a zero at $x = 0$ of order $2n + 1$, can be determined using the Hermite interpolation formula as

$$P_n(x) = (1 + \frac{d}{dx})^{-n-1}.x^n$$

$$Q_n(x) = (-1 + \frac{d}{dx})^{-n-1}.x^n$$

so that

$$R_n(x) = \frac{x^{2n+1}}{n!} \int_0^1 t^n(1-t)^n e^{xt} dt$$

Recurrences defining polynomials $P_n(x)$, $Q_n(x)$, together with their generating functions can be then determined explicitly. Such an explicit determination of Padé approximants is rare, but has interesting number theoretical applications when exists.

We present here some other cases, when the explicit formulas for the recurrences and Pade approximants to other functions of number-theoretical importance are known. In these cases functions satisfy Fuchsian linear differential equations, and then generating functions of Pade approximants are algebraic functions or period of integrals on algebraic varieties (determined by Picard-Fuchs equations). It is difficult to determine explicitly recurrences relating consecutive Padé approximants. One of the existing methods is based on the monodromy considerations [21]. For the Fuchsian linear differential equations monodromy considerations allow to find contiguous relations, i.e. relations between solutions of different o.l.d.e. having the same monodromy group. Contiguous relations in the matrix form [9] are equivalent to the Mahler's matrix recurrence relations (1.1). These contiguous (matrix recurrence) relations are used to determine arithmetic properties of coefficients of Padé approximants and convergence of Padé approximations.

Examples of explicit contiguous relations for functions with simplest monodromy lead to new measures of irrationality for logarithms of rational numbers such as log2, $\pi/\sqrt{3}$ and π.

We will describe now the matrix recurrences for Gauss hypergeometric functions. It is here, when matrix recurrences generate simple three-term linear recurrences satisfied by consecutive Pade approximants to the ratio of two contiguous hypergeometric functions. These three-term recurrences are represented by Gauss's continued fraction expansion.

Gauss contiguous relations for ${}_2F_1\,(a,b;c;x)$ arise, when parameters a,b,c are changed by integers.

The contiguous relations of Gauss can be written down as follows:

$$
\begin{aligned}
F(m+1,\ell,k|z) &= F(m,\ell,k-1|z) + zF(m,\ell,k|z) \\
F(m,\ell+1,k|z) &= F(m,\ell,k-1|z) + (z-1)F(m,\ell,k|z)
\end{aligned}
\tag{2.1}
$$

One can use these recurrence relations to determine explicitly Padé approximants and the remainder function (partial fractions and the error term) in the Gauss continued fraction expansion. Let us consider the near-diagonal Padé approximations to the function $\log(1-\frac{1}{z})$ at $z = \infty$. Then we have:

$$R_n(z) = \frac{1}{2}P_n(z)\log(1-\frac{1}{z}) + Q_n(z)$$

where $R_n(z) = O(z^{-n-1})$ as $|z| \to \infty$, and $P_n(z)$ and $Q_n(z)$ are polynomials of degree n and $n-1$ respectively.

Specialization of initial conditions $F(1,1,k|z)$ gives us three functions: $P_n(z)$, $Q_n(z)$ and $R_n(z)$:

i) If $F_1(1,1,k|z) = \frac{1}{k-2}\{(-z)^{2-k} - (1-z)^{2-k}\}$ for $k \neq 2$ and $F_1(1,1,2|z) = \log(1-1/z)$, then

$$R_n(z) \overset{\text{def}}{=} F_1(n+1,n+1,n+2|z)$$

ii) If $F_2(1,1,k|z) = \delta_{k2}$, then

$$P_n(z) \overset{\text{def}}{=} F_2(n+1,n+1,n+2|z)$$

iii) If $F_3(1,1,k|z) = \frac{1}{k-2}\{(-z)^{2-k} - (1-z)^{2-k}\}$ for $k \neq 2$, $F_3(1,1,2|z) = 0$, then

$$Q_n(z) \overset{\text{def}}{=} F_3(n+1,n+1,n+2|z)$$

The Padé approximants $P_n(z)$, $Q_n(z)$ can be identified with Legendre polynomials

$$P_n(z) = \bar{P}_n(x), \qquad x = 1 - 2z$$

where $\bar{P}_n(x)$ is the Legendre polynomial of the degree n:

$$\bar{P}_n(x) = 2^{-n}(n!)^{-1} \frac{d^n}{dx^n}\{(x^2-1)^n\}$$

Similarly

$$R_n(z) = \bar{Q}_n(x), \qquad x = 1-2z$$

where $\bar{Q}_n(x)$ is a Legendre function of the second kind.

Matrix recurrences (2.1) imply a single three-term linear recurrence

$$(n+1)X_{n+1} - (2n+1)(1-2z)X_n + n\cdot X_{n-1} = 0 \qquad (2.2)$$

satisfied by three sequences $X_n = P_n$, Q_n or R_n. Classical analysis that furnishes the recurrence (2.2), does not give an immediate description of the arithmetic nature of the coefficients of polynomials $P_n(x)$ and $Q_n(x)$. To obtain complete information on coefficients of $P_n(x)$, $Q_n(x)$ one can use previous matrix recurrences. This way one obtains the following main properties of Padé approximants to $\log(1- 1/z)$:

A. Coefficients of polynomial $P_n(z)$ are rational integers;

B. Coefficients of the polynomial $Q_n(z)$ are rational numbers with the common denominator dividing $\ell cm\{1,\dots,n\}$. Here $\ell cm\{1,\dots,n\}$ denotes the least common multiplier of numbers $1,\dots,n$. According to the Prime Number Theorem, this number is asymptotically $\exp\{(1+o(1))n\}$ as $n \to \infty$.

The property $\underline{B}$ is a consequence of an integral representation of $Q_n(z)$ in terms of $P_n(z)$:

$$Q_n(z) = \int_0^1 \frac{P_n(z) - P_n(x_1)}{z - x_1} dx_1$$

For applications of Padé approximations to the measure of irrationality of $\log(1-\frac{1}{z})$ with rational $z \neq 0,1$, we determine the asymptotic behavior of Padé approximants and remainder function for a fixed z and $n \to \infty$. The key elements here are recurrence relations and the following Poincaré lemma on the asymptotic behavior:

LEMMA 2.1. Let

$$\Sigma_{i=0}^{m} a_i(n)X_{n+i} = 0 \tag{2.3}$$

be a linear recurrence with coefficients depending on n such that $a_i(n) \to a_i$ when $n \to \infty$. Suppose the roots of the "limit" characteristic equation $\Sigma_{i=0}^{m} a_i\lambda^i = 0$ are distinct in absolute values: $|\lambda_1| >\ldots> |\lambda_m|$. Then there are m linearly independent solutions $X_n^{(j)}$: $j = 1,\ldots,m$ of (2.3) such that

$$\log|X_n^{(j)}| \sim n \log|\lambda_1|\text{: as } n \to \infty;\ j = 1,\ldots,m$$

and there is only one (up to a multiplicative constant) solution $\bar{X}_n$ of (2.3) such that

$$\log|\bar{X}_n| \sim n \log|\lambda_m| \quad \text{as} \quad n \to \infty$$

The existence of a "dense" sequence of rational approximations to the number θ is the best test of irrationality of θ and is used to determine the measure of irrationality of θ using the following very simple [19] lemma:

LEMMA 2.2. Let us assume that there exist a sequence of rational integers P_n, Q_n such that

$$\left.\begin{array}{l}\log|P_n| \\ \\ \log|Q_n|\end{array}\right\} \sim a \cdot n \quad \text{as} \quad n \to \infty$$

and

$$\log|P_n\theta - Q_n| \sim b \cdot n \quad \text{as} \quad n \to \infty$$

where $b < 0$. Then the number θ is irrational and for any $\epsilon > 0$ and for all rational integers p, q we have

$$|\theta - p/q| > |q|^{a/b-1-\epsilon}$$

provided that $|q| \geq q_0(\epsilon)$.

Applying lemma 2.1 to the recurrence (2.2) we obtain the following asymptotical formulas of Laplace or Riemann [7]:

$$\left.\begin{array}{l}\log|P_n(z)| \\ \\ \log|Q_n(z)|\end{array}\right\} \sim n \cdot \log|2z-1-2\sqrt{z^2-z}|$$

and

$$\log|R_n(z)| \sim n \cdot \log|2z - 1 + 2\sqrt{z^2-z}|$$

as $n \to \infty$. Here the root $\sqrt{z^2-z}$ is chosen in the way that $|2z - 1 + \sqrt{z^2-z}| < |2z - 1 - \sqrt{z^2-z}|$.

Specializing $z = a/b \in \mathbb{Q}$, $z \neq 0,1$ we obtain a sequence of rational approximations $Q_n(a/b)/P_n(a/b)$ to $\log(1-b/a)$. We can use then lemma 2.2 on "dense" approximations. This way we obtain the measure of irrationality for log(1- b/a) for integer a,b provided that $(\sqrt{a-b} - \sqrt{a})^2 < e^{-1}$, see [10].

For example, putting z = -1 and applying lemma 2.2 we obtain the following measure of irrationality of log2:

$$|q \log 2 - p| > |q|^{-3.6221009\ldots} \tag{2.4}$$

for rational integers p,q with $|q| \geq q_0$. Similarly, Gauss's contiguous relations (2.1) or (2.2) and lemma 2.1, 2.2 give the measure of irrationality of $\pi/\sqrt{3}$:

$$|q\frac{\pi}{\sqrt{3}} - p| > |q|^{-7.3099864} \tag{2.5}$$

for $|q| \geq q_1$. The measures of irrationality similar to (2.4), (2.5) were obtained independently by several people [10]-[13], [15].

The possibility to improve the measures of irrationality of logarithms of algebraic numbers are usually connected with the applications of Hermite's Padé approximations to functions $\log^i(1+z)$: i = 1,...,m-1, $m \geq 2$ at z = 0, described in detail in [14], or with Baker's method of linear forms in logarithms of algebraic numbers [16]. Either of these methods do not provide with a sequence of "dense" approximations required by lemma 2.2 and do not give any improvement for the measure of irrationality of numbers log2, $\pi/\sqrt{3}$ or π, etc. We propose a different method, based on Padé-type approximations, that allow us to construct new "dense" sequences of rational approximations to particular values of logarithms.

The new dense sequences of rational approximations to the function log(1-1/z) of Pade-type are chosen in the way that the convergence of approximations is better in the neighborhood of a given point $z = z_0$ away from $z = \infty$. New, Padé-type approximations to the function $\log(1-\frac{1}{z})$, depend now on two integer parameters n and m, $0 \leq m \leq n$. For a given m, and $n \geq m$, these rational approximations have the usual form:

$$R_n(z) = P_n(z) \cdot \log(1 - \frac{1}{z}) - Q_n(z) \tag{2.6}$$

where $P_n(z)$, $Q_n(z)$ are polynomials of degrees at most n, and $R_n(z)$ is regular at $z = \infty$. In complete analogy with the classical theory of Legendre polynomials we have the following arithmetical properties of $P_n(z)$ and $Q_n(z)$:

A. Polynomials $P_n(z)$, $Q_n(z)$ have rational coefficients, and are of degrees $\leqslant n$ in z.

B. $P_n(z) \in \mathbb{Z}[z]$ and the common denominator of the coefficients of the polynomial $Q_n(z)$ is $\ell cm\{1,\ldots,n\}$.

The explicit expressions of $P_n(z)$ (with given $m \leq n$) is particularly simple:

$$P_n(z) = \Sigma_{i=0}^{m} \Sigma_{j=0, n-m \leq i+2j \leq n}^{n-m} \binom{m}{i}\binom{n-m}{j} \times \binom{m}{n-i-2j}(-1)^{n-m-j} z^{n-i-j}(z-1)^{i+j} \tag{2.7}$$

According to the theory of Padé-type approximations, the polynomials $Q_n(z)$ are the adjoint polynomials (polynomials of the second kind):

$$Q_n(z) = \int_0^1 \frac{P_n(z) - P_n(x)}{x - z} dx \tag{2.8}$$

The property B above of Padé-type approximants $P_n(z)$, $Q_n(z)$ is a direct corollary of (2.8).

There are matrix recurrences that lead to scalar recurrences satisfied by $P_n(z)$, $Q_n(z)$, $R_n(z)$. These matrix recurrences are represented in the following form:

$$G(i+1,k,\ell|z) = G(i,k,\ell-2|z) + (2z-1)G(i,k,\ell-1|z)$$
$$+ (z^2-z)G(i,k,\ell|z) \qquad (2.9)$$
$$G(i,k+1,\ell|z) = G(i,k,\ell-2|z) + (z-z^2)G(i,k,\ell|z)$$

Solutions of matrix recurrences (2.9) are completely determined by initial conditions $G(1,1,k|z)$. Initial conditions that determine two sequences $P_n(z)$, $Q_n(z)$ are the following:

i) $G_1(1,1,k|z) = \delta_{k2}$. Then

$$P_n(z) \overset{\text{def}}{=} G_1(m,n-m,n+1|z)$$

ii) $G_2(1,1,k|z) = \frac{1}{k-2}\{(1-z)^{2-k} - (-z)^{2-k}\}$ for $k \neq 2$;

$$G_2(1,1,2|z) = 0$$

Then

$$Q_n(z) \overset{\text{def}}{=} G_2(m,n-m,n+1|z)$$

Padé-type approximants $P_n(z)$, $Q_n(z)$ and the remainder function $R_n(z)$ satisfy a scalar recurrence relation with coefficients that are polynomials in n. This recurrence is, however, not a three-term, but a five-term linear recurrence. Its limit form determine, according to the Poincaré Lemma 2.1, the asymptotics of Padé approximants (corresponding to the largest root of a quartic polynomial) and the asymptotics of the remainder function (corresponding to the root smallest in the absolute value).

This Padé-type scheme can be used at $z = -1$ to improve the measure irrationality of $\log 2$.

We choose a parameter m such that $m = [0.88\cdot n]$ for a sufficiently large integer n. This way we obtain a "dense" sequence of rational approximations Q_n/P_n to log2 such that: a) P_n are rational integers, b) Q_n are rational numbers whose denominators divide $\ell cm\{1,\ldots,n\}$.

Numerically one has

$$\left.\begin{array}{l}\log|P_n| \\ \\ \log|Q_n|\end{array}\right\} \sim 1.5373478\ldots\cdot n$$

and

$$\log|P_n \log 2 - Q_n| \sim -1.77602924\ldots\cdot n$$

as $n \to \infty$.

Hence the application of lemma 2.2 implies the following improvement of the measure of irrationality of log2:

$$|q \cdot \log 2 - p| > |q|^{-3.2696549\ldots} \qquad (2.10)$$

for rational integers p,q with $|q| \geq q_2$.

With different parameters m and n one can slightly improve the "density constants" a and b of sequence of approximations, see lemma 2.2. We obtain new sequences P_n, Q_n of rational numbers satisfying the properties a), b) above such that

$$\left.\begin{array}{l}\log|P_n| \\ \\ \log|Q_n|\end{array}\right\} \sim 1.93902189\ldots\cdot n$$

$$\log|P_n \log 2 - Q_n| \sim -1.93766649\ldots n$$

as $n \to \infty$.

This gives the following improvement for the measure of irrationality of log2, over (2.4) and 2.10:

$$|q\cdot \log 2-p|>|q|^{-3.13400029\ldots}$$

for rational integers p,q with $|q| \geq q_3$.

§3.

New measures of the irrationality of values of inverse trigonometric functions at rational points follow from Padé-type approximations, where apparent singularities are allowed and, consequently, the order of zero of the remainder function is less than the maximal one. We present one example of the Padé-type approximations to the function $(\operatorname{arctg} x)/x$ suited for the best approximation in the neighborhood of the point $x = 1/\sqrt{3}$, in order to study the number $\pi/\sqrt{3}$.

For convenience we will discuss the approximations to the function $\frac{8}{\sqrt{z}} \operatorname{arctg} \frac{1}{\sqrt{z}}$ for $z > 0$. We consider the Padé-type approximations to $\frac{8}{\sqrt{z}} \operatorname{arctg} \frac{1}{\sqrt{z}}$ at $z = \infty$ of the form

$$R_n(z) = Y_n(z)\cdot \frac{8}{\sqrt{z}} \operatorname{arctg} \frac{1}{\sqrt{z}} + X_n(z) \tag{3.1}$$

where the function $R_n(z)$ is regular at $z = \infty$, and $Y_n(z)$, $X_n(z)$ are polynomials in z. The structure of the denominator $Y_n(z)$ in the Padé-type approximation problem we consider is typically represented by a double sum of triple products of binomial coefficients. The simplest expressions of the polynomial $Y_n(z)$ for $n \geq 0$ is the following one:

$$Y_n(z) = \Sigma_{i_1=0}^{3n} \Sigma_{i_2=0,\, i_1+i_2\leq 4n}^{3n} \binom{3n}{i_1}\binom{3n}{i_2} \times \binom{2\{4n-i_1-i_2\}}{4n-i_1-i_2}\cdot(-1)^{i_1+i_2} 4^{i_2} z^{i_1}\cdot\left(\frac{z+1}{4}\right)^{3n-i_1} \tag{3.2}$$

Slightly different expressions for denominators $Y_n(z)$ with a different choice of apparent singularities in the Padé-type approximation scheme (3.1) are presented in [19]. Here, in (3.1), (3.2) for a rational integer z, $z \equiv -1 \pmod 4$, $Y_n(z)$ is a rational integer, while $X_n(z)$ is a rational number whose denominator divides $\ell cm\{1,\dots,4n\}$. These properties are analogous to properties A and B of §2.

Similar to §2 we can present matrix recurrences that determine Padé-type approximations (3.1)-(3.2). The matrix recurrences under investigation depend on parameters n, m, k and z, and correspond to three partial differences. The equations below are the particular cases of contiguous relations when local multiplicities at regular and apparent singularities are increased by one. The multi-branch function we study is denoted by $W(n,m,k|z)$, satisfying the following equations:

$$W(n+1,m,k|z) = W(n,m,k+1|z) - (z+1)W(n,m,k|z)$$

$$W(n,m+1,k|z) = W(n,m,k+1|z) - zW(n,m,k|z) \tag{3.3}$$

$$k(k+1)z(z+1)W(n,m,k-2|z) + \{m(m+1)+(m+1)(n+m)\}W(n,m,k|z)$$

$$= \{m(m+1)(2z+1)+(m+1)((n+m)z+m)\}W(n,m,k-1|z).$$

The specialization of initial conditions for $W(1,\frac{1}{2},0|z)$, $W(1,\frac{1}{2},1|z)$ of the function W in (3.3), determines $Y_n(z)$, $X_n(z)$ and $R_n(z)$ in the following form. E.g., we have

$$Y_n(z) = (-4)^n z^n . W(3n+1,3n+\tfrac{1}{2},-4n-2|z) \tag{3.4}$$

and similar expressions for $X_n(z)$, $R_n(z)$.

Moreover the approximants $Y_n(z)$, $X_n(z)$, as well as $R_n(z)$ in the scheme (3.1)-(3.2) (or its variations) satisfy a single scalar three-term linear recurrence with coefficients that are polynomial in n and z. Instead of presenting this complicated recurrence in the general case, we consider only a specialization to z = 3, when we obtain a "dense" sequence of rational approximations to the number $\frac{4\pi}{3\sqrt{3}}$ which we want to study.

Let now $z = 3$. Then according to (3.2) the expression of the denominator $Y_n \overset{\text{def}}{=} Y_n(3)$ in the sequence of rational approximations X_n/Y_n to $\frac{4\pi}{3\sqrt{3}}$ has the following form:

$$Y_n = \Sigma_{i_1=0}^{3n} \Sigma_{i_2=0,i_1+i_2\leq 4n}^{3n} \binom{3n}{i_1}\binom{3n}{i_2}\binom{2(4n-i_1-i_2)}{4n-i_1-i_2} \times (-1)^{i_1+i_2} \cdot 4^{i_2} \cdot 3^{i_1}. \tag{3.5}$$

The direct analysis of the matrix contiguous relations (3.3) establishes the scalar three-term linear recurrence relation satisfied by each of the sequences X_n and Y_n, that has coefficients polynomial in n

$$\begin{cases} A_2(n)Y_{n+2} + A_1(n)Y_{n+1} + A_0(n)Y_n = 0 \\ A_2(n)X_{n+2} + A_1(n)X_{n+1} + A_0(n)X_n = 0 \end{cases} \tag{3.6}$$

Here $A_0(n)$, $A_1(n)$ and $A_2(n)$ are polynomials in n of degree 9 with integer coefficients and are explicitly the following:

$$A_2(n) = -2^3 \cdot (4n+7) \cdot (4n+5) \cdot (4n+3) \cdot (4n+1) \cdot (2n+3) \cdot (n+2) \times (27279n^3 + 52164n^2 + 31511n + 6046)$$

$$A_1(n) = 3 \cdot (4n+3) \cdot (4n+1) \cdot (15484624281n^7 + 122518066482n^6 + 401859218160n^5 + 706125904254n^4 + 715282318379n^3 + 415975459648n^2 + 128021157420n + 16022087856) \tag{3.7}$$

$$A_0(n) = 2 \cdot 3^3 \cdot (6n+5) \cdot (6n+1) \cdot (3n+2) \cdot (3n+1) \times (2n+1) \cdot (n+1) \cdot (27279n^3 + 134001n^2 + 217676n + 117000)$$

The initial conditions and the first few terms for the solutions X_n and Y_n of the recurrence (3.6.) are the following

$$Y_0 = 1,\ Y_1 = 1250,\ Y_2 = 5915250$$

$$Y_3 = 32189537978,\dots \tag{3.8}$$

$$X_0 = 0,\quad X_1 = 3023,\dots$$

$$\text{and } X_2/Y_2 = 111264499/46007500,\dots$$

The expression (3.2) and its specialization (3.5) for $z = 3$ was chosen from among similar expressions for Padé-type approximations because it can be represetned as a special case of a hypergeometric polynomial closely connected with classical hypergeometric polynomials. It turns out that we can represent the double sum in (3.2) as

$$Y_n(z) = 4^n \cdot \Sigma_{i=0}^{3n} \binom{3n}{i} \cdot \binom{-\frac{1}{2}+3n}{4n-i} \quad (z+1)^{2n-i} \cdot z^i \tag{3.9}$$

or, for $z = 3$,

$$Y_n = \Sigma_{i=0}^{3n} \binom{3n}{i} \cdot \binom{-\frac{1}{2}+3n}{4n-i} 4^{4n-i} \cdot 3^i \tag{3.10}$$

The sequence of "dense" rational approximations X_n/Y_n to $\frac{4\pi}{3\sqrt{3}}$ provides us, according to the Poincaré lemma, with the following system of rational approximations Q_n/P_n to $\pi/\sqrt{3}$:

A. The numbers P_n are rational integers; Q_n are rational numbers, whose denominator divides $\ell cm\{1,\dots,n\}$.

B. The asymptotics of $|P_n|$, $|Q_n|$ and $|P_n \frac{\pi}{\sqrt{3}} - Q_n|$ is determined by the roots of the limit quadratic equation in the following way

$$\left.\begin{array}{l}\log|P_n| \\ \\ \log|Q_n|\end{array}\right\} \sim -1.664392\ldots\cdot n$$

and

$$\log|P_n \frac{\pi}{\sqrt{3}} - Q_n| \sim 2.200669\ldots\cdot n$$

as $n \to \infty$.

Here P_n and Q_n correspond to Y_{4n} and X_{4n}, respectively.

The limit quadratic equation that determines the asymptotics of Y_n and X_n follows from (3.6), (3.7): $2^8x^2 - 3^3\cdot 59\cdot 1069x - 3^5 = 0$.

This implies the following measure of irrationality of $\pi/\sqrt{3}$ presented for the first time in [19]:

$$|q \frac{\pi}{\sqrt{3}} - p| > |q|^{-4.8174417\ldots} \tag{3.11}$$

for arbitrary rational integers p, q with $|q| \geq q_4$.

More complicated Pade-type approximations to $8/\sqrt{z}$ arctg $1/\sqrt{z}$ of the form (3.1), (3.2) provide sequences of rational approximations to $4\pi/3\sqrt{3}$ with better density constants. E.g.:

$$|q \frac{\pi}{\sqrt{3}} - p| > |q|^{-4.792613804\ldots}$$

for rational integers p,q with $|q| \geq q_5$ and (effective) $q_5 > 0$. This is a significant improvement over the exponent $-7.3099\ldots$ in the previous bound (2.5).

New Padé-type approximation schemes for logarithmic functions improve constants in the measure of irrationality of the logarithms of algebraic numbers. For example, let us consider the problem of the best exponent in the measure of irrationality of $\log(a/b)$ for arbitrary integers $a>b\geqslant 1$. Hermite's

Padé approximations to the system of functions $\log^i(1+x)$: $i = 0,1,\ldots,m$ at $x = 0$ for $m > 1$ were completely analyzed by one of the authors in [14]. From general results of [14] we obtained the following:

PROPOSITION 3.1 [20]. Let $a > b \geq 1$ be arbitrary rational integers. Then for arbitrary rational integers p,q we have

$$|q \log(a/b) - p| > |q|^{-(18.567\ldots)\log a}$$

provided that $|q| \geq q_0(a,b)$.

Using new Padé-type approximations this measure of irrationality can be significantly improved and gives constants close to existing conjectures from the theory of linear forms in the logarithms of algebraic numbers.

We want to remark also that Apéry's remarkable measure of the irrationality of π^2 with the exponent -11.85 [15], can be considerably improved:

$$|q\pi^2 - p| > |q|^{-6.4}$$

for arbitrary rational integers p,q with $|q| \geq q_6$.

REFERENCES

[1] K. Mahler, Composito Math. 19 (1968), 95-166.

[2] Ch. Hermite, C. R. Acad. Sci. Paris 77 (1873), 18-24, 74-79, 226-233, 285-293, = Oeuvres, v. III, 150-179.

[3] L. Onsager, Phys. Rev. 65 (1944), 117.

[4] H.P. McKean, P. van Moerbeke, Comm. Pure Appl. Math. 33 (1980), 23-42.

[5] D.V. Chudnovsky, Cargése lectures, in Bifurcation Phenomena in mahtematical physics and related topics, D. Reidel, Boston, 1980, 385-447.

[6] P. van Moerbeke, D. Mumford, Acta Math. 61 (1979), 225.

[7] B. Riemann, Oeuvres Mathematiques, Blanchard, Paris, 1968, 353-363.

[8] C.L. Siegel, Transcendental numbers, Princeton, 1949.

[9] D.V. Chudnovsky. G.V. Chudnovsky, J. Math. Pures. Appl. 61 (1982), 4-16.

[10] G.V. Chudnovsky, C.R. Acad. Sci. Paris 288 (1979), A-607-A-609.

[11] K. Alladi, M.L. Robinson, Lecture Notes Math., v. 751, Springer, 1979, 1-9.

[12] F. Beukers, Lecture Notes Math., Springer, 1981, v. 888, 90-99.

[13] V. Danilov, Math. Zametki 24 (1978), no. 6.

[14] G.V. Chudnovsky. in the Riemann Problem, Complete Integrability and Arithmetic Applications, Lecture Notes Math., Springer, 1982, v. 925, 299-322.

[15] A. Van der Poorten, Math. Intelligencer 1 (1979), 195-203.

[16] Transcendence theory: advances and applications, Academic Press, Longon, 1977.

[17] G.A. Baker, Jr., P. Graves-Morris, Padé approximants, Encycl. Math. v. 13, 14, ed. by G.-C. Rota, Addison-Wesley, 1981.

[18] G.V. Chudnovsky, Festschrift in honor of F. Gursey, Gordon and Breach, 1983.

[19] G.V. Chudnovsky, Proc. Japan Academy v. 58 (1982), 129-133.

[20] G.V. Chudnovsky, C.R. Acad. Sci. Paris, 288 (1979), A-965-A-967.

[21] G.V. Chudnovsky, Lecture Notes Physics, v. 120, Springer, 1980, 103-150.

HARMONIC OSCILLATORS AT LOW ENERGIES

Richard C. Churchill and David Lee*

Hunter College
City University of New York
New York, New York

To study a nonlinear oscillator $m\ddot{x} = -\nabla V(x)$ at low positive energies one can simply write the potential V as a power series and truncate all nonlinear terms. This is the method of *small oscillations*, the simplest example of which is the use of $\ddot{x} = -x$ as an approximation to the pendulum equation $\ddot{x} = -\sin x$. However, once it is realized that higher order terms can be quite influential, this approach has to be abandoned, and replaced by some alternate approximation which hopefully exhibits the predominant nonlinearity in a particularly simple way. The view of this paper is that the normal form of the oscillator provides the appropriate alternative.

This work is an expository account of the use of normal forms in connection with the study of nonlinear harmonic oscillators. It is written at an elementary level, so as to make the ideas accessible to as large an audience as possible, and is self-contained except for some standard definitions and results from the theory of ordinary differential equations, e.g. the variational equation and the existence theorem for flows. Harmonic oscillators are viewed as Hamiltonian systems, but no prior knowledge of such systems is assumed on the part of the reader.

To convert a harmonic oscillator to normal form requires a series of transformations which, although rewarding in result, is tedious in

*Supported by the PSC-CUNY Research Award Program, Grant Number 13719;
Current Affiliation: Department of Computer Science, Columbia University, New York, New York

execution. Fortunately the computations are amenable to symbolic computer manipulation, and we have taken this approach; the FORMAC 73 interpreter was used, and executed on the IBM 3033 equipment of CUNY. The results form the examples, one of which is presented in extreme detail.

In other respects this paper contains little that is new, with the possible exception of the geometrical bias which is everywhere evident. In fact most of the results can be found scattered throughout the literature; here they are simply collected in one place.

Moreover, the paper is not a survey. In fact many important topics of current research, such as the reduced phase space of an oscillator with symmetries, are not even mentioned. Readers wishing to pursue the ideas presented here should consult [1] and [8] for further references.

My thinking as regards normal forms has been strongly influenced by Martin Kummer, both through papers and through personal contact, and I am pleased to acknowledge this fact. I am also grateful to Richard Cushman for helpful conversations over the past few years. Needless to say, errors which may occur in this manuscript are my responsibility alone.

Both Mr. Lee and I would like to thank Samuel Allen for his assistance on the computer aspects of this work.

§1. Elementary Facts Concerning Hamiltonian Systems

Notation

We let $<\ ,\ >$ denote the usual inner product on R^m.

If $F: U \subset R^m \to R$, then F_x will denote the gradient of F (which we regard, along with all elements of R^m, as a column vector), and F_{x_j} will denote the partial derivative of F with respect to the variable x_j. All functions will be assumed (at least) C^3, i.e. three times continuously differentiable.

$J = J_n$ will denote the $2n\times 2n$ skew-symmetric matrix $\begin{pmatrix} 0 & I \\ -I & 0 \end{pmatrix}$, where $I = I_n$ is the $n\times n$ identity matrix. One immediately verifies that

$$J^T = -J = J^{-1} \tag{1.1}$$

where T indicates the transpose. Notice that switching columns k and $k+n$ of J, for all $1 \le k \le n$, gives

$$\det J = (-1)^n \det\begin{pmatrix} I & 0 \\ 0 & -I \end{pmatrix} = (-1)^n(-1)^n = (-1)^{2n} = 1 \tag{1.2}$$

Also note by the skew-symmetry of J that

$$\langle x, Jx\rangle = 0, \qquad x \in R^{2n} \tag{1.3}$$

Hamilton's Equations

A *Hamiltonian system* is a system of ordinary differential equations, defined on an open set $U \subset R^{2n}$, having the form

$$\dot{x} = JH_x \tag{1.4}$$

i.e. $\dot{x} = JH_x(x)$. The function $H:U \to R$ is then called the *Hamiltonian*, or *(total) energy (function)*, and n the number of *degrees of freedom*. Notice that H is unique only up to a constant (assuming U is connected).

If we write $x = (q,p)$, where q and p vary in open subsets of R^n, then (1.4) becomes

$$\dot{q} = H_p, \qquad \dot{p} = H_q \tag{1.5}$$

i.e.

$$\dot{q}_j = \frac{\partial H}{\partial p_j} \qquad \dot{p}_j = -\frac{\partial H}{\partial q_j} \qquad j = 1,\ldots,n \tag{1.6}$$

q and p are called *canonical variables*, and q_j and p_j are said to be *(canonically) conjugate (variables)*. The notation (1.6) is also used, with n replaced by nk, when $q_j = (q_{j1},\ldots,q_{jk})$ and $p_j = (p_{j1},\ldots,p_{jk})$ are vector variables assuming values in open subsets of R^k. In that case (1.6) is an abbreviation for

$$\dot{q}_{ji} = \frac{\partial H}{\partial p_{ji}} \qquad \dot{p}_{ji} = -\frac{\partial H}{\partial q_{ji}} \qquad i = 1,\ldots,k \qquad j = 1,\ldots,n \tag{1.7}$$

and we refer to (1.6) as *vector notation* for (1.7), or simply as *vector notation*.

Suppose r represents one of p_j or q_j in (1.6). If H is periodic in r then r is called an *angular* or *cyclic variable*. If $\partial H/\partial r \equiv 0$, then r is *ignorable*.

To see the relationship of Hamiltonian systems to Classical Mechanics let $n > 0$ and $k > 0$ be integers, let $m_1,\ldots,m_n$ be positive constants, and let $(1/2m)$ be the block diagonal matrix with j^{th}-block $(1/2m_j)I_k$, where I_k is the $k\times k$ identity matrix. If H has the form

$$(1.8) \qquad H(q,p) = \langle(1/2m)p,p\rangle + V(q)$$

then in vector notation the associated equations are

$$(1.9) \qquad \dot{q}_j = (1/m_j)p_j \quad \dot{p}_j = -V_{q_j} \qquad j = 1,\ldots,n$$

i.e.

$$(1.10) \qquad m_j\ddot{q}_j = -V_{q_j}$$

precisely Newton's equations for n particles, of masses $m_1,\ldots,m_n$, moving in R^k subject to a force with potential function V. Following the custom in mechanics we call q_j the jth*-position variable*, and $p_j = m_j\dot{q}_j$ the jth*-momentum variable*.

Integrals

Let $V \subset U \subset R^m$ be open, $f:U \to R^m$ and $G:V \to R$. G is a *local integral* of the ordinary differential equation

$$(1.11) \qquad \dot{x} = f(x)$$

if G is constant along each solution of this equation, i.e. if for any solution $x = x(t)$ of (1.11) we have $(d/dt)G(x(t)) \equiv 0$. If G is such and $V = U$, then G is a *global integral*, or simply an *integral*, of (1.11). Be aware that some authors do not distinguish between local and global integrals.

For f, G and $x = x(t)$ as in the previous paragraph the chain-rule gives

$$\frac{d}{dt}G(x) = \langle G_x, \dot{x} \rangle = \langle G_x, f \rangle \tag{1.12}$$

and so G is a local integral of (1.11) if and only if G_x is perpendicular to f at each $x \in V$. But of course G_x is perpendicular to the level surfaces of G; hence f must then be tangent to such surfaces. In other words, the geometrical requirement for G to be a local integral of (1.11) is that solution curves of this equation lie on level surfaces of G, as in Figure 1. The existence of a local integral thus reduces the dimension of a given equation.

Now suppose (1.11) is Hamiltonian, i.e. has the form (1.4). Then (1.12) and (1.3), with H_x replacing x, show

$$\frac{d}{dt}H(x) = \langle H_x, JH_x \rangle \equiv 0 \tag{1.13}$$

This establishes the following result.

1.14 Theorem (Conservation of Energy): *The Hamiltonian for a Hamiltonian system is always an integral of the associated equations.*

Level surfaces of the Hamiltonian are called *energy surfaces* of the corresponding system, and the value of the Hamiltonian along a particular solution is called the *energy* of that solution.

If $V \subset U \subset R^{2n}$ are open and $F,G:V \to R$, define the *Poisson bracket* $\{F,G\}:V \to R$ of F and G by

$$\{F,G\} = \langle F_x, JG_x \rangle \tag{1.15}$$

i.e. $\{F,G\}(x) = \langle F_x(x), JG_x(x) \rangle$, $x \in V$. If $x = x(t)$ satisfies (1.4) then (1.12) gives

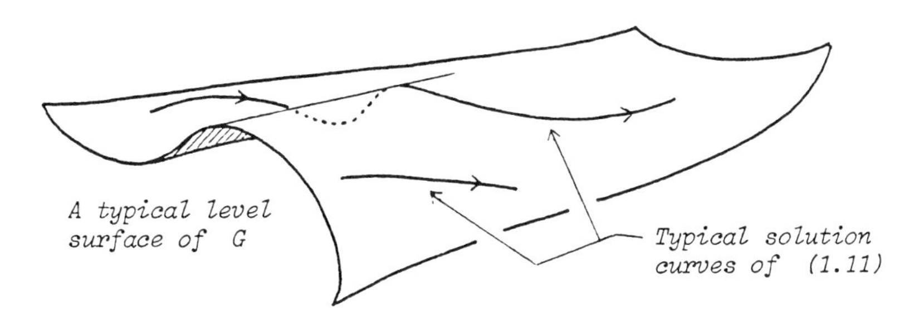

FIGURE 1

$$\frac{d}{dt}G(x) = \{G,H\} \tag{1.16}$$

from which we obtain the following result.

1.17 Proposition: *$G:V \to R$ is a local integral of (1.4) if and only if $\{G,H\} = 0$.*

For computational purposes the important properties of the Poisson bracket are

$$(1.18)\begin{cases} \text{(a)} & \{F,G\} \equiv -\{G,F\} \\ \text{(b)} & \{E,\{F,G\}\} + \{G,\{E,F\}\} + \{F,\{G,E\}\} \equiv 0 \end{cases}$$

where of course $E:V \to R$, (b) being called the *Jacobi identity*. (The usual cross product on R^3 satisfies similar identities.) These relations will certainly be familiar to anyone even mildly acquainted with Lie algebras, but we do not make this assumption on the part of the reader.

Integrable Systems

Using the notation of the previous section call a collection of k local integrals $G^{(j)}:V \to R$ of (1.11) *independent* if $\{G_x^{(j)}(x)\}_{j=1}^{k}$ is a linearly independent set for all x in some dense open subset of V. (Some authors require this condition on all of V, but this is rarely encountered in practice.) The Hamiltonian system (1.4) is *integrable* if there are n independent integrals $G^{(j)}:U \to R$ which are in *involution*, i.e.

$$\{G^{(i)},G^{(j)}\} \equiv 0, \quad 1 \le i,j \le n \tag{1.19}$$

Integrability is a rare but desirable phenomenon.

Canonical Transformations

A (real) $2n\times 2n$ matrix M is *symplectic* if

$$M^T JM = J \tag{1.20}$$

notice from (1.1) that J satisfies this equation. Since $\det M = \det M^T$ and $\det J = 1$ (see (1.2)), (1.20) implies

$$\det M = \pm 1 \tag{1.21}$$

(in fact one can show, although we will not, that $\det M = 1$). In particular, a symplectic matrix must be nonsingular. Multiplying (1.20) on the left by $(M^T)^{-1} = (M^{-1})^T$, and on the right by M^{-1}, we find

$$(M^{-1})^T J M^{-1} = J \tag{1.22}$$

and so M^{-1} is symplectic whenever M has this property. Also, if N is such then $(NM)^T J(NM) = M^T(N^T JN)M = M^T JM = J$. Since $I = I_{2n}$ is obviously symplectic, we conclude that such matrices form a group. This is the *(real) symplectic group*, denoted $Sp(n,R)$.

Notice that (1.20) holds if and only if $M^T = -JM^{-1}J$, which in turn is true if and only if

$$MJM^T = J \tag{1.23}$$

In short, (1.20) and (1.23) are equivalent. In particular, M is symplectic if and only if M^T is symplectic.

We could just as well have considered complex matrices in the foregoing discussion. In that case we would have obtained the *(complex) symplectic group* $Sp(n,C)$.

1.24 Examples of Symplectic Matrices

(a) For a 2×2 matrix M one computes easily that $M^T JM = (\det M)J$. Therefore, *a 2×2 matrix M is symplectic if and only if $\det M = 1$.*

(b) If N and S are $n\times n$ matrices, with N nonsingular and S symmetric, then $M = \begin{pmatrix} N & NS \\ 0 & (N^T)^{-1} \end{pmatrix}$ is symplectic. The proof is a straightforward computation using (1.20).

Let $V \subset R^{2n}$ be open, let $U \subset R^{2n}$ be open, and let $f: V \to U$. Then f is called a *canonical transformation* if the Jacobian matrix $\frac{df}{dy}(y)$

is symplectic at each $y \in V$. Notice from (1.22) and (1.23) that this implies

(1.25) $$(\frac{df}{dy})^{-1} J((\frac{df}{dy})^{-1})^T = J$$

at each point of V. Regarding $x = f(y)$ as a change of variables in (1.4) we have $\frac{df}{dy}(y)\dot{y} = \dot{x} = JH_x(x) = JH_x(f(y)) = J((\frac{df}{dy})^{-1})^T(\frac{df}{dy})^T H_x(f(y))$ $= J((\frac{df}{dy})^{-1})^T (H \circ f)_y(y)$, the last equality by the chain rule (which is transposed because of our column vector convention for the gradient). (1.25) then gives $\dot{y} = (\frac{df}{dy})^{-1} J((\frac{df}{dy})^{-1})^T (H \circ f)_y(y) = J(H \circ f)_y(y)$, i.e.

(1.26) $$\dot{y} = J(H \circ f)_y(y)$$

This result is summarized in the following statement.

1.27 Theorem: *A canonical transformation carries a Hamiltonian system into a Hamiltonian system, with the new Hamiltonian simply being the composition of the original Hamiltonian with the transformation.*

The theorem explains the importance of canonical transformations in the study of Hamilton's equations.

1.28 Examples of Canonical Transformations

(a) For any nonsingular $n \times n$ matrix N the transformation $q = Nu$, $p = (N^T)^{-1}v$ is canonical. Indeed, let $S = 0$ in 1.24(b).

(b) The transformation $q_j = \sqrt{2I_j}\cos\theta_j$, $p_j = \sqrt{2I_j}\sin\theta_j$, $I_j > 0$, is canonical, as the reader can easily verify.

The transformation (b) can be useful if one needs angular variables in a given Hamiltonian.

Let φ^t be the flow of (1.4), i.e. if $x_0 \in U$ then $\gamma(t) = \varphi^t(x_0)$ will be the unique solution of (1.4) satisfying $\gamma(0) = x_0$.

1.29 Theorem: φ^t *is canonical for each* t. *More precisely, if for each* $t \in R$ *we let* $U_t = \{ x \in U : \varphi^t(x)$ *is defined* $\}$, *then* $\varphi^t : U_t \to U$ *is a canonical transformation.*

Proof: Let H_{xx} denote the Hessian matrix $(\frac{\partial^2 H}{\partial x_i \partial x_j})$, and write φ_x^t for $\frac{d}{dx}\varphi^t$. The trick is to recall that φ_x^t satisfies the variational equation $\dot{z} = JH_{xx}(\varphi^t(x))z$, $s(0) = I$ $(= I_{2n})$, of (1.4), i.e. $\dot{\varphi}_x^t = JH_{xx}\varphi_x^t$, $\varphi_x^0(x) = I$, and so by (1.1) we have $(\dot{\varphi}_x^t)^T = -(\varphi_x^t)^T H_{xx} J$. But then $\frac{d}{dt}(\varphi_x^t)^T J \varphi_x^t = (\varphi_x^t)^T J \dot{\varphi}_x^t + (\dot{\varphi}_x^t)^T J \varphi_x^t = (\varphi_x^t)^T JJH_{xx}\varphi_x^t - (\varphi_x^t)^T H_{xx} JJ\varphi_x^t = -(\varphi_x^t)^T H_{xx}\varphi_x^t + (\varphi_x^t)^T H_{xx}\varphi_x^t = 0$. This shows $g(t) = (\varphi_x^t)^T J \varphi_x^t$ is constant, and since $\varphi_x^0 = I$ we conclude $g(t) = g(0) = J$ as desired. Q.E.D.

As a corollary we see from (1.21) that

$$\det \varphi_x^t \equiv 1 \tag{1.30}$$

Indeed, $h(t) = \det \varphi_x^t$ is continuous in t, and $\varphi_x^0 = I$.

1.31 Liouville's Theorem: *The flow* φ^t *is volume preserving, i.e.*

$$\int_{\varphi^t(C)} dV = \int_C dV$$

for any 2n-cell $C \subset U \subset R^{2n}$.

Proof: By the change of variables formula we have

$$\int_{\varphi^t(C)} dy = \int_C |\det \varphi_x^t| dx$$

and so the result follows from (1.30). Q.E.D.

Volume preserving flows are also called *incompressible*.

Time and Parameter-dependent Hamiltonians

Let $U \subset R^{2n}$ and $I \subset R$ be open, let x vary in U and t in I, and let $H: I \times U \to R$. The differential equation

(1.32) $$\dot{x} = JH_x$$

i.e. $\dot{x} = JH_x(t,x)$, is then called a *(time dependent) Hamiltonian system*. Alternatively, suppose $U \subset R^{2n}$ and $W \subset R^k$ are open, let λ vary in W, and let $H:W\times U \to R$. In this case the differential equation (1.32), now meaning $\dot{x} = JH_x(\lambda,x)$, is called a *parameter dependent Hamiltonian system*, with λ being the *parameter*. Finally, if $H:W\times I\times U \to R$ then (1.32), i.e. $\dot{x} = JH_x(\lambda,t,x)$, is called a *time and parameter-dependent Hamiltonian system*. In all cases H is called the *Hamiltonian (function)*.

Now suppose (1.32) is time and/or parameter dependent and let η^t be the associated time and/or parameter dependent flow, i.e. if $(\lambda_0,t_0,x_0) \in W\times I\times U$, then $\gamma(t) = \eta^t(\lambda_0,t_0,x_0)$ is the unique solution of (1.32), with λ fixed at λ_0, satisfying $\gamma(t_0) = x_0$.

1.33 Theorem: *As a mapping from U into U, η^t is canonical for each t. More precisely, if for $t \in I$ we let $U_t = \{ x \in U : \eta^t(\lambda_0,t_0,x_0)$ is defined $\}$, then $\eta^t:U_t \to U$ is a canonical transformation.*

The proof is easily adapted from that of Theorem 1.29, using the appropriate variational equation.

1.34 Corollary: *If $n = 1$, so $U \subset R^2$, the mapping $\eta^t:U_t \to U$ is area-preserving.*

Proof: Here we can argue as in Liouville's Theorem 1.31, or use Example 1.24(a). Q.E.D.

§2. Harmonic Oscillators

Normalizing the Quadratic Terms

Consider a Hamiltonian of the form

(2.1) $$H(q,p) = \tfrac{1}{2}\langle Ap,p\rangle + \tfrac{1}{2}\langle Bq,q\rangle + \cdots$$

where A and B are symmetric $n\times n$ matrices, with A positive definite, and where the dots denote higher order terms. Notice this form includes the Hamiltonians (1.8) of classical mechanics for which $V(0) = 0$ and $V_q(0) = 0$, in which case $B = V_{qq}(0)$ is the Hessian matrix.

Given (2.1), construct matrices C and Q according to the following recipe:

(2.2)
- (a) Choose a nonsingular $n \times n$ matrix C such that $C^T AC = I$; then
- (b) Let E denote the symmetric matrix $C^{-1}B(C^{-1})^T$, and choose an orthogonal $n \times n$ matrix Q which diagonalizes E, i.e. $Q^T EQ = \text{diag}\{k_1, \ldots, k_n\}$.

That such choices are possible is well-known from elementary linear algebra; the k_j are real since E is symmetric.

2.3 Lemma: *In the above notation,* $k_1, \ldots, k_n$ *are the roots of* $\det(B-\lambda A^{-1}) = 0$, *and all are positive if and only if* B *is positive definite.*

Proof: $k_1, \ldots, k_n$ are obviously the roots of $\det(Q^T EQ-\lambda I) = 0$. But E and $Q^T EQ$ are similar, since $Q^T = Q^{-1}$, and so $k_1, \ldots, k_n$ must also be the roots of $\det(E-\lambda I) = 0$. Taking inverses in (2.2a) we obtain $I = C^{-1}A^{-1}(C^{-1})^T$, hence $E-\lambda I = C^{-1}B(C^{-1})^T - \lambda C^{-1}A^{-1}(C^{-1})^T = C^{-1}(B-\lambda A^{-1})(C^{-1})^T$, and the first assertion now follows from the product rule for determinants. Since B and $Q^T EQ$ are equivalent, both have the same number of positive eigenvalues (see any book on elementary linear algebra), and this gives the second assertion. Q.E.D.

2.4 Theorem: *For* C *and* Q *as above the substitution*

$$q = (C^{-1})^T Qu, \quad p = CQv \tag{1}$$

is canonical, and transforms (2.1) *into*

$$\hat{H}(u,v) = \sum_{j=0}^{n} \tfrac{1}{2}(v_j^2 + k_j u_j^2) + \cdots \tag{2}$$

where the dots denote higher order terms.

Proof: (1) has the form $q = Nu$, $p = (N^T)^{-1}v$, where $N = (C^{-1})^T Q$, and so the first statement follows from Example 1.28(a). As for the second, we have $\langle ACQv, CQv\rangle = \langle Q^T C^T ACQv, v\rangle = \langle Q^{-1}Qv, v\rangle = |v|^2$ and $\langle BNu, Nu\rangle = \langle N^T BNu, u\rangle = \langle Q^T C^{-1} B(C^{-1})Qu, u\rangle = \langle Q^T EQu, u\rangle = \Sigma k_j u_j^2$ as asserted. Q.E.D.

2.5 Theorem: *Suppose* $k_j > 0$ *in* (2) *of Theorem 2.4, and write* $k_j = \omega_j^2$, $\omega_j > 0$, $j = 1,\ldots,n$. *Then the substitution*

$$u_j = \omega_j^{-\frac{1}{2}} q_j, \quad v_j = \omega_j^{\frac{1}{2}} p_j \tag{1}$$

is canonical, and transforms (2) *of Theorem 2.4 into*

$$H(q,p) = \sum_{j=1}^{n} (\omega_j/2)(q_j^2 + p_j^2) + \cdots \tag{2}$$

where the dots denote higher order terms.

By Lemma 2.3 the condition $k_j > 0$ for all j is equivalent to B in (2.1) being positive definite. Notice that the H, q, and p in the statement above are not the H, q, and p of (2.1).

Proof: That (1) is canonical follows from Example 1.28(a), and the remaining assertion is obvious. Q.E.D.

2.6 Example

Consider the Hamiltonian

$$(1/2m_1)p_1^2 + (1/2m_2)p_2^2 + \tfrac{1}{2}(b_{11}q_1^2 + 2b_{12}q_1q_2 + b_{22}q_2^2) + \cdots \tag{1}$$

where $m_1 > 0$, $m_2 > 0$. This two-degree of freedom example is of the form (2.1), with

$$A = \begin{pmatrix} 1/m_1 & 0 \\ 0 & 1/m_2 \end{pmatrix}, \quad B = \begin{pmatrix} b_{11} & b_{12} \\ b_{12} & b_{22} \end{pmatrix}$$

Here

$$\det(B-\lambda A^{-1}) = m_1m_2\lambda^2-(b_{11}m_2+b_{22}m_1)\lambda+(b_{11}b_{22}-b_{12}^2)$$

and so

$$(2) \qquad k_1,k_2 = \frac{1}{2m_1m_2}((b_{11}m_2+b_{22}m_1)\pm\{(b_{11}m_2-b_{22}m_1)^2+4m_1m_2b_{12}^2\}^{\frac{1}{2}})$$

in particular, *$k_1 = k_2$ if and only if $b_{11}m_2 = b_{22}m_1$ and $b_{12} = 0$.* For the matrix C of (2.2a) we can take

$$C = \begin{pmatrix} m_1^{\frac{1}{2}} & 0 \\ 0 & m_2^{\frac{1}{2}} \end{pmatrix}$$

and one then easily computes the matrix $E = C^{-1}B(C^{-1})^T$ of (2.2b) to be

$$E = \begin{pmatrix} \alpha & \beta \\ \beta & \gamma \end{pmatrix}, \quad \alpha = b_{11}/m_1, \quad \beta = b_{12}/\sqrt{m_1m_2}\ , \quad \gamma = b_{22}/m_2.$$

If $\beta = 0$ set $Q = I$; otherwise define

$$\rho = ((\alpha-k_1)^2+\beta^2)^{\frac{1}{2}} > 0$$

and set

$$Q = \rho^{-1}\begin{pmatrix} \beta & \alpha-k_1 \\ k_1-\alpha & \beta \end{pmatrix}$$

Assuming $\beta \neq 0$ and using the relations $k_1+k_2 = \alpha+\gamma$ and $k_1k_2 = \alpha\gamma-\beta^2$, which follow since k_1,k_2 are the eigenvalues of E, one easily checks that $Q^TEQ = \mathrm{diag}\{k_1,k_2\}$. Moreover, a simple calculation shows that the transformation (1) of Theorem 2.4 in this case is given by

$$q = \rho^{-1}\begin{pmatrix} \beta/\sqrt{m_1} & (\alpha-k_1)/\sqrt{m_1} \\ (k_1-\alpha)/\sqrt{m_2} & \beta/\sqrt{m_2} \end{pmatrix}u, \qquad p = \rho^{-1}\begin{pmatrix} \beta\sqrt{m_1} & (\alpha-k_1)\sqrt{m_1} \\ (k_1-\alpha)\sqrt{m_2} & \beta\sqrt{m_2} \end{pmatrix}v$$

In summary, *this last transformation is canonical, and will convert (1) to the form (2) of Theorem 2.4, with k_1 and k_2 given by (2).* Of course if k_1 and k_2 are positive, we can then apply Theorem 2.5 so as to further transform (1), arriving at the form (2) of that statement.

A *harmonic oscillator*, or simply an *oscillator*, is a Hamiltonian system defined on an open neighborhood $U \subset R^{2n}$ of the origin, with Hamiltonian $H:U \to R$ having the form

$$H(q,p) = \sum_{j=1}^{n}(\omega_j/2)(q_j^2+p_j^2) + \cdots \tag{2.7}$$

Here the dots denote terms of order at least three in q_j and/or p_j, and if such terms are actually present the oscillator is called *nonlinear*; otherwise it is *linear*. The constants $\omega_1,\ldots,\omega_n$ are the *(characteristic, relative, or fundamental) frequencies* of the system, and one speaks of *nonresonance* in connection with (2.7) if these numbers are independent over the rationals, i.e. if $\Sigma r_j\omega_j = 0$ for rational r_j implies each $r_j = 0$. If this is not the case, then one speaks of *resonance*.

Theorems 2.4 and 2.5 show that (2.1) can be converted to a harmonic oscillator, by means of a linear canonical transformation, if B is positive definite. This is a special case $(S = \begin{pmatrix} A & 0 \\ 0 & B \end{pmatrix})$ of the following result.

2.8 Theorem: *Let* $U \subset R^{2n}$ *be an open neighborhood of the origin, and let* $H:U \to R$ *have the form*

$$H(x) = \tfrac{1}{2}\langle Sx,x\rangle + \cdots$$

where S *is a positive definite symmetric matrix and the dots denote higher order terms. Then there is a linear canonical transformation which converts* H *to a harmonic oscillator with positive frequencies.*

For a proof see ([1], pp. 492, 494-5), and for an example see [4].

Complex Notation

Both for theoretical and practical purposes it is convenient to treat harmonic oscillators (and many other Hamiltonian systems as well) in terms of the complex "variables"

$$z_j = q_j+ip_j, \quad \bar{z}_j = q_j-ip_j, \quad j = 1,\ldots,n \tag{2.9}$$

From the relations

$$(2.10) \qquad q_j = \tfrac{1}{2}(z_j+\bar{z}_j), \quad p_j = -\tfrac{1}{2}i(z_j-\bar{z}_j)$$

it is clear that any real-valued polynomial in q,p, i.e. in $q_1,\ldots,q_n$, $p_1,\ldots,p_n$, can be expressed as a real-valued polynomial in $z,\bar{z}$, i.e. in $z_1,\ldots,z_n,\bar{z}_1,\ldots,\bar{z}_n$, and conversely.

2.11 Example

For any real numbers λ and δ we have

$$\begin{aligned}
&\lambda q_1^3 + \delta q_1 q_2^2 = \\
&(\lambda/8)z_1^3 + (3\lambda/8)z_1^2\bar{z}_1 + (3\lambda/8)z_1\bar{z}_1^2 + (\lambda/8)\bar{z}_1^3 \\
&+ (\delta/8)z_1 z_2^2 + (\delta/4)z_1 z_2\bar{z}_2 + (\delta/8)z_1\bar{z}_2^2 \\
&+ (\delta/8)z_2^2\bar{z}_1 + (\delta/4)z_2\bar{z}_1\bar{z}_2 + (\delta/8)\bar{z}_1\bar{z}_2^2
\end{aligned}$$

Now introduce the operators

$$(2.12) \qquad \frac{\partial}{\partial z_j} = \tfrac{1}{2}\left(\frac{\partial}{\partial q_j} - i\frac{\partial}{\partial p_j}\right), \quad \frac{\partial}{\partial \bar{z}_j} = \tfrac{1}{2}\left(\frac{\partial}{\partial q_j} + i\frac{\partial}{\partial p_j}\right), \quad j = 1,\ldots,n$$

and notice that if a Hamiltonian H is written in terms of $z,\bar{z}$ then Hamilton's equations simply become

$$(2.13) \qquad \dot{z}_j = -2i\frac{\partial H}{\partial \bar{z}_j}, \quad j = 1,\ldots,n$$

we abbreviate this as

$$(2.14) \qquad \dot{z} = -2i\frac{\partial H}{\partial \bar{z}}$$

Also, if F and G in (1.15) are written in terms of $z,\bar{z}$, then (2.12) easily implies

$$(2.15) \qquad \{F,G\} = 2i\sum_{j=1}^n\left(\frac{\partial F}{\partial \bar{z}_j}\frac{\partial G}{\partial z_j} - \frac{\partial F}{\partial z_j}\frac{\partial G}{\partial \bar{z}_j}\right)$$

Finally, if for obvious reasons we write

(2.16) $$z_j\bar{z}_j = |z_j|^2$$

when working with the variables $z,\bar{z}$, then the harmonic oscillator (2.7) becomes

(2.17) $$H(z,\bar{z}) = \sum_{j=1}^{n}(\omega_j/2)|z_j|^2 + \cdots$$

with associated equations

(2.18) $$\dot{z}_j = -i\omega_j z_j + \cdots, \quad j = 1,\ldots,n$$

The operators (2.12) obey the usual rules for partial differentiation when applied to polynomials in $z,\bar{z}$, wherein $z_1,\ldots,z_n,\bar{z}_1,\ldots,\bar{z}_n$ are imagined as $2n$ independent variables. As examples check that

$$\frac{\partial}{\partial z_1}(z_1 z_2^2\bar{z}_1 + 3z_1 z_2\bar{z}_2) = z_2^2\bar{z}_1 + 3z_2\bar{z}_2$$

and

$$\frac{\partial}{\partial \bar{z}_2}(z_1 z_2\bar{z}_2 + z_1\bar{z}_1 + \bar{z}_1^3) = z_1 z_2 + 3\bar{z}_2^2$$

Also compare (2.18) with (2.14), using H as in (2.17).

Stretching Variables

Consider (for simplicity) an analytic harmonic oscillator

(2.19) $$\begin{cases} H(z,\bar{z}) = H_2(z,\bar{z}) + H_3(z,\bar{z}) + H_4(z,\bar{z}) + \cdots \\ H_2(z,\bar{z}) = \sum_{j=1}^{n}(\omega_j/2)|z_j|^2 \end{cases}$$

defined on an open neighborhood $U \subset R^{2n}$ of the origin. At the same time consider the associated parameter-dependent Hamiltonian

(2.20) $$H_\varepsilon(z,\bar{z}) = H_2(z,\bar{z}) + \varepsilon H_3(z,\bar{z}) + \varepsilon^2 H_4(z,\bar{z}) + o(\varepsilon^2)$$

in which the homogeneous polynomial $H_d(z,\bar{z})$ agrees for all $d \geq 2$ with that in (2.19).

2.21 Theorem: *For any* $\varepsilon > 0$ *the Hamiltonian system associated with (2.19) is equivalent to that associated with (2.20) by means of the (noncanonical) substitution*

$$(1) \qquad z \to \varepsilon z$$

In fact, solutions of (2.20) of energy $h > 0$ *correspond, via (1), to (reparameterized) solutions of (2.19) of energy* $\varepsilon^2 h$.

This is verified by a straightforward calculation.

The point of the result is that we can study (2.19) at small positive energies by studying (2.20) when $\varepsilon > 0$ is small. But this latter system can be regarded as a small perturbation of the linear oscillator $H_2(z,\bar{z})$, and so our next task should be to study the linear case.

The use of (1) in Theorem 2.21 is called *stretching variables*. It is also coming to be known as *blowing up the singularity 0*, due to the fact that (2.20) makes sense when $\varepsilon = 0$, even though (1) does not (the terminology is borrowed from algebraic geometry).

Linear Oscillators

Here we consider the linear oscillator

$$(2.22) \qquad H(z,\bar{z}) = \textstyle\sum_{j=1}^{n} (\omega_j/2)|z_j|^2$$

with associated equations

$$(2.23) \qquad \dot{z}_j = -i\omega_j z_j, \quad j = 1,\ldots,n$$

and associated flow

$$(2.24) \qquad \sigma^t(z) = (e^{-i\omega_1 t}z_1,\ldots,e^{-i\omega_n t}z_n)$$

The following result is trivially verified.

2.25 Theorem: *The Hamiltonian system (2.23) is integrable. Indeed, a collection of* n *independent integrals in involution is provided by the functions*

$$(1) \qquad G_j : z \to |z_j|^2, \quad j = 1,\ldots,n$$

Continuing with the above notation, we claim that if $c = (c_1,\ldots,c_n) \in R^n$, $c_j \geq 0$, then the set

$$(2.26) \qquad T_c = \{ z \in R^{2n} : G_j(z) = c_j, \quad j = 1,\ldots,n \}$$

is a torus of dimension at most n. To see this first note that the projection of T_c into the z_j-plane is $\{ z_j : |z_j|^2 = c_j \}$, obviously a circle if $c_j > 0$ and the origin if $c_j = 0$; the assertion follows by observing that T_c can be identified with the cartesian product of these projections. Since the G_j are integrals of (2.23), we conclude that solutions of that equation are constrained to lie on tori. Moreover, by varying $c = (c_1,\ldots,c_n)$ we see that R^{2n} is stratified (or "foliated') by these σ^t-invariant tori. In fact such behavior is the case in any integrable system, i.e. the domain must decompose into tori and/or cylinders which are invariant under the associated flow ([2], p. 279).

Now recall that conservation of energy guarantees that the energy surfaces of (2.22), obviously quadratic surfaces, are also σ^t-invariant, hence must also be foliated by invariant tori. We can visualize the case $n = 2$, when $\omega_1 > 0$ and $\omega_2 > 0$, as follows. First, in

$$(2.27) \qquad H(z,\bar{z}) = (\omega_1/2)|z_1|^2 + (\omega_2/2)|z_2|^2, \quad \omega_1 > 0, \quad \omega_2 > 0,$$

write $z_1 = re^{i\theta}$, so that the constraint $H = h$, which obviously defines an ellipsoid, reduces to $\omega_1 r^2 + \omega_2 |z_2|^2 = 2h$. As a consequence z_2 determines r on $H = h$, and we may use θ and z_2, where

$$(2.28) \qquad -\pi < \theta < \pi \quad \text{and} \quad |z_2|^2 < 2h/\omega_2$$

as coordinates. However, we can also view (2.28) as defining coordinates for the "double solid cone" illustrated in Figure 2. More specifically, if p is as indicated (it is assumed in the interior of the upper cone), then the coordinates (θ, z_2) of p are obtained as follows: z_2 is the projection of p into the plane, q is obtained by drawing the line segment from 0 through z_2 to the circle, and θ is then obtained by drawing the line segment from q through p to the

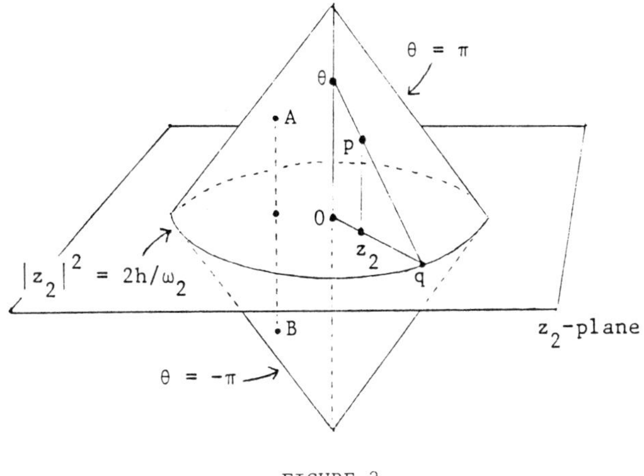

FIGURE 2

vertical axis. Notice in the picture that any two points on the "boundary" (i.e. where $\theta = \pm\pi$) must be "identified" (i.e. considered the same) if both have the same z_2-coordinate, e.g. A and B.

The inner vertical axis and outer planar circle in Figure 2 represent the *normal modes* of (2.23) of energy h, i.e. the orbits of energy h having initial conditions of the form $(z_1,0)$ and $(0,z_2)$. The remaining orbits of energy h are shown winding around their containing tori in Figure 3; from (2.24) one can see that such orbits close only in the resonance case, and will be dense within the containing tori in the nonresonance case.

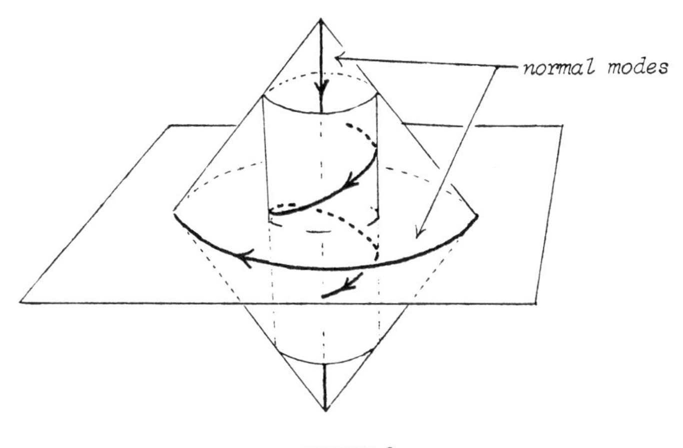

FIGURE 3

It is important to realize that Figure 3 can be "untwisted' by means of the substitution

$$(2.29)\qquad \begin{cases} z_1 = \frac{1}{\sqrt{\omega_1}}(2L-\omega_2|w|^2)^{\frac{1}{2}} ie^{-i\omega_1\rho} \\ z_2 = we^{-i\omega_2\rho} \end{cases}$$

which we view as defining a transformation

$$(2.30)\qquad (\rho+iL,w) \to (z_1,z_2)$$

Notice that w may simply be regarded as a rotation of z_2 by $i\omega_2\rho$.

2.31 Proposition: *The substitution* (2.29) *is canonical, and is one-one on the domain* $-\pi/\omega_1 < \rho < \rho/\omega_2$, $L > 0$, $|w|^2 < 2L/\omega_2$.

Proof: The first statement is a straightforward verification. As for the second, first notice from (2.29) that

$$(1)\qquad \omega_1|z_1|^2 + \omega_2|z_2|^2 = 2L, \qquad |z_2|^2 = |w|^2$$

hence $2L - \omega_2|w|^2 = 2L - \omega_2|z_2|^2 = \omega_1|z_1|^2$, and so

$$(2)\qquad z_1 = |z_1| ie^{-i\omega_1\rho}$$

Given z_1 and z_2 we can determine L from (1), a unique $\rho \in (-(\pi/\omega_1),\rho/\omega_1]$ from (2), and a unique w such that $|w|^2 < 2L/\omega_2$ from $w = z_2e^{i\omega_2\rho}$. Q.E.D.

From Proposition 2.31 and Theorem 1.27 we see that (2.29) converts (2.22) to

$$(2.32)\qquad H(\rho,\mathrm{Re}\ w,L,\mathrm{Im}\ w) = L$$

for which the associated equations are simply

$$(2.33)\qquad \begin{array}{ll} \dot{\rho} = 1 & \dot{L} = 0 \\ (\mathrm{Re}\ w)^{\cdot} = 0, & (\mathrm{Im}\ w)^{\cdot} = 0 \end{array}$$

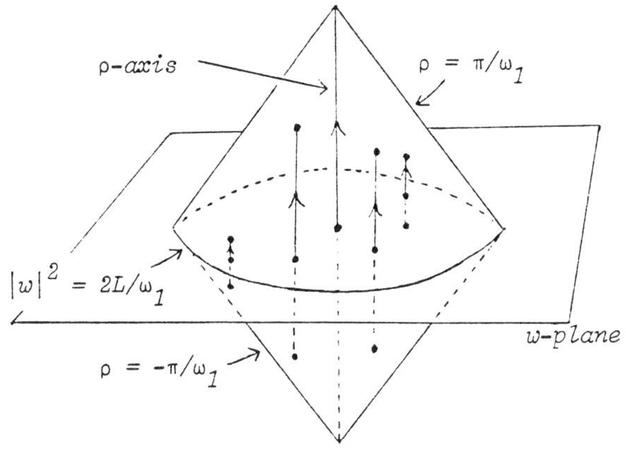

FIGURE 4

If we mimic Figure 2 by viewing w and ρ as coordinates on a double solid cone, then from (2.33) we see that Figure 3 becomes Figure 4, which is certainly much simpler. However, *unless* $\omega_1 = \omega_2 = 1$, *the identifications used with Figure 2 must be adjusted in Figure 4.* Indeed, without such adjustments all orbits would be closed even in the nonresonance case.

§3. Transforming Harmonic Oscillators into Normal Form

Normal Forms

Consider a harmonic oscillator

$$(3.1)\quad \begin{cases} \text{(a)}\quad H(z,z) = H_2(z,z) + H_3(z,z) + \cdots + H_s(z,z) + \cdots \\ \text{(b)}\quad H_2(z,z) = \sum_{j=1}^{n} (\omega_j/2)|z_j|^2 \end{cases}$$

defined on some open neighborhood $U \subset R^{2n}$ of the origin, where $H_d(z,\bar{z})$ is a homogeneous polynomial of degree d in $z_1,\ldots,z_n,\bar{z}_1,\ldots,\bar{z}_n$, $3 \leq d \leq s$, and where the omitted terms in (a) are of order at least $s+1$. If

$$(3.2)\qquad \{H_2,H_r\} \equiv 0 \qquad r = 3,\ldots,s$$

then we say (3.1a) is in *normal form through terms of order* s (that (3.2) holds for $r = 2$ follows from (1.18a)).

Since

$$\frac{\partial H_2}{\partial \bar{z}_j} = (\omega_j/2)z_j, \quad \frac{\partial H_2}{\partial z_j} = (\omega_j/2)\bar{z}_j$$

(2.15) shows the equivalence of (3.2) with

$$\sum_{j=1}^{n} i\omega_j(z_j\frac{\partial H_r}{\partial z_j} - \bar{z}_j\frac{\partial H_r}{\partial \bar{z}_j}) \tag{3.3}$$

or, defining a differential operator D by

$$D = i\sum_{j=1}^{n}\omega_j(z_j\frac{\partial}{\partial z_j} - \bar{z}_j\frac{\partial}{\partial \bar{z}_j}) \tag{3.4}$$

with

$$DH_r \equiv 0, \quad r = 3,\ldots,s \tag{3.5}$$

In terms of the variables $q_j = \frac{1}{2}(z_j+\bar{z}_j)$ and $p_j = -\frac{1}{2}i(z_j-\bar{z}_j)$, (3.4) is simply

$$D = \sum_{j=1}^{n}\omega_j(q_j\frac{\partial}{\partial p_j} - p_j\frac{\partial}{\partial q_j}) \tag{3.6}$$

One can also express (3.2) in terms of Lie derivatives (i.e. as $L_{X_2}X_r \equiv 0$, where $X_j = J(H_j)_x$), but we do not assume familiarity with this notion.

If

$$z^k\bar{z}^\ell = z_1^{k_1}\cdots z_n^{k_n}\bar{z}_1^{\ell_1}\cdots\bar{z}_n^{\ell_n} \tag{3.7}$$

is any monomial, then one can easily check that

$$Dz^k\bar{z}^\ell = \sum_{j=1}^{n} i\omega_j(k_j-\ell_j)z^k\bar{z}^\ell \tag{3.8}$$

which we abbreviate as

$$Dz^k\bar{z}^\ell = i\langle\omega,k-\ell\rangle z^k\bar{z}^\ell \tag{3.9}$$

The following result is then immediate.

3.10 Proposition: *For* $z^k\bar{z}^\ell$ *as in (3.7),*

$$Dz^k\bar{z}^\ell = 0 \tag{1}$$

if and only if

(2) $$\langle\omega,k-\ell\rangle = \Sigma\omega_j(k_j-\ell_j) = 0$$

and if this is not the case then

(3) $$z^k\bar{z}^\ell = D(-i\langle\omega,k-\ell\rangle^{-1}z^k\bar{z}^\ell)$$

3.11 Remark: Notice that if $cz^{k'}\bar{z}^{\ell'}$ is such that $Dcz^{k'}\bar{z}^{\ell'} = 0$, then $D(-i\langle\omega,k-\ell\rangle^{-1}z^k\bar{z}^\ell + cz^{k'}\bar{z}^{\ell'})$ also equals $z^k\bar{z}^\ell$. In other words, the solution of $DX = z^k\bar{z}^\ell$ given by (3) of Proposition 3.10 need not be unique.

The Transformation Algorithm

Our approach uses a "Lie series" method, but one can also use "generating function" techniques (e.g. see [8]).

Assume that (3.1) is in normal form through terms of order $r-1$, $4 \leq r \leq s$, or that $r = 3$, and define a new function K^r according to the following recipe:

(3.12)
(a) Write $H_r(z,\bar{z}) = \Sigma_N c_{k\ell}z^k\bar{z}^\ell + \Sigma_R c_{k\ell}z^k\bar{z}^\ell = H_r^N(z,\bar{z}) + H_r^R(z,\bar{z})$, where the first sum is over all (vectors) k,ℓ such that $\langle\omega,k-\ell\rangle = 0$, and the second is over all remaining terms. Then

(b) Define

$$K^r(z,\bar{z}) = \Sigma_R i\langle\omega,k-\ell\rangle^{-1}c_{k\ell}z^k\bar{z}^\ell$$

Notice from (3) of Proposition 3.10 that

(3.13) $$DK^r = -H_r^R$$

and from (1) and (2) of Proposition 3.10 that

(3.14) $$DH_r^N = 0$$

3.15 Theorem: *Using the notation above let* ρ^t *be the flow generated by* $\dot{z} = -2i(\partial K^r/\partial\bar{z})$. *Then there is a neighborhood* $V \subset U$ *of the origin such that* $\rho^t: V \to U$ *is defined for* $|t| < 2$, *and each of these mappings is canonical. Moreover, if we let* $\rho = \rho^1 : V \to U$, *then* $H\circ\rho$ *will be in normal form through terms of order* r, *will agree with* H *up through terms of order* $r-1$, *and will have* $H_r^N(z,\bar{z})$ *as* r^{th}*-term. In fact*

$$(1) \qquad H\circ\rho = H + \{H,K^r\} + \frac{1}{2!}\{\{H,K^r\},K^r\} + \cdots$$

where the omitted terms are as one would guess. For any $k \geq 3$ *we can also write* (1) *as*

$$(2) \qquad \begin{cases} H\circ\rho = H + \sum_{j=2}^{k}\{H_j,K^r\}_{j+r-2} + \frac{1}{2!}\sum_{j=2}^{k}\{\{H_j,K^r\},K^r\}_{j+2(r-2)} \\ \qquad + \frac{1}{3!}\sum_{j=2}^{k}\{\{\{H_j,K^r\},K^r\},K^r\}_{j+3(r-2)} + \cdots \end{cases}$$

where the subscripts on the brackets denote the order in $z,\bar{z}$ *of the homogeneous polynomials defined by those brackets, and where the omitted terms have order at least* $2+4(r-2) > r$.

The form (2) is useful when the final goal is to transform a given Hamiltonian into normal form through terms of order k.

Proof: The homogeneity of K^r implies that the origin is an equilibrium point, hence $\rho^t(0) \equiv 0$, $t \in R$, and since the domain of a flow is always open the existence of V follows. That ρ^t is canonical is seen from Theorem 1.29.

For $|t| < 2$ and $x \in V$ define $F(t) = H\circ\rho^t = H\circ\rho^t(x)$, and by Taylor's theorem we have

$$F(t) = F(0) + F'(0)t + \frac{1}{2!}F''(0)t^2 + \cdots$$

But by (1.16), (1.18) and the (obvious) linearity of the Poisson bracket we have

$$F'(t) = \frac{d}{dt}H\circ\rho^t = \{H,K^r\} = \Sigma\{H_j,K^r\}_{j+r-2} + \cdots$$

and

$$F''(t) = \frac{d}{dt}\{H,K^r\} = \{\{H,K^r\},K^r\}$$

etc., and (1) and (2) follow.

Finally, since $j+\ell(r-2) \geq r$ for $\ell \geq 1$, with equality only when $j = 2$, $\ell = 1$ and $r = 3$, of all terms preceded by a summation sign in (2) only $\{H_2,K^r\}$ has order less than or equal r, and that order is r. Moreover, by (3.2), (3.5) and (3.13) we have $\{H_2,K^r\} = DK^r = -H_r^R$. Therefore, the terms in (1) and (2) of order at most r are

$$H_2 + \cdots + H_{r-1} + H_r^N + H_r^R - H_r^R$$
$$= H_2 + \cdots + H_{r-1} + H_r^N,$$

and this is in normal form through terms of order r by hypothesis and by (3.14). Q.E.D.

At this point we should mention an alternate notation used in conjunction with (1) of Theorem 3.15. If we define $ad_{K^r}^0 H = H$, $ad_{K^r}H = \{H,K^r\}$, and $ad_{K^r}^m H = \{ad_{K^r}^{m-1}H,K^r\}$, $m > 1$, then we can write (1) in that statement as

$$\text{(3.16)} \qquad H\circ\rho = \sum_{m=1}^{\infty}(1/m!)ad_{K^r}^m H$$

(assuming H is analytic). This should explain why $H\circ\rho$ is sometimes written

$$\text{(3.17)} \qquad (\exp ad_{K^r})H$$

The "ad" is an abbreviation for "adjoint representation," and is borrowed from the vocabulary of Lie algebras (where a minus sign is often included with the definition).

We should also point out why the "normalized term" $H_r^N(z,\bar{z})$ is often called the *averaged term*, or *average, of* H_r *with respect to the linearized flow.*

3.18 Theorem: *If* $\omega_j > 0$ *in* (3.1b), $j = 1,\ldots,n$, *then in the notation of* (3.12) *we have*

$$\text{(1)} \qquad H_r^N(z,\bar{z}) = \lim_{T\to\infty}\int_0^T H_r(\sigma^t(z,\bar{z}))dt$$

where σ^t *is the flow of the "linearized equation of* (3.1) *about* $z = 0$," *i.e.* $\dot{z} = -2i(\partial H_2/\partial\bar{z})$.

Proof: (3.14) shows $\frac{d}{dt}H_r^N(\sigma^t) = \{H_r^N, H_2\} = -\{H_2, H_r^N\} = -DH_r^N = 0$, and so $\frac{1}{T}\int^T H_r^N = H_r^N(z,\bar{z})$. Similarly, by choosing K^r as in (3.12), (3.13) gives $\frac{d}{dt}K^r(\sigma^t) = H_r^R$, hence $\frac{1}{T}\int^T H_r^R = \frac{1}{T}K^r$. But $\omega_j > 0$ implies that the level surfaces $H = h$ are ellipsoids, hence that $K^r(\sigma^t)$ is bounded, and so $\frac{1}{T}K^r \to 0$. The result now follows since $H_r = H_r^N + H_r^R$. Q.E.D.

A Detailed Example

We will apply the algorithm (3.12) successively to the *Hénon-Heiles Hamiltonian*

$$H(q,p) = \tfrac{1}{2}(q_1^2+p_1^2) + \tfrac{1}{2}(q_2^2+p_2^2) + \lambda q_1^2 + \delta q_1 q_2^2 \tag{3.19}$$

so as to convert the problem into normal form through terms of order six. Here λ and δ are parameters, with $\lambda = 1/3$ and $\delta = -1$ being the values originally studied [5]. In fact by a simple rescaling it is easy to see that only the values +1, 0 and -1 need be considered for δ, but the extra parameter should make the computations easier to follow.

We first convert (3.19) to $z,\bar{z}$ notation. Using Example 2.11, the result is seen to be

$$\left\{\begin{aligned} H(z,\bar{z}) &= H_2 + H_3, \\ H_2 &= \tfrac{1}{2}|z_1|^2 + \tfrac{1}{2}|z_2|^2, \\ H_3 &= (\lambda/8)z_1^3 + (3\lambda/8)z_1^2\bar{z}_1 + (3\lambda/8)z_1\bar{z}_1^2 + (\lambda/8)\bar{z}_1^3 \\ &\quad + (\delta/8)z_1 z_2^2 + (\delta/4)z_1 z_2\bar{z}_2 + (\delta/8)z_1 z_2^2 \\ &\quad + (\delta/8)z_2^2\bar{z}_1 + (\delta/4)z_2\bar{z}_1\bar{z}_2 + (\delta/8)\bar{z}_1\bar{z}_2^2 \end{aligned}\right. \tag{3.20}$$

Conversion Through Third Order

From (3.12a) we compute that $H_3^R = H_3$, $H_3^N = 0$, and from (3.12b) that

$$\begin{aligned} K^3 = {} & i(\lambda/24)z_1^3 + i(3\lambda/8)z_1^2\bar{z}_1 - i(3\lambda/8)z_1\bar{z}_1^2 + i(\delta/24)z_1 z_2^2 + i(\delta/4)z_1 z_2\bar{z}_2 \\ & -i(\lambda/24)\bar{z}_1^3 + i(\delta/8)z_2^2\bar{z}_1 - i(\delta/4)z_2\bar{z}_1\bar{z}_2 - i(\delta/24)\bar{z}_1\bar{z}_2^2 \end{aligned}$$

Using (2.15) and (2) of Theorem 3.15 we then compute

$$\text{(3.21)} \qquad \mathrm{Hop} = H(2) + H(3) + H(4) + H(5) + H(6) + \cdots$$

where

$$H(2) = \tfrac{1}{2}|z_1|^2 + \tfrac{1}{2}|z_2|^2$$

$$H(3) = 0$$

$$\begin{aligned} H(4) = {} & (3\lambda^2/32)z_1^4 - (3\lambda^2/8)z_1^3\bar z_1 - (15\lambda^2/16)z_1^2\bar z_1^2 + (\lambda\delta/16)z_1^2z_2^2 + (\delta^2/24)z_1^2z_2^2 \\ & -(\lambda\delta/8)z_1^2z_2\bar z_2 - (\delta^2/12)z_1^2z_2\bar z_2 + (\lambda\delta/16)z_1^2\bar z_2^2 - (\delta^2/8)z_1^2\bar z_2^2 \\ & -(3\lambda^2/8)z_1\bar z_1^3 - (\lambda\delta/8)z_1z_2^2\bar z_1 - (\delta^2/12)z_1z_2^2\bar z_1 - (3\lambda\delta/4)z_1z_2\bar z_1\bar z_2 \\ & -(\delta^2/6)z_1z_2\bar z_1\bar z_2 - (\lambda\delta/8)z_1\bar z_1\bar z_2^2 - (\delta^2/12)z_1\bar z_1\bar z_2^2 + (3\lambda^2/32)\bar z_1^4 \\ & +(\lambda\delta/16)z_2^2\bar z_1^2 - (\delta^2/8)z_2^2\bar z_1^2 - (\lambda\delta/8)z_2\bar z_1^2\bar z_2 - (\delta^2/12)z_2\bar z_1^2\bar z_2 \\ & +(\lambda\delta/16)\bar z_1^2\bar z_2^2 + (\delta^2/24)\bar z_1^2\bar z_2^2 + (\delta^2/96)z_2^4 - (\delta^2/24)z_2^3\bar z_2 - (5\delta^2/48)z_2^2\bar z_2^2 \\ & (\delta^2/24)z_2\bar z_2^3 + (\delta^2/96)\bar z_2^4 \end{aligned}$$

$$\begin{aligned} H(5) = {} & (\lambda^3/4)z_1^5 - (\lambda^3/4)z_1^4\bar z_1 + 2\lambda^3 z_1^3\bar z_1^2 + (\lambda^2\delta/6)z_1^3z_2^2 + (\lambda\delta^2/9)z_1^3z_2^2 \\ & +(\delta^3/27)z_1^3z_2^2 - (\lambda\delta^2/9)z_1^3z_2\bar z_2 - (\lambda\delta^2/9)z_1^3\bar z_2^2 + (\delta^3/9)z_1^3\bar z_2^2 \\ & +2\lambda^3 z_1^2\bar z_1^3 - (\lambda^2\delta/6)z_1^2z_2^2\bar z_1 - (\delta^3/27)z_1^2z_2^2\bar z_1 + (5\lambda^2\delta/3)z_1^2z_2\bar z_1\bar z_2 \\ & +(\lambda\delta^2/3)z_1^2z_2\bar z_1\bar z_2 + (4\delta^3/27)z_1^2z_2\bar z_1\bar z_2 - (\lambda^2\delta/3)z_1^2\bar z_1\bar z_2^2 \\ & +(2\lambda\delta^2/3)z_1^2\bar z_1\bar z_2^2 + (\delta^3/27)z_1^2\bar z_1\bar z_2^2 - (\lambda^3/4)z_1\bar z_1^4 - (\lambda^2\delta/3)z_1z_2^2\bar z_1^2 \\ & +(2\lambda\delta^2/3)z_1z_2^2\bar z_1^2 + (\delta^3/27)z_1z_2^2\bar z_1^2 + (5\lambda^2\delta/3)z_1z_2\bar z_1^2\bar z_2 \\ & +(\lambda\delta^2/3)z_1z_2\bar z_1^2\bar z_2 + (4\delta^3/27)z_1z_2\bar z_1^2\bar z_2 - (\lambda^2\delta/6)z_1\bar z_1^2\bar z_2^2 \\ & -(\delta^3/27)z_1\bar z_1^2\bar z_2^2 + (\lambda\delta^2/36)z_1z_2^4 + (\delta^3/27)z_1z_2^4 - (\delta^3/27)z_1z_2^3\bar z_2 \\ & +(\lambda\delta^2/3)z_1z_2^2\bar z_2^2 + (\delta^3/9)z_1z_2^2\bar z_2^2 - (\lambda\delta^2/9)z_1z_2\bar z_2^3 + (5\delta^3/27)z_1z_2\bar z_2^3 \\ & -(\lambda\delta^2/36)z_1\bar z_2^4 + (\lambda^3/4)\bar z_1^5 - (\lambda\delta^2/9)z_2^2\bar z_1^3 + (\delta^3/9)z_2^2\bar z_1^3 \\ & -(\lambda\delta^2/9)z_2\bar z_1^3\bar z_2 + (\lambda^2\delta/6)\bar z_1^3\bar z_2^2 + (\lambda\delta^2/9)\bar z_1^3\bar z_2^2 + (\delta^3/27)\bar z_1^3\bar z_2^2 \\ & -(\lambda\delta^2/36)z_2^4\bar z_1 - (\lambda\delta^2/9)z_2^3\bar z_1\bar z_2 + (5\delta^3/27)z_2^3\bar z_1\bar z_2 + (\lambda\delta^2/3)z_2^2\bar z_1\bar z_2^2 \\ & +(\delta^3/9)z_2^2\bar z_1\bar z_2^2 - (\delta^3/27)z_2\bar z_1\bar z_2^3 + (\lambda\delta^2/36)\bar z_1\bar z_2^4 + (\delta^3/27)\bar z_1\bar z_2^4 \end{aligned}$$

$$\begin{aligned}
H(6) = {} & (3\lambda^4/8)z_1^6 - (39\lambda^4/32)z_1^5\bar{z}_1 - (3\lambda^4/4)z_1^4\bar{z}_1^2 + (9\lambda^3\delta/32)z_1^4z_2^2 \\
& +(5\lambda^2\delta^2/32)z_1^4z_2^2 + (\lambda\delta^3/12)z_1^4z_2^2 + (\delta^4/72)z_1^4z_2^2 - (11\lambda^3\delta/32)z_1^4z_2\bar{z}_2 \\
& -(3\lambda^2\delta^2/16)z_1^4z_2\bar{z}_2 - (\lambda\delta^3/24)z_1^4z_2\bar{z}_2 - (\delta^4/36)z_1^4z_2\bar{z}_2 \\
& -(\lambda^3\delta/32)z_1^4\bar{z}_2^2 - (5\lambda^2\delta^2/32)z_1^4\bar{z}_2^2 + (11\lambda\delta^3/72)z_1^4\bar{z}_2^2 - (\delta^4/24)z_1^4\bar{z}_2^2 \\
& -(69\lambda^4/16)z_1^3\bar{z}_1^3 - (7\lambda^3\delta/16)z_1^3z_2^2\bar{z}_1 - (13\lambda^2\delta^2/24)z_1^3z_2^2\bar{z}_1 \\
& -(\lambda\delta^3/9)z_1^3z_2^2\bar{z}_1 - (\delta^4/27)z_1^3z_2^2\bar{z}_1 - (7\lambda^3\delta/8)z_1^3z_2\bar{z}_1\bar{z}_2 \\
& +(5\lambda^2\delta^2/24)z_1^3z_2\bar{z}_1\bar{z}_2 - (\lambda\delta^3/12)z_1^3z_2\bar{z}_1\bar{z}_2 - (\delta^4/27)z_1^3z_2\bar{z}_1\bar{z}_2 \\
& +(7\lambda^3\delta/16)z_1^3\bar{z}_1\bar{z}_2^2 - (19\lambda\delta^3/36)z_1^3\bar{z}_1\bar{z}_2^2 - (3\lambda^4/4)z_1^2\bar{z}_1^4 \\
& +(3\lambda^3\delta/4)z_1^2z_2^2\bar{z}_1^2 - (25\lambda^2\delta^2/24)z_1^2z_2^2\bar{z}_1^2 + (\lambda\delta^3/8)z_1^2z_2^2\bar{z}_1^2 \\
& -(\delta^4/108)z_1^2z_2^2\bar{z}_1^2 - (81\lambda^3\delta/16)z_1^2z_2\bar{z}_1^2\bar{z}_2 - (5\lambda^2\delta^2/24)z_1^2z_2\bar{z}_1^2\bar{z}_2 \\
& -(3\lambda\delta^3/4)z_1^2z_2\bar{z}_1^2\bar{z}_2 - (\delta^4/54)z_1^2z_2\bar{z}_1^2\bar{z}_2 + (3\lambda^3\delta/4)z_1^2\bar{z}_1^2\bar{z}_2^2 \\
& -(25\lambda^2\delta^2/24)z_1^2\bar{z}_1^2\bar{z}_2^2 + (\lambda\delta^3/8)z_1^2\bar{z}_1^2\bar{z}_2^2 - (\delta^4/108)z_1^2\bar{z}_1^2\bar{z}_2^2 + (\lambda^2\delta^2/16)z_1^2z_2^4 \\
& +(\lambda\delta^3/16)z_1^2z_2^4 + (\delta^4/24)z_1^2z_2^4 - (5\lambda^2\delta^2/48)z_1^2z_2^3\bar{z}_2 - (7\lambda\delta^3/36)z_1^2z_2^3\bar{z}_2 \\
& -(2\delta^4/27)z_1^2z_2^3\bar{z}_2 - (13\lambda^2\delta^2/48)z_1^2z_2^2\bar{z}_2^2 - (\lambda\delta^3/48)z_1^2z_2^2\bar{z}_2^2 - (\delta^4/54)z_1^2z_2^2\bar{z}_2^2 \\
& +(3\lambda^2\delta^2/16)z_1^2z_2\bar{z}_2^3 - (\lambda\delta^3/24)z_1^2z_2\bar{z}_2^3 - (\delta^4/6)z_1^2z_2\bar{z}_2^3 - (\lambda^2\delta^2/24)z_1^2\bar{z}_2^4 \\
& +(\lambda\delta^3/9)z_1^2\bar{z}_2^4 - (\delta^4/24)z_1^2\bar{z}_2^4 - (39\lambda^4/32)z_1\bar{z}_1^5 + (7\lambda^3\delta/16)z_1z_2^2\bar{z}_1^3 \\
& -(19\lambda\delta^3/36)z_1z_2^2\bar{z}_1^3 - (7\lambda^3\delta/8)z_1z_2\bar{z}_1^3\bar{z}_2 + (5\lambda^2\delta^2/24)z_1z_2\bar{z}_1^3\bar{z}_2 \\
& -(\lambda\delta^3/12)z_1z_2\bar{z}_1^3\bar{z}_2 - (\delta^4/27)z_1z_2\bar{z}_1^3\bar{z}_2 - (7\lambda^3\delta/16)z_1\bar{z}_1^3\bar{z}_2^2 \\
& -(13\lambda^2\delta^2/24)z_1\bar{z}_1^3\bar{z}_2^2 - (\lambda\delta^3/9)z_1\bar{z}_1^3\bar{z}_2^2 - (\delta^4/27)z_1\bar{z}_1^3\bar{z}_2^2 - (\lambda^2\delta^2/96)z_1z_2^4\bar{z}_1 \\
& -(7\lambda\delta^3/72)z_1z_2^4\bar{z}_1 - (\delta^4/24)z_1z_2^4\bar{z}_1 + (11\lambda^2\delta^2/24)z_1z_2^3\bar{z}_1\bar{z}_2 \\
& -(35\lambda\delta^3/72)z_1z_2^3\bar{z}_1\bar{z}_2 + (\delta^4/27)z_1z_2^3\bar{z}_1\bar{z}_2 - (83\lambda^2\delta^2/48)z_1z_2^2\bar{z}_1\bar{z}_2^2 \\
& -(31\delta^4/108)z_1z_2^2\bar{z}_1\bar{z}_2^2 + (11\lambda^2\delta^2/24)z_1z_2\bar{z}_1\bar{z}_2^3 - (35\lambda\delta^3/72)z_1z_2\bar{z}_1\bar{z}_2^3 \\
& +(\delta^4/27)z_1z_2\bar{z}_1\bar{z}_2^3 - (\lambda^2\delta^2/96)z_1\bar{z}_1\bar{z}_2^4 - (7\lambda\delta^3/72)z_1\bar{z}_1\bar{z}_2^4 - (\delta^4/24)z_1\bar{z}_1\bar{z}_2^4 \\
& +(3\lambda^4/8)\bar{z}_1^6 - (\lambda^3\delta/32)z_2^2\bar{z}_1^4 - (5\lambda^2\delta^2/32)z_2^2\bar{z}_1^4 + (11\lambda\delta^3/72)z_2^2\bar{z}_1^4 \\
& -(\delta^4/24)z_2^2\bar{z}_1^4 - (11\lambda^3\delta/32)z_2\bar{z}_1^4\bar{z}_2 - (3\lambda^2\delta^2/16)z_2\bar{z}_1^4\bar{z}_2 - (\lambda\delta^3/24)z_2\bar{z}_1^4\bar{z}_2 \\
& -(\delta^4/36)z_2\bar{z}_1^4\bar{z}_2 + (9\lambda^3\delta/32)\bar{z}_1^4\bar{z}_2^2 + (5\lambda^2\delta^2/32)\bar{z}_1^4\bar{z}_2^2 + (\lambda\delta^3/12)\bar{z}_1^4\bar{z}_2^2 \\
& +(\delta^4/72)\bar{z}_1^4\bar{z}_2^2 - (\lambda^2\delta^2/24)\bar{z}_1^2\bar{z}_2^4 + (\lambda\delta^3/9)z_2^4\bar{z}_1^2 - (\delta^4/24)z_2^4\bar{z}_1^2 \\
& +(3\lambda^2\delta^2/16)z_2^3\bar{z}_1^2\bar{z}_2 - (\lambda\delta^3/24)z_2^3\bar{z}_1^2\bar{z}_2 - (\delta^4/6)z_2^3\bar{z}_1^2\bar{z}_2 - (13\lambda^2\delta^2/48)z_2^2\bar{z}_1^2\bar{z}_2^2 \\
& -(\lambda\delta^3/48)z_2^2\bar{z}_1^2\bar{z}_2^2 - (\delta^4/54)z_2^2\bar{z}_1^2\bar{z}_2^2 - (5\lambda^2\delta^2/48)z_2\bar{z}_1^2\bar{z}_2^3 - (7\lambda\delta^3/36)z_2\bar{z}_1^2\bar{z}_2^3
\end{aligned}$$

$$
\begin{aligned}
&-(2\delta^4/27)z_2\bar{z}_1^2\bar{z}_2^3 + (\lambda^2\delta^2/16)\bar{z}_1^2\bar{z}_2^4 + (\lambda\delta^3/16)\bar{z}_1^2\bar{z}_2^4 + (\delta^4/24)\bar{z}_1^2\bar{z}_2^4 \\
&+(\lambda\delta^3/288)z_2^6 + (\delta^4/288)z_2^6 + (\lambda\delta^3/288)z_2^5\bar{z}_2 - (7\delta^4/432)z_2^5\bar{z}_2 \\
&+(11\lambda\delta^3/288)z_2^4\bar{z}_2^2 - (19\delta^4/864)z_2^4\bar{z}_2^2 - (7\lambda\delta^3/48)z_2^3\bar{z}_2^3 - (\delta^4/216)z_2^3\bar{z}_2^3 \\
&+(11\lambda\delta^3/288)z_2^2\bar{z}_2^4 - (19\delta^4/864)z_2^2\bar{z}_2^4 + (\lambda\delta^3/288)z_2\bar{z}_2^5 - (7\delta^4/432)z_2\bar{z}_2^5 \\
&+(\lambda\delta^3/288)\bar{z}_2^6 + (\delta^4/288)\bar{z}_2^6
\end{aligned}
$$

Conversion Through Fourth Order

To avoid subscripting we write the Hamiltonian (3.21) as H, and compute from (3.12) that

$$
\begin{aligned}
K^4 = {} & i(3\lambda^2/128)z_1^4 - i(3\lambda^2/16)z_1^3\bar{z}_1 + i(\lambda\delta/64)z_1^2z_2^2 + i(\delta^2/96)z_1^2z_2^2 \\
&-i(\lambda\delta/16)z_1^2z_2\bar{z}_2 - i(\delta^2/24)z_1^2z_2\bar{z}_2 + i(3\lambda^2/16)z_1\bar{z}_1^3 - i(\lambda\delta/16)z_1z_2^2\bar{z}_1 \\
&-i(\delta^2/24)z_1z_2^2\bar{z}_1 + i(\lambda\delta/16)z_1\bar{z}_1\bar{z}_2^2 + i(\delta^2/24)z_1\bar{z}_1\bar{z}_2^2 - i(3\lambda^2/128)\bar{z}_1^4 \\
&+i(\lambda\delta/16)z_2\bar{z}_1^2\bar{z}_2 + i(\delta^2/24)z_2\bar{z}_1^2\bar{z}_2 - i(\lambda\delta/64)\bar{z}_1^2\bar{z}_1^2 - i(\delta^2/96)\bar{z}_1^2\bar{z}_2^2 \\
&+i(\delta^2/384)z_2^4 - i(\delta^2/48)z_2^3\bar{z}_2 + i(\delta^2/48)z_2\bar{z}_2^3 - i(\delta^2/384)\bar{z}_2^4.
\end{aligned}
$$

Now write the Hop of (3.21) as H, and again use (2.15) and (2) of Theorem 3.15. With the ρ now being that associated with K^4, we compute

$$
\text{(3.22)} \qquad H\circ\rho = H(2) + H(3) + H(4) + H(5) + H(6) + \cdots
$$

where

$$
H(2) = \tfrac{1}{2}|z_1|^2 + \tfrac{1}{2}|z_2|^2
$$

$$
H(3) = 0
$$

$$
\begin{aligned}
H(4) = {} & -(15\lambda^2/16)z_1^2\bar{z}_1^2 + (\lambda\delta/16)z_1^2\bar{z}_2^2 - (\delta^2/8)z_1^2\bar{z}_2^2 - (3\lambda\delta/4)z_1z_2\bar{z}_1\bar{z}_2 \\
&-(\delta^2/6)z_1z_2\bar{z}_1\bar{z}_2 + (\lambda\delta/16)z_2^2\bar{z}_1^2 - (\delta^2/8)z_2^2\bar{z}_1^2 - (5\delta^2/48)z_2^2\bar{z}_2^2
\end{aligned}
$$

$$
\begin{aligned}
H(5) = {} & (\lambda^3/4)z_1^5 - (\lambda^3/4)z_1^4\bar{z}_1 + 2\lambda^3z_1^3\bar{z}_1^2 + (\lambda^2\delta/6)z_1^3z_2^2 + (\lambda\delta^2/9)z_1^3z_2^2 \\
&+(\delta^3/27)z_1^3z_2^2 - (\lambda\delta^2/9)z_1^3z_2\bar{z}_2 - (\lambda\delta^2/9)z_1^3\bar{z}_2^2 + (\delta^3/9)z_1^3\bar{z}_2^2 + 2\lambda^3z_1^2\bar{z}_1^3
\end{aligned}
$$

$$-(\lambda^2\delta/6)z_1^2z_2^2\bar{z}_1 - (\delta^3/27)z_1^2z_2^2\bar{z}_1 + (5\lambda^2\delta/3)z_1^2z_2\bar{z}_1\bar{z}_2 + (\lambda\delta^2/3)z_1^2z_2\bar{z}_1\bar{z}_2$$

$$+(4\delta^3/27)z_1^2z_2\bar{z}_1\bar{z}_2 - (\lambda^2\delta/3)z_1^2\bar{z}_1\bar{z}_2^2 + (2\lambda\delta^2/3)z_1^2\bar{z}_1\bar{z}_2^2 + (\delta^3/27)z_1^2\bar{z}_1\bar{z}_2^2$$

$$-(\lambda^3/4)z_1\bar{z}_1^4 - (\lambda^2\delta/3)z_1z_2^2\bar{z}_1^2 + (2\lambda\delta^2/3)z_1z_2^2\bar{z}_1^2 + (\delta^3/27)z_1z_2^2\bar{z}_1^2$$

$$+(5\lambda^2\delta/3)z_1z_2\bar{z}_1^2\bar{z}_2 + (\lambda\delta^2/3)z_1z_2\bar{z}_1^2\bar{z}_2 + (4\delta^3/27)z_1z_2\bar{z}_1^2\bar{z}_2$$

$$-(\lambda^2\delta/6)z_1\bar{z}_1^2\bar{z}_2^2 - (\delta^3/27)z_1\bar{z}_1^2\bar{z}_2^2 + (\lambda\delta^2/36)z_1z_2^4 + (\delta^3/27)z_1z_2^4$$

$$-(\delta^3/27)z_1z_2^3\bar{z}_2 + (\lambda\delta^2/3)z_1z_2^2\bar{z}_2^2 + (\delta^3/9)z_1z_2^2\bar{z}_2^2 - (\lambda\delta^2/9)z_1z_2\bar{z}_2^3$$

$$+(5\delta^3/27)z_1z_2\bar{z}_2^3 - (\lambda\delta^2/36)z_1\bar{z}_2^4 + (\lambda^3/4)\bar{z}_1^5 - (\lambda\delta^2/9)z_2^2\bar{z}_1^3$$

$$+(\delta^3/9)z_2^2\bar{z}_1^3 - (\lambda\delta^2/9)z_2\bar{z}_1^3\bar{z}_2 + (\lambda^2\delta/6)\bar{z}_1^3\bar{z}_2^2 + (\lambda\delta^2/9)\bar{z}_1^3\bar{z}_2^2$$

$$+(\delta^3/27)\bar{z}_1^3\bar{z}_2^2 - (\lambda\delta^2/36)z_1^4\bar{z}_1 - (\lambda\delta^2/9)z_2^3\bar{z}_1\bar{z}_2 + (5\delta^3/27)z_2^3\bar{z}_1\bar{z}_2$$

$$+(\lambda\delta^2/3)z_2^2\bar{z}_1\bar{z}_2^2 + (\delta^3/9)z_2^2\bar{z}_1\bar{z}_2^2 - (\delta^3/27)z_2\bar{z}_1\bar{z}_2^3 + (\lambda\delta^2/36)\bar{z}_1\bar{z}_2^4$$

$$+(\delta^3/27)\bar{z}_1\bar{z}_2^4$$

$$H(6) = (87\lambda^4/256)z_1^6 - (111\lambda^4/128)z_1^5\bar{z}_1 - (471\lambda^4/256)z_1^4\bar{z}_1^2 + (33\lambda^3\delta/128)z_1^4z_2^2$$

$$+(35\lambda^2\delta^2/256)z_1^4z_2^2 + (5\lambda\delta^3/64)z_1^4z_2^2 + (7\delta^4/576)z_1^4z_2^2 - (13\lambda^3\delta/64)z_1^4z_2\bar{z}_2$$

$$-(21\lambda^2\delta^2/128)z_1^4z_2\bar{z}_2 - (\lambda\delta^3/32)z_1^4z_2\bar{z}_2 - (5\delta^4/288)z_1^4z_2\bar{z}_2$$

$$-(11\lambda^3\delta/256)z_1^4\bar{z}_2^2 - (3\lambda^2\delta^2/128)z_1^4\bar{z}_2^2 + (19\lambda\delta^3/144)z_1^4\bar{z}_2^2$$

$$-(\delta^4/16)z_1^4\bar{z}_2^2 - (705\lambda^4/128)z_1^3\bar{z}_1^3 - (11\lambda^3\delta/32)z_1^3z_2^2\bar{z}_1 - (71\lambda^2\delta^2/192)z_1^3z_2^2\bar{z}_1$$

$$-(5\lambda\delta^3/72)z_1^3z_2^2\bar{z}_1 - (13\delta^4/432)z_1^3z_2^2\bar{z}_1 - (235\lambda^3\delta/128)z_1^3z_2\bar{z}_1\bar{z}_2$$

$$-(49\lambda^2\delta^2/384)z_1^3z_2\bar{z}_1\bar{z}_2 - (3\lambda\delta^3/32)z_1^3z_2\bar{z}_1\bar{z}_2 - (59\delta^4/864)z_1^3z_2\bar{z}_1\bar{z}_2$$

$$+(41\lambda^3\delta/128)z_1^3\bar{z}_1\bar{z}_2^2 - (7\lambda^2\delta^2/64)z_1^3\bar{z}_1\bar{z}_2^2 - (41\lambda\delta^3/72)z_1^3\bar{z}_1\bar{z}_2^2$$

$$-(\delta^4/72)z_1^3\bar{z}_1\bar{z}_2^2 - (471\lambda^4/256)z_1^2\bar{z}_1^4 + (255\lambda^3\delta/256)z_1^2z_2^2\bar{z}_1^2$$

$$-(1109\lambda^2\delta^2/768)z_1^2z_2^2\bar{z}_1^2 - (\lambda\delta^3/192)z_1^2z_2^2\bar{z}_1^2 - (19\delta^4/1728)z_1^2z_2^2\bar{z}_1^2$$

$$-(45\lambda^3\delta/8)z_1^2z_2\bar{z}_1^2\bar{z}_2 - (251\lambda^2\delta^2/384)z_1^2z_2\bar{z}_1^2\bar{z}_2 - (27\lambda\delta^3/32)z_1^2z_2\bar{z}_1^2\bar{z}_2$$

$$-(43\delta^4/864)z_1^2z_2\bar{z}_1^2\bar{z}_2 + (255\lambda^3\delta/256)z_1^2\bar{z}_1^2\bar{z}_2^2 - (1109\lambda^2\delta^2/768)z_1^2\bar{z}_1^2\bar{z}_2^2$$

$$-(\lambda\delta^3/192)z_1^2\bar{z}_1^2\bar{z}_2^2 - (19\delta^4/1728)z_1^2\bar{z}_1^2\bar{z}_2^2 + (15\lambda^2\delta^2/256)z_1^2z_2^4$$

$$+(7\lambda\delta^3/128)z_1^2z_2^4 + (11\delta^4/288)z_1^2z_2^4 - (11\lambda^2\delta^2/192)z_1^2z_2^3\bar{z}_2$$

$$-(41\lambda\delta^3/288)z_1^2z_2^3\bar{z}_2 - (23\delta^4/432)z_1^2z_2^3\bar{z}_2 - (355\lambda^2\delta^2/768)z_1^2z_2^2\bar{z}_2^2$$

$$-(95\lambda\delta^3/768)z_1^2z_2^2\bar{z}_2^2 - (151\delta^4/3456)z_1^2z_2^2\bar{z}_2^2 + (5\lambda^2\delta^2/32)z_1^2z_2\bar{z}_2^3$$

$$-(37\lambda\delta^3/384)z_1^2z_2\bar{z}_2^3 - (109\delta^4/576)z_1^2z_2\bar{z}_2^3 - (5\lambda^2\delta^2/192)z_1^2\bar{z}_2^4$$

$$+(205\lambda\delta^3/2304)z_1^2\bar{z}_2^4 - (19\delta^4/384)z_1^2\bar{z}_2^4 - (111\lambda^4/128)z_1\bar{z}_1^5$$

$$
\begin{aligned}
&+(41\lambda^3\delta/128)z_1z_2^2\bar{z}_1^3 - (7\lambda^2\delta^2/64)z_1z_2^2\bar{z}_1^3 - (41\lambda\delta^3/72)z_1z_2^2\bar{z}_1^3 \\
&-(\delta^4/72)z_1z_2^2\bar{z}_1^3 - (235\lambda^3\delta/128)z_1z_2\bar{z}_1^3\bar{z}_2 - (49\lambda^2\delta^2/384)z_1z_2\bar{z}_1^3\bar{z}_2 \\
&-(3\lambda\delta^3/32)z_1z_2\bar{z}_1^3\bar{z}_2 - (59\delta^4/864)z_1z_2\bar{z}_1^3\bar{z}_2 - (11\lambda^3\delta/32)z_1\bar{z}_1^3\bar{z}_2 \\
&-(71\lambda^2\delta^2/192)z_1\bar{z}_1^3\bar{z}_2^2 - (5\lambda\delta^3/72)z_1\bar{z}_1^3\bar{z}_2^2 - (13\delta^4/432)z_1\bar{z}_1^3\bar{z}_2^2 \\
&-(7\lambda^2\delta^2/384)z_1z_2^4\bar{z}_1 - (41\lambda\delta^3/576)z_1z_2^4\bar{z}_1 - (\delta^4/36)z_1z_2^4\bar{z}_1 \\
&+(197\lambda^2\delta^2/384)z_1z_2^3\bar{z}_1 - (695\lambda\delta^3/1152)z_1z_2^3\bar{z}_1 - (65\delta^4/1728)z_1z_2^3\bar{z}_1\bar{z}_2 \\
&-(691\lambda^2\delta^2/384)z_1z_2^2\bar{z}_1\bar{z}_2^2 - (5\lambda\delta^3/32)z_1z_2^2\bar{z}_1\bar{z}_2^2 - (311\delta^4/864)z_1z_2^2\bar{z}_1 \\
&+(197\lambda^2\delta^2/384)z_1z_2\bar{z}_1\bar{z}_2^3 - (695\lambda\delta^3/1152)z_1z_2\bar{z}_1\bar{z}_2^3 - (65\delta^4/1728)z_1z_2\bar{z}_1\bar{z}_2^3 \\
&-(7\lambda^2\delta^2/384)z_1\bar{z}_1\bar{z}_2^4 - (41\lambda\delta^3/576)z_1\bar{z}_1\bar{z}_2^4 - (\delta^4/36)z_1\bar{z}_1\bar{z}_2^4 \\
&+(87\lambda^4/256)\bar{z}_1^6 - (11\lambda^3\delta/256z_2^2\bar{z}_1^4 - (3\lambda^2\delta^2/128)z_2^2\bar{z}_1^4 + (19\lambda\delta^3/144)z_2^2\bar{z}_1^4 \\
&-(\delta^4/16)z_2^2\bar{z}_1^4 - (13\lambda^3\delta/64)z_2\bar{z}_1^4\bar{z}_2 - (21\lambda^2\delta^2/128)z_2\bar{z}_1^4\bar{z}_2 \\
&-(\lambda\delta^3/32)z_2\bar{z}_1^4\bar{z}_2 - (5\delta^4/288)z_2\bar{z}_1^4\bar{z}_2 + (33\lambda^3\delta/128)\bar{z}_1^4\bar{z}_2^2 \\
&+(35\lambda^2\delta^2/256)\bar{z}_1^4\bar{z}_2^2 + (5\lambda\delta^3/64)\bar{z}_1^4\bar{z}_2^2 + (7\delta^4/576)\bar{z}_1^4\bar{z}_2^2 - (5\lambda^2\delta^2/192)z_2^4\bar{z}_1^2 \\
&+(205\lambda\delta^3/2304)z_2^4\bar{z}_1^2 - (19\delta^4/384)z_2^4\bar{z}_1^2 + (5\lambda^2\delta^2/32)z_2^3\bar{z}_1^2\bar{z}_2 \\
&-(37\lambda\delta^3/384)z_2^3\bar{z}_1^2\bar{z}_2 - (109\delta^4/576)z_2^3\bar{z}_1^2\bar{z}_2 - (355\lambda^2\delta^2/768)z_2^2\bar{z}_1^2\bar{z}_2^2 \\
&-(95\lambda\delta^3/768)z_2^2\bar{z}_1^2\bar{z}_2^2 - (151\delta^4/3456)z_2^2\bar{z}_1^2\bar{z}_2^2 - (11\lambda^2\delta^2/192)z_2\bar{z}_1^2\bar{z}_2^3 \\
&-(41\lambda\delta^3/288)z_2\bar{z}_1^2\bar{z}_2^3 - (23\delta^4/432)z_2\bar{z}_1^2\bar{z}_2^3 + (15\lambda^2\delta^2/256)\bar{z}_1^2\bar{z}_2^4 \\
&+(7\lambda\delta^3/128)\bar{z}_1^2\bar{z}_2^4 + (11\delta^4/288)\bar{z}_1^2\bar{z}_2^4 + (\lambda\delta^3/288)z_2^6 + (7\delta^4/2304)z_2^6 \\
&+(\lambda\delta^3/288)z_2^5\bar{z}_2 - (41\delta^4/3456)z_2^5\bar{z}_2 + (11\lambda\delta^3/288)z_2^4\bar{z}_2^2 \\
&-(245\delta^4/6912)z_2^4\bar{z}_2^2 - (7\lambda\delta^3/48)z_2^3\bar{z}_2^3 - (67\delta^4/3456)z_2^3\bar{z}_2^3 \\
&+(11\lambda\delta^3/288)z_2^2\bar{z}_2^4 - (245\delta^4/6912)z_2^2\bar{z}_2^4 + (\lambda\delta^3/288)z_2\bar{z}_2^5 \\
&-(41\delta^4/3456)z_2\bar{z}_2^5 + (\lambda\delta^3/288)\bar{z}_2^6 + (7\delta^4/2304)\bar{z}_2^6
\end{aligned}
$$

Conversion Through Fifth and Sixth Order

To continue avoiding subscripts we now write (3.22) as H, and compute from (3.12) that

$$
\begin{aligned}
K^5 = {} & i(\lambda^3/20)z_1^5 - i(\lambda^3/12)z_1^4\bar{z}_1 + 2i\lambda^3z_1^3\bar{z}_1^2 + i(\lambda^2\delta/30)z_1^3z_2^2 + i(\lambda\delta^2/45)z_1^3z_2^2 \\
&+i(\delta^3/135)z_1^3z_2^2 - i(\lambda\delta^2/27)z_1^3z_2\bar{z}_2 - i(\lambda\delta^2/9)z_1^3\bar{z}_2^2 + i(\delta^3/9)z_1^3\bar{z}_2^2 \\
&-2i\lambda^3z_1^2\bar{z}_1^3 - i(\lambda^2\delta/18)z_1^2z_2^2\bar{z}_1 - i(\delta^3/81)z_1^2z_2^2\bar{z}_1 + i(5\lambda^2\delta/3)z_1^2z_2\bar{z}_1\bar{z}_2
\end{aligned}
$$

$$
\begin{aligned}
&+i(\lambda\delta^2/3)z_1^2z_2\bar{z}_1\bar{z}_2 + i(4\delta^3/27)z_1^2z_2\bar{z}_1\bar{z}_2 + i(\lambda^2\delta/3)z_1^2\bar{z}_1\bar{z}_2^2 \\
&-i(2\lambda\delta^2/3)z_1^2\bar{z}_1\bar{z}_2^2 - i(\delta^3/27)z_1^2\bar{z}_1\bar{z}_2^2 + i(\lambda^3/12)z_1\bar{z}_1^4 - i(\lambda^2\delta/3)z_1z_2^2\bar{z}_1^2 \\
&+i(2\lambda\delta^2/3)z_1z_2^2\bar{z}_1^2 + i(\delta^3/27)z_1z_2^2\bar{z}_1^2 - i(5\lambda^2\delta/3)z_1z_2\bar{z}_1^2\bar{z}_2 \\
&-i(\lambda\delta^2/3)z_1z_2\bar{z}_1^2\bar{z}_2 - i(4\delta^3/27)z_1z_2\bar{z}_1^2\bar{z}_2 + i(\lambda^2\delta/18)z_1\bar{z}_1^2\bar{z}_2^2 \\
&+i(\delta^3/81)z_1\bar{z}_1^2\bar{z}_2^2 + i(\lambda\delta^2/180)z_1z_2^4 + i(\delta^3/135)z_1z_2^4 - i(\delta^3/81)z_1z_2^3\bar{z}_2 \\
&+i(\lambda\delta^2/3)z_1z_2^2\bar{z}_2^2 + i(\delta^3/9)z_1z_2^2\bar{z}_2^2 + i(\lambda\delta^2/9)z_1z_2\bar{z}_2^3 - i(5\delta^3/27)z_1z_2\bar{z}_2^3 \\
&+i(\lambda\delta^2/108)z_1\bar{z}_2^4 - i(\lambda^3/20)\bar{z}_1^5 + i(\lambda\delta^2/9)z_2^2\bar{z}_1^3 - i(\delta^3/9)z_2^2\bar{z}_1^3 \\
&+i(\lambda\delta^2/27)z_2\bar{z}_1^3\bar{z}_2 - i(\lambda^2\delta/30)\bar{z}_1^3\bar{z}_2^2 - i(\lambda\delta^2/45)\bar{z}_1^3\bar{z}_2^2 - i(\delta^3/135)\bar{z}_1^3\bar{z}_2^2 \\
&-i(\lambda\delta^2/108)z_2^4\bar{z}_1 - i(\lambda\delta^2/9)z_2^3\bar{z}_1\bar{z}_2 + i(5\delta^3/27)z_2^3\bar{z}_1\bar{z}_2 - i(\lambda\delta^2/3)z_2^2\bar{z}_1\bar{z}_2^2 \\
&-i(\delta^3/9)z_2^2\bar{z}_1\bar{z}_2^2 + i(\delta^3/81)z_2\bar{z}_1\bar{z}_2^3 - i(\lambda\delta^2/180)\bar{z}_1\bar{z}_2^4 - i(\delta^3/135)\bar{z}_1\bar{z}_2^4
\end{aligned}
$$

Now write the $H\circ\rho$ of (3.22) as H, and again use (2.15) and (2) of Theorem 3.15. With the ρ now being that associated with K^5, we compute

$$
(3.23) \qquad H\circ\rho = H(2) + H(3) + H(4) + H(5) + H(6) + \cdots
$$

where

$$
H(2) = \tfrac{1}{2}|z_1|^2 + \tfrac{1}{2}|z_2|^2
$$

$$
H(3) = 0,
$$

$$
\begin{aligned}
H(4) = &-(15\lambda^2/16)z_1^2\bar{z}_1^2 + (\lambda\delta/16)z_1^2\bar{z}_2^2 - (\delta^2/8)z_1^2\bar{z}_2^2 - (3\lambda\delta/4)z_1z_2\bar{z}_1\bar{z}_2 \\
&-(\delta^2/6)z_1z_2\bar{z}_1\bar{z}_2 + (\lambda\delta/16)z_2^2\bar{z}_1^2 - (\delta^2/8)z_2^2\bar{z}_1^2 - (5\delta^2/48)z_2^2\bar{z}_2^2
\end{aligned}
$$

$$
H(5) = 0
$$

$$
\begin{aligned}
H(6) = &(87\lambda^4/256)z_1^6 - (111\lambda^4/128)z_1^5\bar{z}_1 - (471\lambda^4/256)z_1^4\bar{z}_1^2 + (33\lambda^3\delta/128)z_1^4z_2^2 \\
&+(35\lambda^2\delta^2/256)z_1^4z_2^2 + (5\lambda\delta^3/64)z_1^4z_2^2 + (7\delta^4/576)z_1^4z_2^2 - (13\lambda^3\delta/64)z_1^4z_2\bar{z}_2 \\
&-(21\lambda^2\delta^2/128)z_1^4z_2\bar{z}_2 - (\lambda\delta^3/32)z_1^4z_2\bar{z}_2 - (5\delta^4/288)z_1^4z_2\bar{z}_2 \\
&-(11\lambda^3\delta/256)z_1^4\bar{z}_2^2 - (3\lambda^2\delta^2/128)z_1^4\bar{z}_2^2 + (19\lambda\delta^3/144)z_1^4\bar{z}_2^8 \\
&-(\delta^4/16)z_1^4\bar{z}_2^2 - (705\lambda^4/128)z_1^3\bar{z}_1^3 - (11\lambda^3\delta/32)z_1^3z_2^2\bar{z}_1 - (71\lambda^2\delta^2/192)z_1^3z_2^2\bar{z}_1 \\
&-(5\lambda\delta^3/72)z_1^3z_2^2\bar{z}_1 - (13\delta^4/432)z_1^3z_2^2\bar{z}_1 - (235\lambda^3\delta/128)z_1^3z_2\bar{z}_1\bar{z}_2
\end{aligned}
$$

$$
\begin{aligned}
&-(49\lambda^2\delta^2/384)z_1^3z_2\bar{z}_1\bar{z}_2 - (3\lambda\delta^3/32)z_1^3z_2\bar{z}_1\bar{z}_2 - (59\delta^4/864)z_1^3z_2\bar{z}_1\bar{z}_2 \\
&+(41\lambda^3\delta/128)z_1^3\bar{z}_1\bar{z}_2^2 - (7\lambda^2\delta^2/64)z_1^3\bar{z}_1\bar{z}_2^2 - (41\lambda\delta^3/72)z_1^3\bar{z}_1\bar{z}_2^2 \\
&-(\delta^4/72)z_1^3\bar{z}_1\bar{z}_2^2 - (471\lambda^4/256)z_1^2\bar{z}_1^2 + (255\lambda^3\delta/256)z_1^2z_2^2\bar{z}_1^2 \\
&-(1109\lambda^2\delta^2/768)z_1^2z_2^2\bar{z}_1^2 - (\lambda\delta^3/192)z_1^2z_2^2\bar{z}_1^2 - (19\delta^4/1728)z_1^2z_2^2\bar{z}_1^2 \\
&-(45\lambda^3\delta/8)z_1^2z_2\bar{z}_1^2\bar{z}_2 - (251\lambda^2\delta^2/384)z_1^2z_2\bar{z}_1^2\bar{z}_2 - (27\lambda\delta^3/32)z_1^2z_2\bar{z}_1^2\bar{z}_2 \\
&-(43\delta^4/864)z_1^2z_2\bar{z}_1^2\bar{z}_2 + (255\lambda^3\delta/256)z_1^2\bar{z}_1^2\bar{z}_2^2 - (1109\lambda^2\delta^2/768)z_1^2\bar{z}_1^2\bar{z}_2^2 \\
&-(\lambda\delta^3/192)z_1^2\bar{z}_1^2\bar{z}_2^2 - (19\delta^4/1728)z_1^2\bar{z}_1^2\bar{z}_2^2 + (15\lambda^2\delta^2/256)z_1^2z_2^4 \\
&+(7\lambda\delta^3/128)z_1^2z_2^4 + (11\delta^4/288)z_1^2z_2^4 - (11\lambda^2\delta^2/192)z_1^2z_2^3\bar{z}_2 \\
&-(41\lambda\delta^3/288)z_1^2z_2^3\bar{z}_2 - (23\delta^4/432)z_1^2z_2^3\bar{z}_2 - (355\lambda^2\delta^2/768)z_1^2z_2^2\bar{z}_2^2 \\
&-(95\lambda\delta^3/768)z_1^2z_2^2\bar{z}_2^2 - (151\delta^4/3456)z_1^2z_2^2\bar{z}_2^2 + (5\lambda^2\delta^2/32)z_1^2z_2\bar{z}_2^3 \\
&-(37\lambda\delta^3/384)z_1^2z_2\bar{z}_2^3 - (109\delta^4/576)z_1^2z_2\bar{z}_2^3 - (5\lambda^2\delta^2/192)z_1^2\bar{z}_2^4 \\
&+(205\lambda\delta^3/2304)z_1^2\bar{z}_2^4 - (19\delta^4/384)z_1^2\bar{z}_2^4 - (111\lambda^4/128)z_1\bar{z}_1^5 \\
&+(41\lambda^3\delta/128)z_1z_2^2\bar{z}_1^3 - (7\lambda^2\delta^2/64)z_1z_2^2\bar{z}_1^3 - (41\lambda\delta^3/72)z_1z_2^2\bar{z}_1^3 \\
&-(\delta^4/72)z_1z_2^2\bar{z}_1^3 - (235\lambda^3\delta/128)z_1z_2\bar{z}_1^3\bar{z}_2-(49\lambda^2\delta^2/384)z_1z_2\bar{z}_1^3\bar{z}_2 \\
&-(3\lambda\delta^3/32)z_1z_2\bar{z}_1^3\bar{z}_2 - (59\delta^4/864)z_1z_2\bar{z}_1^3\bar{z}_2 - (11\lambda^3\delta/32)z_1\bar{z}_1^3\bar{z}_2^2 \\
&-(71\lambda^2\delta^2/192)z_1\bar{z}_1^3\bar{z}_2^2 - (5\lambda\delta^3/72)z_1\bar{z}_1^3\bar{z}_2^2 - (13\delta^4/432)z_1\bar{z}_1^3\bar{z}_2^2 \\
&-(7\lambda^2\delta^2/384)z_1z_2^4\bar{z}_1 - (41\lambda\delta^3/576)z_1z_2^4\bar{z}_1 - (\delta^4/36)z_1z_2^4\bar{z}_1 \\
&+(197\lambda^2\delta^2/384)z_1z_2^3\bar{z}_1\bar{z}_2 - (695\lambda\delta^3/1152)z_1z_2^3\bar{z}_1\bar{z}_2 - (65\delta^4/1728)z_1z_2^3\bar{z}_1\bar{z}_2 \\
&-(691\lambda^2\delta^2/384)z_1z_2^2\bar{z}_1\bar{z}_2^2 - (5\lambda\delta^3/32)z_1z_2^2\bar{z}_1\bar{z}_2^2 - (311\delta^4/864)z_1z_2^2\bar{z}_1\bar{z}_2^2 \\
&+(197\lambda^2\delta^2/384)z_1z_2\bar{z}_1\bar{z}_2^3 - (695\lambda\delta^3/1152)z_1z_2\bar{z}_1\bar{z}_2^3 - (65\delta^4/1728)z_1z_2\bar{z}_1\bar{z}_2^3 \\
&-(7\lambda^2\delta^2/384)z_1\bar{z}_1\bar{z}_2^4 - (41\lambda\delta^3/576)z_1\bar{z}_1\bar{z}_2^4 - (\delta^4/36)z_1\bar{z}_1\bar{z}_2^4 \\
&+(87\lambda^4/256)\bar{z}_1^6 - (11\lambda^3\delta/256)z_2^2\bar{z}_1^4 - (3\lambda^2\delta^2/128)z_2^2\bar{z}_1^4 \\
&+(19\lambda\delta^3/144)z_2^2\bar{z}_1^4 - (\delta^4/16)z_2^2\bar{z}_1^4 - (13\lambda^3\delta/64)z_2\bar{z}_1^4\bar{z}_2 \\
&-(21\lambda^2\delta^2/128)z_2\bar{z}_1^4\bar{z}_2 - (\lambda\delta^3/32)z_2\bar{z}_1^4\bar{z}_2 - (5\delta^4/288)z_2\bar{z}_1^4\bar{z}_2 \\
&+(33\lambda^3\delta/128)\bar{z}_1^4\bar{z}_2^2 + (35\lambda^2\delta^2/256)\bar{z}_1^4\bar{z}_2^2 + (5\lambda\delta^3/64)\bar{z}_1^4\bar{z}_2^2 \\
&+(7\delta^4/576)\bar{z}_1^4\bar{z}_2^2 - (5\lambda^2\delta^2/192)z_2^4\bar{z}_1^2 + (205\lambda\delta^3/2304)z_2^4\bar{z}_1^2 \\
&-(19\delta^4/384)z_2^4\bar{z}_1^2 + (5\lambda^2\delta^2/32)z_2^3\bar{z}_1^2\bar{z}_2 - (37\lambda\delta^3/384)z_2^3\bar{z}_1^2\bar{z}_2 \\
&-(109\delta^4/576)z_2^3\bar{z}_1^2\bar{z}_2 - (355\lambda^2\delta^2/768)z_2^2\bar{z}_1^2\bar{z}_2^2 - (95\lambda\delta^3/768)z_2^2\bar{z}_1^2\bar{z}_2^2 \\
&-(151\delta^4/3456)z_2^2\bar{z}_1^2\bar{z}_2^2 - (11\lambda^2\delta^2/192)z_2\bar{z}_1^2\bar{z}_2^3 - (41\lambda\delta^3/288)z_2\bar{z}_1^2\bar{z}_2^3
\end{aligned}
$$

$$
\begin{aligned}
&-(23\delta^4/432)z_2\bar{z}_1^2\bar{z}_2^3 + (15\lambda^2\delta^2/256)\bar{z}_1^2\bar{z}_2^4 + (7\lambda\delta^3/128)\bar{z}_1^2\bar{z}_2^4 \\
&+(11\delta^4/288)\bar{z}_1^2\bar{z}_2^4 + (\lambda\delta^3/288)z_2^6 + (7\delta^4/2304)z_2^6 + (\lambda\delta^3/288)z_2^5\bar{z}_2 \\
&-(41\delta^4/3456)z_2^5\bar{z}_2 + (11\lambda\delta^3/288)z_2^4\bar{z}_2^2 - (245\delta^4/6912)z_2^4\bar{z}_2^2 \\
&-(7\lambda\delta^3/48)z_2^3\bar{z}_2^3 - (67\delta^4/3456)z_2^3\bar{z}_2^3 + (11\lambda\delta^3/288)z_2^2\bar{z}_2^4 \\
&-(245\delta^4/6912)z_2^2\bar{z}_2^4 + (\lambda\delta^3/288)z_2\bar{z}_2^5 - (41\delta^4/3456)z_2\bar{z}_2^5 \\
&+(\lambda\delta^3/288)\bar{z}_2^6 + (7\delta^4/2304)\bar{z}_2^6
\end{aligned}
$$

Conversion to sixth order in now relatively simple, since we are not interested in seventh order terms. Indeed, if we use Theorem 3.14 to transform (3.23) to normal form through terms of order six, then the resulting Hamiltonian must be the one just written above, except that all terms of $H(6)$ in H_6^R must be removed. In particular, we do not have to compute K^6. Removing these "bad" terms we find that the Hamiltonian (3.19), when converted using Theorem 3.14 to normal form through terms of order six, is

$$
(3.24) \qquad H(2) + H(3) + H(4) + H(5) + H(6) + \cdots
$$

(here $H(6)$ has a different meaning than in (3.23)), where

$$
H(2) = \tfrac{1}{2}|z_1|^2 + \tfrac{1}{2}|z_2|^2
$$

$$
H(3) = 0
$$

$$
\begin{aligned}
H(4) = &-(15\lambda^2/16)z_1^2\bar{z}_1^2 + (\lambda\delta/16)z_1^2\bar{z}_2^2 - (\delta^2/8)z_1^2\bar{z}_2^2 - (3\lambda\delta/4)z_1z_2\bar{z}_1\bar{z}_2 \\
&-(\delta^2/6)z_1z_2\bar{z}_1\bar{z}_2 + (\lambda\delta/16)z_2^2\bar{z}_1^2 - (\delta^2/8)z_2^2\bar{z}_1^2 - (5\delta^2/48)z_2^2\bar{z}_2^2
\end{aligned}
$$

$$
H(5) = 0
$$

$$
\begin{aligned}
H(6) = &-(705\lambda^4/128)z_1^3\bar{z}_1^3 + (41\lambda^3\delta/128)z_1^3\bar{z}_1\bar{z}_2^2 - (7\lambda^2\delta^2/64)z_1^3\bar{z}_1\bar{z}_2^2 \\
&-(41\lambda\delta^3/72)z_1^3\bar{z}_1\bar{z}_2^2 - (\delta^4/72)z_1^3\bar{z}_1\bar{z}_2^2 - (45\lambda^3\delta/8)z_1^2z_2\bar{z}_1^2\bar{z}_2 \\
&-(251\lambda^2\delta^2/384)z_1^2z_2\bar{z}_1^2\bar{z}_2 - (27\lambda\delta^3/32)z_1^2z_2\bar{z}_1^2\bar{z}_2 - (43\delta^4/864)z_1^2z_2\bar{z}_1^2\bar{z}_2 \\
&+(5\lambda^2\delta^2/32)z_1^2z_2\bar{z}_2^3 - (37\lambda\delta^3/384)z_1^2z_2\bar{z}_2^3 - (109\delta^4/576)z_1^2z_2\bar{z}_2^3 \\
&+(41\lambda^3\delta/128)z_1z_2^2\bar{z}_1^3 - (7\lambda^2\delta^2/64)z_1z_2^2\bar{z}_1^3 - (41\lambda\delta^3/72)z_1z_2^2\bar{z}_1^3 \\
&-(\delta^4/72)z_1z_2^2\bar{z}_1^3 - (691\lambda^2\delta^2/384)z_1z_2^2\bar{z}_1\bar{z}_2^2 - (5\lambda\delta^3/32)z_1z_2^2\bar{z}_1\bar{z}_2^2 \\
&-(311\delta^4/864)z_1z_2^2\bar{z}_1\bar{z}_2^2 + (5\lambda^2\delta^2/32)z_2^3\bar{z}_1^2\bar{z}_2 - (37\lambda\delta^3/384)z_2^3\bar{z}_1^2\bar{z}_2 \\
&-(109\delta^4/576)z_2^3\bar{z}_1^2\bar{z}_2 - (7\lambda\delta^3/48)z_2^3\bar{z}_2^3 - (67\delta^4/3456)z_2^3\bar{z}_2^3
\end{aligned}
$$

Summary

Gathering coefficients in this last computation, we arrive at the following result: By repeated use of Theorem 3.15, we can convert the Hamiltonian

$$H(q,p) = \tfrac{1}{2}(q_1^2+p_1^2) + \tfrac{1}{2}(q_2^2+p_2^2) + \lambda q_1^3 + \delta q_1 q_2^2 \tag{3.25}$$

into normal form through sixth order terms, obtaining

$$H(2) + H(3) + H(4) + H(5) + H(6) + \cdots \tag{3.26}$$

where the omitted terms are of higher order, and where

$$H(2) = \tfrac{1}{2}|z_1|^2 + \tfrac{1}{2}|z_2|^2$$

$$H(3) = 0$$

$$H(4) = -\frac{15\lambda^2}{16}|z_1|^4 - \frac{1}{12}(9\lambda\delta+2\delta^2)|z_1|^2|z_2|^2 - \frac{5\delta^2}{48}|z_2|^4 + \frac{1}{16}(\lambda\delta-2\delta^2)z_1^2\bar{z}_2^2 + \frac{1}{16}(\lambda\delta-2\delta^2)z_2^2\bar{z}_1^2$$

$$H(5) = 0$$

$$\begin{aligned} H(6) = &-\frac{705\lambda^4}{128}|z_1|^6 - \frac{1}{3,456}(19,440\lambda^3\delta+2,259\lambda^2\delta^2+2,916\lambda\delta^3+172\delta^4)|z_1|^4|z_2|^2 \\ &-\frac{1}{3,456}(6,219\lambda^2\delta^2+540\lambda\delta^3+1,244\delta^4)|z_1|^2|z_2|^4 \\ &-\frac{1}{3,456}(504\lambda\delta^3+67\delta^4)|z_2|^6 \\ &+\frac{1}{1,152}(369\lambda^3\delta-126\lambda^2\delta^2-656\lambda\delta^3-16\delta^4)z_1^3\bar{z}_1\bar{z}_2^2 \\ &+\frac{1}{1,152}(369\lambda^3\delta-126\lambda^2\delta^2-656\lambda\delta^3-16\delta^4)z_1z_2^2\bar{z}_1^3 \\ &+\frac{1}{1,152}(180\lambda^2\delta^2-111\lambda\delta^3-218\delta^4)z_1^2z_2\bar{z}_2^3 \\ &+\frac{1}{1,152}(180\lambda^2\delta^2-111\lambda\delta^3-218\delta^4)z_2^3\bar{z}_1^2\bar{z}_2 \end{aligned}$$

In particular, for the values $\lambda = 1/3$, $\delta = -1$ originally studied [5], the terms $H(j)$ of (3.26) become

$$H(2) = \tfrac{1}{2}|z_1|^2 + \tfrac{1}{2}|z_2|^2$$

$$H(3) = 0$$

$$H(4) = -\frac{5}{48}|z_1|^4 + \frac{1}{12}|z_1|^2|z_2|^2 - \frac{5}{48}|z_2|^4 - \frac{7}{48}z_1^2\bar{z}_2^2 - \frac{7}{48}z_2^2\bar{z}_1^2$$

$$H(5) = 0$$

$$H(6) = -\frac{235}{3,456}|z_1|^6 + \frac{47}{128}|z_1|^4|z_2|^2 - \frac{65}{128}|z_1|^2|z_2|^4 + \frac{101}{3,456}|z_2|^6 + \frac{175}{1,152}z_1^3\bar{z}_1\bar{z}_2^2 + \frac{175}{1,152}z_1z_2^2\bar{z}_1^3 - \frac{161}{1,152}z_1^2z_2\bar{z}_2^3 - \frac{161}{1,152}z_2^3\bar{z}_1^2\bar{z}_2$$

We should remark that this last result, at least through fourth order, is precisely what is obtained if generating function techniques are used in place of Theorem 3.15, e.g. see [3] (where an additional factor of ½, which is irrelevant for the results of that paper, was obtained when this author made the calculation of the $H(4)$ term, thereby making it agree with the $H(4)$ term shown above).

Further Two-degree of Freedom Examples

Here we present the results obtained when Theorem 3.25 is used to convert the indicated Hamiltonians into normal form through fourth order terms.

Two-one Resonance

For the Hamiltonian

$$H(q,p) = (q_1^2+p_1^2) + \tfrac{1}{2}(q_2^2+p_2^2) + q_1q_2^2 \tag{3.27}$$

the result is

$$H(2) = |z_1|^2 + \tfrac{1}{2}|z_2|^2$$

$$H(3) = \frac{1}{8}z_1\bar{z}_2^2 + \frac{1}{8}z_2^2\bar{z}_1$$

$$H(4) = -\frac{1}{32}|z_1|^2|z_2|^2 - \frac{9}{28}|z_2|^4$$

Three-one Resonance

For the Hamiltonian

$$H(q,p) = \frac{3}{2}(q_1^2+p_1^2) + \frac{1}{2}(q_2^2+p_2^2) + q_1^2q_2^2 \tag{3.28}$$

the result is

$$H(2) = \frac{3}{2}|z_1|^2 + \frac{1}{2}|z_2|^2$$

$$H(3) = 0$$

$$H(4) = \frac{1}{4}|z_1|^2|z_2|^2$$

Minus one-one Resonance

For the Hamiltonian

$$(3.29) \qquad H(q,p) = -\frac{1}{2}(q_1^2+p_1^2) + \frac{1}{2}(q_2^2+p_2^2) + \frac{1}{3}q_1^3 - q_1q_2^2$$

the result is

$$H(2) = -\frac{1}{2}|z_1|^2 + \frac{1}{2}|z_2|^2$$

$$H(3) = 0$$

$$H(4) = \frac{5}{48}|z_1|^4 - \frac{5}{12}|z_1|^2|z_2|^2 + \frac{5}{48}|z_2|^4 - \frac{5}{48}z_1^2z_2^2 - \frac{5}{48}\bar{z}_1^2\bar{z}_2^2$$

Nonresonance

For the Hamiltonian

$$(3.30) \qquad H(q,p) = \frac{\alpha}{2}(q_1^2+p_1^2) + \frac{1}{2}(q_2^2+p_2^2) + \lambda q_1^3 + \delta q_1q_2^2$$

where α is irrational, the result is

$$H(2) = \frac{\alpha}{2}|z_1|^2 + \frac{1}{2}|z_2|^2$$

$$H(3) = 0$$

$$H(4) = -\frac{15\lambda^2}{16\alpha}|z_1|^4 + \{(\delta^2/(2\alpha^2-8)) - (3\lambda\delta/4\alpha)\}|z_1|^2|z_2|^2$$
$$- \{(\alpha\delta^2/(16\alpha^2-64)) + (\delta^2/8\alpha)\}|z_2|^4$$

Three-degree of Freedom Examples

Here we list three examples of the use of Theorem 3.15 to convert three-degree of freedom harmonic oscillators into norm form. The first we give through sixth order terms, and the remaining two through fourth order terms.

One-one-one resonance

For the Hamiltonian

$$\text{(3.31)} \quad H(q,p) = \tfrac{1}{2}(q_1^2+p_1^2) + \tfrac{1}{2}(q_2^2+p_2^2) + \tfrac{1}{2}(q_3^2+p_3^2) + q_1^2(q_2+q_3)$$

the result is

$$H(2) = \tfrac{1}{2}|z_1|^2 + \tfrac{1}{2}|z_2|^2 + \tfrac{1}{2}|z_3|^2$$

$$H(3) = 0$$

$$\begin{aligned} H(4) = &-\tfrac{5}{24}|z_1|^4 - \tfrac{1}{8}z_1^2\bar{z}_2^2 - \tfrac{1}{4}z_1^2\bar{z}_2\bar{z}_3 - \tfrac{1}{8}z_1^2\bar{z}_3^2 - \tfrac{1}{6}|z_1|^2|z_2|^2 - \tfrac{1}{6}z_1z_2\bar{z}_1\bar{z}_3 \\ &-\tfrac{1}{6}z_1z_3\bar{z}_1\bar{z}_2 - \tfrac{1}{6}|z_1|^2|z_3|^2 - \tfrac{1}{8}z_2^2\bar{z}_1^2 - \tfrac{1}{4}z_2z_3\bar{z}_1^2 - \tfrac{1}{8}z_3^2\bar{z}_1^2 \end{aligned}$$

$$H(5) = 0$$

$$\begin{aligned} H(6) = &-\tfrac{67}{864}|z_1|^6 - \tfrac{109}{288}z_1^3\bar{z}_1\bar{z}_2^2 - \tfrac{109}{144}z_1^3\bar{z}_1\bar{z}_2\bar{z}_3 - \tfrac{109}{288}z_1^3\bar{z}_1\bar{z}_3^2 - \tfrac{311}{432}|z_1|^4|z_2|^2 \\ &-\tfrac{311}{432}z_1^2z_2\bar{z}_1^2\bar{z}_3 - \tfrac{311}{432}z_1^2z_3\bar{z}_1^2\bar{z}_2 - \tfrac{311}{432}|z_1|^4|z_3|^2 - \tfrac{1}{72}z_1^2z_2\bar{z}_2^3 \\ &-\tfrac{1}{24}z_1^2z_2\bar{z}_2^2\bar{z}_3 - \tfrac{1}{24}z_1^2z_2\bar{z}_2\bar{z}_3^2 - \tfrac{1}{72}z_1^2z_2\bar{z}_3^3 - \tfrac{1}{72}z_1^2z_3\bar{z}_2^3 - \tfrac{1}{24}z_1^2z_3\bar{z}_2^2\bar{z}_3 \\ &-\tfrac{1}{24}z_1^2z_3\bar{z}_2\bar{z}_3^2 - \tfrac{1}{72}z_1^2z_3\bar{z}_3^3 - \tfrac{109}{288}z_1z_2^2\bar{z}_1^3 - \tfrac{109}{144}z_1z_2z_3\bar{z}_1^3 - \tfrac{109}{288}z_1z_3^2\bar{z}_1^3 \\ &-\tfrac{43}{864}|z_1|^2|z_2|^4 - \tfrac{43}{432}z_1z_2^2\bar{z}_1\bar{z}_2\bar{z}_3 - \tfrac{43}{864}z_1z_2^2\bar{z}_1\bar{z}_2^2 - \tfrac{43}{432}z_1z_2z_3\bar{z}_1\bar{z}_2^2 \\ &-\tfrac{43}{216}|z_1|^2|z_2|^2|z_3|^2 - \tfrac{43}{432}z_1z_2z_3\bar{z}_1\bar{z}_3^2 - \tfrac{43}{864}z_1z_3^2\bar{z}_1\bar{z}_2^2 \\ &-\tfrac{43}{432}z_1z_3^2\bar{z}_1\bar{z}_2\bar{z}_3 - \tfrac{43}{864}|z_1|^2|z_3|^4 - \tfrac{1}{72}z_2^3\bar{z}_1^2\bar{z}_2 - \tfrac{1}{72}z_2^3\bar{z}_1^2\bar{z}_3 \\ &-\tfrac{1}{24}z_2^2z_3\bar{z}_1^2\bar{z}_2 - \tfrac{1}{24}z_2^2z_3\bar{z}_1^2\bar{z}_3 - \tfrac{1}{24}z_2z_3^2\bar{z}_1^2\bar{z}_2 - \tfrac{1}{24}z_2z_3^2\bar{z}_1^2\bar{z}_3 \\ &-\tfrac{1}{72}z_3^3\bar{z}_1^2\bar{z}_2 - \tfrac{1}{72}z_3^3\bar{z}_1^2\bar{z}_3 \end{aligned}$$

One-minus one-one resonance

For the Hamiltonian

$$\text{(3.32)} \quad H(q,p) = \tfrac{1}{2}(q_1^2+p_1^2) - \tfrac{1}{2}(q_2^2+p_2^2) + \tfrac{1}{2}(q_3^2+p_3^2) + q_1^2(q_2+q_3)$$

the result is

$$H(2) = \tfrac{1}{2}|z_1|^2 - \tfrac{1}{2}|z_2|^2 + \tfrac{1}{2}|z_3|^2$$

$$H(3) = 0$$

$$H(4) = -\frac{1}{8}|z_1|^4 - \frac{1}{4}z_1^2 z_2 \bar{z}_3 - \frac{1}{8}z_1^2 \bar{z}_3^2 - \frac{1}{6}|z_1|^2|z_2|^2 - \frac{1}{6}z_1 z_2 z_3 \bar{z}_1$$
$$- \frac{1}{6}z_1 \bar{z}_1 \bar{z}_2 \bar{z}_3 - \frac{1}{6}|z_1|^2|z_3|^2 - \frac{1}{8}\bar{z}_1^2 \bar{z}_2^2 - \frac{1}{4}z_3 \bar{z}_1^2 \bar{z}_2 - \frac{1}{8}z_3^2 \bar{z}_1^2$$

Two-three-five resonance

For the Hamiltonian

$$H(q,p) = (q_1^2+p_1^2) + \frac{3}{2}(q_2^2+p_2^2) + \frac{5}{2}(q_3^2+p_3^2) + q_1^2(q_2+q_3) \tag{3.33}$$

the result is

$$H(2) = |z_1|^2 + \frac{3}{2}|z_2|^2 + \frac{5}{2}|z_3|^2$$

$$H(3) = 0$$

$$H(4) = -\frac{47}{630}|z_1|^4 - \frac{1}{7}|z_1|^2|z_2|^2 + \frac{1}{9}|z_1|^2|z_3|^2$$

§4. Applications of Normal Forms

Isoenergetic Reduction

Here we present a technical trick relating to Hamiltonian systems (not only harmonic oscillators) which will be connected with normal forms in the following section.

Consider a Hamiltonian $H(q,p)$ defined on some open set $U \subset R^{2n}$ containing the origin, and assume that on some energy surface $H = h_0$ we have

$$\frac{\partial H}{\partial p_1} \neq 0 \tag{4.1}$$

By the implicit function theorem we can then write p_1, near each point of $H = h_0$, as a function of the remaining variables, say

$$p_1 = -T(q_1,\ldots,q_n,p_2,\ldots,p_n,h) \tag{4.2}$$

(the minus sign is a notational convenience). It follows that

$$H(q_1,\ldots,q_n,-T,p_2,\ldots,p_n) = h \tag{4.3}$$

and differentiating with respect to $s = q_j$ and p_j, $j = 1,\dots,n$, then gives $H_s - H_{p_1}T_s = 0$, i.e.

$$(4.4) \qquad T_{q_j} = H_{q_j}/H_{p_1}, \quad T_{p_j} = H_{p_j}/H_{p_1}$$

But in the system associated with H we have $\dot{q}_1 = H_{p_1}$, and so (4.1) guarantees that we can replace the time variable by q_1. Writing $'$ for d/dq_1 and using (4.4) we thus find, for $j = 2,\dots,n$, that

$$q_j' = \dot{q}_j/H_{p_1} = H_{p_j}/H_{p_1} = T_{p_j}$$
$$p_j' = \dot{p}_j/H_{p_1} = -H_{q_j}/H_{p_1} = -T_{q_j}$$

i.e.

$$(4.5) \qquad q_j' = T_{p_j}, \quad p_j' = -T_{q_j}, \qquad j = 2,\dots,n$$

Notice this is a time (= q_1) and parameter (= h) dependent Hamiltonian with Hamiltonian function T.

In summary, we have shown that, near each point of $H = h_0$, *solutions of (4.5), for which* $h = h_0$, *are reparametrizations of solutions of energy* h_0 *of* $\dot{q} = H_p$, $\dot{p} = -H_q$. Any technique which involves changing the time variable for the purpose of stydying solutions of a Hamiltonian at a fixed energy is called *isoenergetic reduction*, amd this is a example.

We can apply isoenergetic reduction as above to a parameter-dependent Hamiltonian $H(q,p;\varepsilon)$ of the form

$$(4.6) \qquad H(q,p;\varepsilon) = p_1 + \varepsilon^{s-2}F(q_2,\dots,q_n,p_1,\dots,p_n) + O(\varepsilon^{s-1})$$

(the absence of q_1 from F is intentional), where $s \geq 3$ and where H is defined on some open set $U\times(-\delta,\delta) \subset R^{2n}\times R$. Here

$$H_{p_1} = 1 + \varepsilon^{s-2}F_{p_1} + O(\varepsilon^{s-1})$$

and so (4.1) holds for all $\varepsilon > 0$ sufficiently small. A straightforward calculation then shows

$$(4.7) \qquad T = -h_0 + \varepsilon^{s-2}F(q_2,\dots,q_n,h_0,p_2,\dots,p_n) + O(\varepsilon^{s-1})$$

and so here (4.5) becomes

$$(4.8) \qquad \begin{cases} q_j' = \varepsilon^{s-2}F_{p_j} + O(\varepsilon^{s-1}) \\ p_j' = -\varepsilon^{s-2}F_{q_j} + O(\varepsilon^{s-1}) \end{cases} \qquad j = 2,\dots,n$$

where the final terms may involve q_1 even though F does not.

In connection with (4.8) consider the *truncated system*

$$(4.9) \qquad \begin{cases} q_j' = \varepsilon^{s-2}F_{p_j} \\ p_j' = -\varepsilon^{s-2}F_{q_j}, \end{cases} \qquad j = 2,\dots,n$$

which by rescaling time is equivalent to

$$(4.10) \qquad \begin{cases} q_j' = F_{p_j} \\ p_j' = -F_{q_j} \end{cases} \qquad j = 2,\dots,n$$

We can study solutions of (4.8) by adjoining $q_1' = 1$ to those equations, thereby making the system autonomous, and studying the resulting flow. We could also do this with (4.9), and use the second flow as an approximation to the first. However, notice that the second flow is much simpler. Indeed, from (4.7) we see that the system (4.9) is already autonomous, and is in reality defined on a subset $V \subset R^{2n-2}$. The second flow is thus a product flow on $V \times R$, which in the second factor is simply a uniform motion.

Normal Forms in the case of Two Degrees of Freedom

We will apply the previous considerations to harmonic oscillators of two degrees of freedom, but first we need some preliminary work.

We consider a harmonic oscillator

$$(4.11) \qquad H(z,\bar{z}) = (\omega_1/2)|z_1|^2 + (\omega_2/2)|z_2|^2 + H_s(z,\bar{z}) + \cdots$$

where $\omega_1 > 0$, $\omega_2 > 0$, H_s is a sum of monomials of degree $s \geq 3$, and the omitted terms are of order greater than s. Notice that if

ω_1/ω_2 is rational (the resonance case), then by rescaling time we may assume ω_1 and ω_2 are relatively prime integers or else both 1, and this we do. In any case, a typical monomial of H_s will have the form

$$(4.12) \qquad cz^k\bar{z}^\ell = cz_1^{k_1}z_2^{k_2}\bar{z}_1^{\ell_1}\bar{z}_2^{\ell_2}$$

where $c \in C$ and where

$$(4.13) \qquad 3 \leq k_1+k_2+\ell_1+\ell_2 = s$$

By Proposition 3.10 such a monomial will be in H_s^N if and only if

$$(4.14) \qquad \omega_1(k_1-\ell_1) + \omega_2(k_2-\ell_2) = 0$$

4.15 Theorem: *Assume the notation above.*

(a) *(The nonresonance case) If ω_1/ω_2 is irrational, then (4.11) is in normal form through terms of order s if and only if H_s can be written as a polynomial in the "variables" $N_j = |z_j|^2$, $j = 1,2$.*

(b) *(The resonance case) If ω_1/ω_2 is rational (and so as above ω_1 and ω_2 may be assumed relatively prime integers or else both 1), then (4.11) is in normal form through terms of order s if and only if H_s can be written as a polynomial in the "variables" N_1, N_2 (see (a)), $M_1 = Re\ z_1^{\omega_2}\bar{z}_2^{\omega_1}$ and $M_2 = Im\ z_1^{\omega_2}\bar{z}_2^{\omega_1}$.*

Although special cases have been studied, the generalization of this result to more than two degrees of freedom does not seem to be known.

Proof: (a) If ω_1/ω_2 is irrational, then (4.14) can hold if and only if $k_j = \ell_j$, $j = 1,2$, and this immediately gives (a).

(b) Suppose (4.12) is in H_s^N. Since ω_1 and ω_2 are relatively prime integers or else both 1, we can find integers u and v such that

$$(1) \qquad \omega_1 u + \omega_2 v = 1$$

But set

(2) $$m = (k_1-\ell_1)v - (k_2-\ell_2)u$$

If $m \neq 0$, then solving (4.14) and (1) for ω_1 and ω_2 we find that

(3) $$k_1-\ell_1 = m\omega_2, \quad k_2-\ell_2 = -m\omega_1$$

whereas if $m = 0$ than solving (4.14) and (2) for $k_1-\ell_1$ and $k_2-\ell_2$ we again obtain (3). For $m \geq 0$ it follows that

$$\begin{aligned} z^k\bar{z}^\ell &= z_1^{k_1}z_2^{k_2}\bar{z}_1^{\ell_1}\bar{z}_2^{\ell_2} \\ &= z_1^{\ell_1+m\omega_2}z_2^{k_2}\bar{z}_1^{\ell_1}\bar{z}_2^{k_2+m\omega_1} \\ &= |z_1|^{2\ell_1}|z_2|^{2k_2}(z_1^{\omega_2}\bar{z}_2^{\omega_1})^m \end{aligned}$$

whereas for $m < 0$ a similar computation gives

$$z^k\bar{z}^\ell = |z_1|^{2k_1}|z_2|^{2\ell_2}(\bar{z}_1^{\omega_2}z_2^{\omega_1})^{-m}$$

In other words, in the two respective cases we have

$$z^k\bar{z}^\ell = N_1^{\ell_1}N_2^{k_2}(M_1+iM_2)^m$$

$$z^k\bar{z}^\ell = N_1^{k_1}N_2^{\ell_2}(M_1-iM_2)^{-m}$$

Since these are polynomials in the desired variables, the forward implication of (b) is established; the reverse implication is clear from (4.14). Q.E.D.

Now assume (4.11) is in normal form through terms of order s; we have seen that this can always be accomplished by using Theorem 3.15. By stretching variables we can then achieve the form

(4.16) $$H_\varepsilon(z,\bar{z}) = (\omega_1/2)|z_1|^2+(\omega_2/2)|z_2|^2+\varepsilon^{s-2}H_s(z,\bar{z})+O(\varepsilon^{s-1})$$

and, as in Theorem 2.21, we can use this to study solutions of $\dot{z} = -2i(\partial H/\partial\bar{z})$ at small positive energies.

Assuming $\omega_1 > 0$ and $\omega_2 > 0$, Theorem 4.15 guarantees that we can write

(4.17) $$H_s = P(N_1, N_2, M_1, M_2)$$

where P is a polynomial, and where M_1 and M_2 occur only in the case of resonance. But now observe that when composed with the canonical transformation (2.29), the functions N_1, N_2, M_1 and M_2 become independent of ρ. In other words, if we apply the substitution (2.29), then H_s becomes a function of $w, \bar{w}$ and L alone. This is precisely the situation in (4.6) (with $q_1 = \rho$, $p_1 = L$, $F = H_s$), and so here the F in (4.7) can be written

(4.18) $$F = F(w, \bar{w}, h_0)$$

while (4.8), (4.9) and (4.10) become

(4.19) $$w' = -2i\varepsilon^{s-2}\frac{\partial F}{\partial \bar{w}} + O(\varepsilon^{s-1})$$

(4.20) $$w' = -2i\varepsilon^{s-2}\frac{\partial F}{\partial \bar{w}}$$

and

(4.21) $$w' = -2i\frac{\partial F}{\partial \bar{w}}$$

respectively.

We regard F, and thus the Hamiltonian system (4.21), as being defined on the domain $|w|^2 < 2h_0/\omega_2$. (This is purely for aesthetics. In fact the implicit function theorem might not guarantee such a large domain, but the heurestic arguments which follow can be made rigorous without such an assumption.) Notice the orbits of (4.21) then coincide with the level curves of F. If we adjoin $\rho' = 1$ to that equation, where $|\rho| < \pi/2\omega_1$, then as before we can regard w and ρ as coordinates on a double solid cone, and we can view the product flow within this figure. It is geometrically evident (see Figure 5) that critical points of F on the disc correcpond to periodic orbits of the product flow, provided we assume resonance. Alternatively, and again in the case of resonance, periodic orbits of the disc flow should correspond to invariant tori of the product flow (see Figure 6).

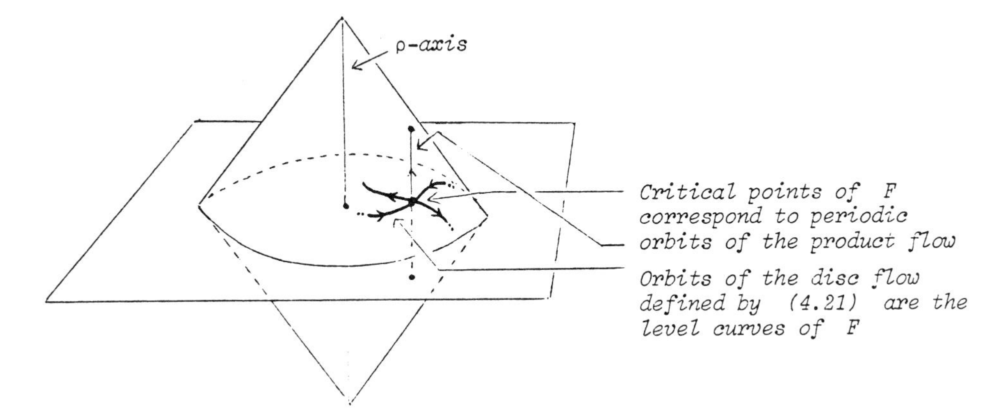

FIGURE 5

By using the implicit function theorem, M. Kummer used the first idea (Figure 5) to prove the existence of periodic solutions of (4.19), hence of (4.11) at low positive energies, from the existence of nondegenerate critical points of F in the disc [6]. As he pointed out, and as would be expected from Theorem 3.18, this is related to the averaging method of Reeb and Moser [7]. Moreover, Kummer also gave criteria for the critical points which guarantee the stability type of the associated periodic solutions: These orbits are elliptic or hyperbolic according as the determinant of the Hessian matrix at the critical point is positive or negative. Alternatively, by using so-called "action-angle variables" M. Braun used the second

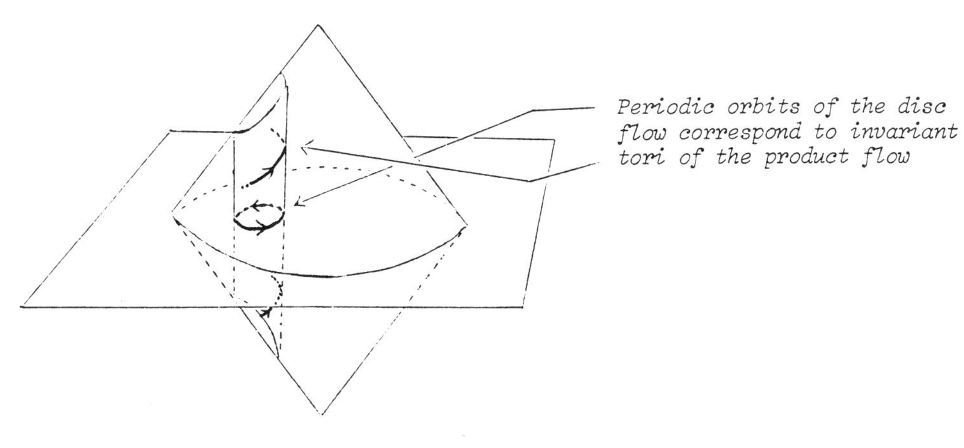

FIGURE 6

idea (Figure 6) to prove the existence of invariant tori in (4.19), hence in (4.11) at low positive energies; the key tool was Moser's twist theorem [3].

As a concrete example consider (3.19) with $\lambda = 1/3$ and $\delta = -1$, in which case the results following (3.46) give

$$(4.22)\quad \begin{cases} H_s(z,\bar{z}) = H_4(z,\bar{z}) \\ = -\frac{1}{48}(5|z_1|^4 + 5|z_2|^4 - 4|z_1|^2|z_2|^2 + 14\mathrm{Re}\; z_1^2\bar{z}_2^2) \\ = -\frac{1}{48}(5N_1^2 + 5N_2^2 - 4N_1N_2 + 14(M_1^2 - M_2^2)) \end{cases}$$

Here one computes $F = F(w,\bar{w},h_0)$ to be

$$(4.23)\qquad F(w,\bar{w},h_0) = -\frac{5}{12}h_0 + \frac{7}{12}(2h_0 - |w|^2)(\mathrm{Re}\; w)^2$$

which has level curves appearing as in Figure 7. The analogue of Figure 6 is therefore Figure 4-4. We thus have a clear picture of the "predominant" nonlinearity, at low energies, of this particular Hamiltonian system.

We should point out that both Braun and Kummer applied their ideas to this particular example; in fact Braun's work was motivated by it.

Reduction to a Mapping

It is easy to see that the flows illustrated in Figures 5 and 6 can be studied by means of a disc mapping $\phi : D \to D$, where $D = \{ w \in C : |w|^2 < 2h_0/\omega_2 \}$. Indeed, we simply pick a point $p \in D$, follow the orbit through p until it next intersects D, and call the point of intersection $\phi(p)$ (see Figure 9). This mapping obviously reflects the nature of the flow, e.g. fixed points correspond to periodic orbits. Moreover, by a variation of Corollary 1.34 ϕ is area-preserving. The study of the harmonic oscillator (4.11) at low energies is thus reduced to the study of an area-preserving homeomorphism of a disc.

In fact the Hénon-Heiles Hamiltonian was first studied by means of such a mapping (although not quite the one above), but using numerical methods [5]. The observed picture, at low energies, was essentially Figure 7 (at least to first order), which we have now explained

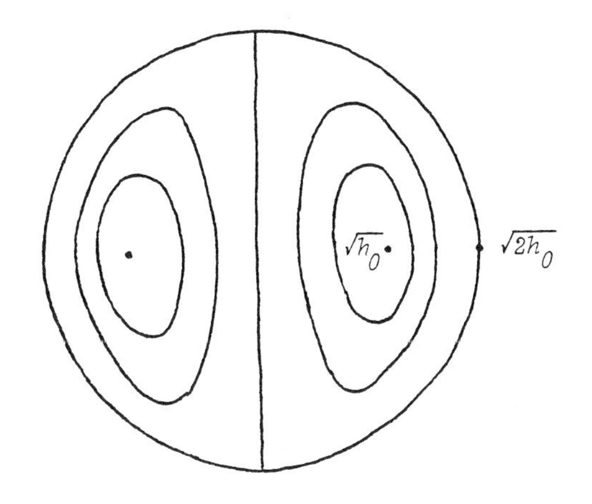

FIGURE 7

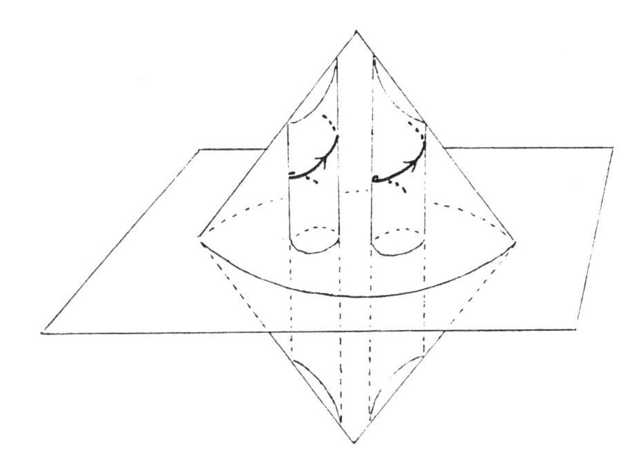

FIGURE 8

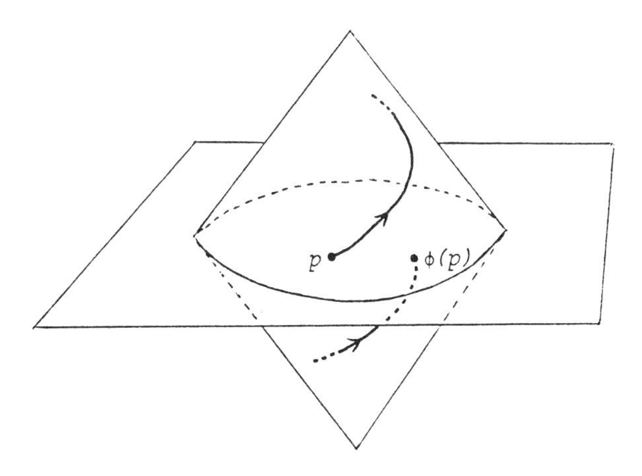

FIGURE 9

rather completely. (The similarity is far more striking if fourth order terms are added to F.) What remains to be explained in that problem, *at a rigorous level*, is why the picture becomes so chaotic as the energy is increased. A discussion in this direction is given in [8].

References

[1] R. Abraham and J. Marsden, *Foundations of Mechanics, 2nd ed.*, Benjamin/Cummings, Reading, Mass., 1978.

[2] V. I. Arnold, *Mathematical Methods of Classical Mechanics*, Springer-Verlag, New York, 1978.

[3] M. Braun, On the applicability of the third integral of motion, *J. Differential Equations* 13 (1973), 300-318.

[4] N. Burgoyne and R. Cushman, Normal forms for real linear Hamiltonian systems with purely imaginary eigenvalues, *Celestial Mechanics* 8 (1974), 435-443.

[5] M. Hénon and C. Heiles, The applicability of the third integral of motion; some numerical experiments, *Astronom. J.* 69 (1964) 73-79.

[6] M. Kummer, On resonant nonlinear coupled oscillators with two equal frequencies, *Commun. Math. Phys.* 48 (1976), 53-79.

[7] J. Moser, Regularization of Kepler's problem and the averaging method on a manifold, *Comm. Pure Appl. Math.* 23 (1970), 609-636.

[8] J. Moser, "Lectures on Hamiltonian Systems," Memoirs Amer. Math. Soc., No. 81 (1968), 1-60.

THE QUANTIZATION OF A CLASSICALLY ERGODIC SYSTEM

Martin C. Gutzwiller

IBM Thomas J. Watson Research Center
Yorktown Heights, New York

1. INTRODUCTION

As a physicist I am mainly concerned about solving particular problems, rather than putting together a general theory. Therefore,I want to explain first what I am looking for. There is no guarantee that I will have provided an answer by the time this presentation is completed, but you will at least understand my motivation and the peculiar mix of rigor and speculation which characterizes a physicists work as opposed to that of a mathematician.

GENERAL PROBLEM. Can classical mechanics be used as an approximation to quantum mechanics ?

Most contemporary physicists who have received a standard education will immediately answer with a resounding yes. They will quote the derivation of energy levels in the hydrogen atom by Niels Bohr in 1912 as the great success in modifying classical mechanics in order to yield a quantum mechanical result.

When pressed for further examples, they will mention the harmonic oscillator and its many variations. Then they will talk about the great difficultyof the classical three-body problem, and that it is really much simpler to solve Schroedinger's equation. In their opinion, it might just be a waste of time to worry about the connection between classical and quantal mechanics. Whatever subleties might emerge from such an endeavor, can hardly be of any practical significance.

Many things can be said in rebuttal. They range from the philosophical to the purely empirical. Surely, it is not an accident that classical mechanics was discovered first, and the obvious reason is that classical mechanics is simpler not only conceptually, but also technically, that is mathematically. More importantly, our thinking seems better adapted to the classical interpretation of physics, to the point where understanding implies to a great extent a translation into classical terms. It is not clear at all whether scientists and the rest of humanity will ever work themselves out of this predicament. Meanwhile the classical approximation to quantum physics has become crucial to the progress of certain areas, in particular high energy physics in the last seven years where the traditional perturbation theory was found to lead nowhere.

Why then has it taken so long for the classical approximation to quantum mechanics to be appreciated? The answer is really quite simple: most classical mechanical systems are badly behaved, they have a stochastic element built into them, and are therefore difficult to understand. The traditional courses and textbooks in classical mechanics ignore completely this aspect, and most physicists as well as mathematicians are unaware of it. Yet it is precisely this feature which leads to statistical mechanics and from there to thermodynamics, fields with which physicists are generally familiar. The connection is made by the ergodic hypothesis whose general validity is in doubt, but which must be correct sufficiently often to be taken for granted whenever it is needed. Thus, a great gap exists in most physicists minds. It is acknowledged when scientists address themselves to statistical mechanics, but it is not at all realized with respect to quantum mechanics. Obviously. if the ergodic hypothesis is so prevalent, one has to cope with systems where it is correct. Otherwise one is forced to leave aside many interesting situations.

Einstein[1] was the first and for his whole life the only scientist to be aware of this problem with respect to quantum mechanics. He wrote a crucial paper on this topic in 1917 which was completely ignored for forty years. He was at the time at the height of his career. and served as president of the German Physical Society in whose transactions this article was published. But I have found no reference to it in any of the great expositions of quantum mechanics in the twenties, or thirties by people like Sommerfeld, Born, Pauli, etc. Even the fifty page article on Einstein's contributions to quantum mechanics in the Reviews of Modern Physics on the occasion of the 100th birthday in 1979, by Abraham Pais,[2] mentions this work only in the bibliography, but does not even allude to it in the text.

At the time of Einstein's paper, quantum mechanics was in its infancy, and it took another eight years before it found its definitive form in Heisenberg's operator formalism and Schroedinger's partial differential equation. But Einstein tried to establish to what extent it could be applied in its rudimentary form which was based on classical mechanics. In his usual low keyed and intuitive style, he managed to make two important statements. In modern terminology they can be paraphrased as follows. The quantization conditions of Sommerfeld can be applied only when the classical trajectories lie on tori in phase space of dimensionality not exceeding the number of degrees of freedom. If a typical trajectory, however, fills up a volume in phasespace of higher dimension, no sensible quantization condition can be formulated. While

Einstein then went ahead and gave a particularly appealing description of the quantization conditions in the first case, he limited himself to general remarks about the second case. He insisted on the fact that the second case is the more frequently occurring, and thereby challenged his fellow physicists to find a method to go from classical to quantum mechanics when one could not take advantage of special situations. This challenge will be taken up in this paper.

2. ERGODIC VERSUS INTEGRABLE SYSTEMS

The problem will now be formulated more technically. Consider a Hamiltonian system which has only two degrees of freedom. Its momentum is given by a vector (u,v) and its position by a vector (x,y). There is a Hamiltonian H(u,v,x,y) from which the equations of motion follow in the usual manner. The value of H is constant along any particular trajectory, and this value will generally be designated by E, the total energy for that trajectory. While the space of momenta and positions is 4-dimensional, any particular trajectory can be found in the 3-dimensional subspace which is defined by the condition H(u,v,x,y) = E. Sometimes it happens that a trajectory lies on a subspace of dimensionality two, equal to the number of degrees of freedom. Such subspaces generally turn out to be tori, and in exceptional cases such a torus reduces to a simple loop. It is important that these tori be smooth. Surfaces which are topologically pathological, will not be considered. It may well be that some part of non-vanishing volume in phase space is covered by these 2-dimensional tori, while the remaining part does not have any or at least very few of them in a measure theoretic sense. There seems to be no general rules which would tell whether any particular Hamiltonian leads to many tori in phase space or not. The situation may change drastically as one considers 3-dimensional subspaces of different energy. A system whose phase space, at least for some fixed energy, is covered by tori, will be called integrable, for that energy. A particle moving in a spherically symmetric potential is an integrable system at all energies.

A mechanical system will be called ergodic, if in the energy range of interest a typical trajectory fills up a nonvanishing subspace on its "energy surface" $H(u,v,x,y) = E$. The words "typical", "filling up" and "nonvanishing" are always referred to the natural Liouville measure in phase space or in an energy surface. This definition is not the conventional one, but it serves our purpose. It includes, in particular, all the cases where some parts of the phase space still have some tori of dimension two. Such a mixture was already envisaged by Einstein, and constitutes the generic situation, although for the purposes of detailed discussion the pure situations (only tori or no tori at all) may be the only ones for the time being. One should keep in mind that the mixed case may present problems of much greater difficulty than either one of the pure cases. Even the description of the known mixed cases is still quite incomplete. There has been a lot of numerical exploration, but no analytic scheme has been found as yet to model the transition region from the tori to the volume without tori.

The two examples which will be discussed in this paper are of the pure variety. In both examples a typical trajectory covers all of the available energy surface which corresponds to the usual definition of ergodic behavior. There is an example for the mathematicians, the geodesics on a surface of constant negative curvature, and an example for physicists, the motion of an electron around the donor impurity in a semiconductor. The former has been known for a long time, it was first discussed by Hadamard[3] in 1898, and by many distinguished Mathematicians since. Even the connection between the classical and quantum system has been established since 25 years, although it had not been interpreted in this manner. This was done by Selberg[4] in his celebrated formula which constitutes at the present the prime example of the kind which Einstein tried to work out. On the one hand the geodesics are well understood as an ergodic mechanical system, and on the other hand Selberg's trace formula relates the spectrum of the Laplacian to the lengths of the closed (periodic) geodesics of a closed, compact, orientable surface of constant negative curvature.

The motion of an electron around a donor impurity in a semiconductor was first proposed by the author.[5] Its purely ergodic behavior was discovered through extensive numerical work,[6] and then conjectured on the basis of a number of mathematical statements which were proved by the author[7], and independently and differently by Devaney[8] in a number of papers. The transition to quantum mechanics was performed with the help of certain general formulas which the author extracted from Feynman's Path integral formulation of quantum mechanics. This procedure is quite heuristic, but it gives a result which is eminently reasonable and contains Selberg's trace formula as a special case.

The bulk of this paper will consist first in explaining this general procedure. Then the ergodic behavior of the anisotropic Kepler problem will be discussed. This name was chosen for the motion of the electron around a donor impurity, because the main feature in it is indeed a seemingly minor modification of the ordinary Kepler problem. (It is fascinating that the latter should be the most regular, least ergodic of all mechanical systems and yet become ergodic so easily.) Finally, the ergodic behavior of the trajectories in the anisotropic Kepler problem (AKP) will be used in order to find approximate values for the energy levels of the quantum mechanical AKP. This is the first time such a calculation has been carried out for an ergodic system. One might think that Selberg's trace formula could be used in order to obtain the spectrum of the Laplacian from the knowledge of all the closed geodesics. The latter are very difficult to enumerate, however, and the mathematicians have rather concluded in the opposite sense, that is from the knowledge of the spectrum to the distribution of the closed geodesics. An exception is the paper of Randol[9)] on the occurrence of small eigenvalues of the Laplacian.

It looks as if the main technical problem in this whole program is the effective enumeration and integration of all periodic orbits in an ergodic system. Once this is done, the transition to quantum mechanics is relatively straightforward. It is interesting to note that the study of the periodic orbits always seemed to be a central goal of mathematicians, although it was never explained, what further results would follow from it.

3. CLASSICAL VERSUS QUANTAL MECHANICS

In classical mechanics the momentum (u,v) and the position (x,y) are real vectors which depend on time t. This time dependence follows from the equations of motion,

$$\frac{du}{dt} = -\frac{\partial H}{\partial x}, \quad \frac{dv}{dt} = -\frac{\partial H}{\partial y}$$
$$\frac{dx}{dt} = \frac{\partial H}{\partial u}, \quad \frac{dy}{dt} = \frac{\partial H}{\partial v} \tag{1}$$

A particular trajectory can be characterized by the initial conditions, i.e. the values of the momentum and the position at a given time t_0.

In quantum mechanics the momentum (u,v) and the position (x,y) are operators in a Hilbert space. They satisfy the commutation rules

$$[x,y] = 0, \quad [u,x] = -i\hbar, \quad [u,y] = 0$$
$$[u,v] = 0, \quad [v,x] = 0, \quad [v,y] = -i\hbar \tag{2}$$

where $\hbar$ is Planck's constant divided by 2π. In many cases it is convenient to realize the Hilbert space as the set of complex valued functions of the position, $\Psi(x,y)$, which change with time according as the Schroedinger equation

$$i\hbar \frac{\partial \Psi}{\partial t} = H(u,v,x,y) \cdot \Psi \tag{3}$$

$H(u,v,x,y)$ is now interpreted as an operator. The so called stationary states have the time dependence $\exp(-i\omega t)$, while their position dependence is obtained from the eigenvalue equation

$$H(u,v,x,y)\Psi = E\Psi \tag{4}$$

where $E = \hbar\omega$ is then called an energy level of the mechanical system.

4. THE TRADITIONAL PROCEDURE

The vector Ψ in Hilbert space is called "wave function" by the physicists. They think of it as a function of the position (x,y) and the time t. Consequently, they write more explicitly $\Psi(x,y,t)$. The operators x and y are simply multiplications of Ψ with x and y, while the momentum operators u and v become the differentiations

$$u\Psi = \frac{\hbar}{i} \frac{\partial \Psi}{\partial x}, \quad v\Psi = \frac{\hbar}{i} \frac{\partial \Psi}{\partial y} \tag{5}$$

In order to make the connection with classical mechanics the wave function is written explicitly as a complex number with the real and positive amplitude A(x,y,t) and the phase function S(x,y,t) in the following form

$$\Psi = A(x,y,t)\exp(\frac{i}{\hbar}S(x,y,t)) \qquad (6)$$

Notice that Planck's quantum $\hbar$ occurs in a very specific manner, so that this peculiar Ansatz will in general not lead to a solution of the Schroedinger equation (3). It can be seen, however, by inserting (6) into (3), that the correction terms vanish with the second power of $\hbar$. The Ansatz (6) is the lowest approximation, if we expand in powers of $\hbar$.

The condition to be satisfied by the phase function S(x,y,t) turns out to be the Hamilton-Jacobi equation

$$H(\frac{\partial S}{\partial x},\frac{\partial S}{\partial y},x,y) + \frac{\partial S}{dt} = 0 \qquad (7)$$

This is a first order partial differential equation, and the equations for its characteristics are just the equations of motion (1). The change of the phase function as one proceeds along a characteristic, i.e. along a trajectory, follows from the additional first order ordinary differential equation

$$\frac{dS}{dt} = u\,\frac{dx}{dt} + v\,\frac{dy}{dt} - H \qquad (8)$$

If one is interested in stationary states, the Hamilton-Jacobi equation simplifies, because the time dependence of S(x,y,t) becomes linear

$$S(x,y,t) = S_0(x,y,E) - Et \qquad (9)$$

E is the constant value of H for the special subset of trajectories which go into the phase function.

The main problem is now: can one construct sensible and smooth functions $S(x,y,t)$ or $S_0(x,y,E)$?

In order to find an answer, one has to look more closely at the relation between the phase function and the trajectories from which it is constructed. Since the phase function depends only on the position coordinates, but not on the momentum, one has to determine first of all the momentum (or equivalently the velocity) of the particular trajectory which passes through a given position (x,y) at the time t, or at the energy E in the case of stationary states. The formula is simply

$$u = \frac{\partial S}{\partial x}, \quad v = \frac{\partial S}{\partial y} \tag{10}$$

At a given time t (or given energy E) the phase function defines, therefore, a 2-dimensional manifold in the 4-dimensional phase space of (u,v,x,y). If a particular trajectory was used in the construction of the phase function, and we know that it passes through the neighborhood of the position (x,y), then we can immediately find its momentum in that neighborhood through the formula (10). The question arises: given any fixed trajectory and knowing that it passes through the neighborhood of (x,y), does it always have the same momentum as it goes through that neighborhood?

5. A FEW SIMPLE CASES OF TRAJECTORIES

Consider what happens in the three following cases. Each case results from a Hamiltonian which consists of the sum of the kinetic energy $\frac{1}{2}(uu + vv)$ and the potential energy $V(x,y)$. We will also use the abbreviation $r = (xx + yy)^{1/2}$.

a) Coulomb potential: $V(x,y) = -1/r$. Every trajectory of negative energy E is an ellipse around the origin, the famous Kepler ellipse. Clearly, every time this trajectory passes through a given position (x,y), it has the same momentum, as shown in figure 1.

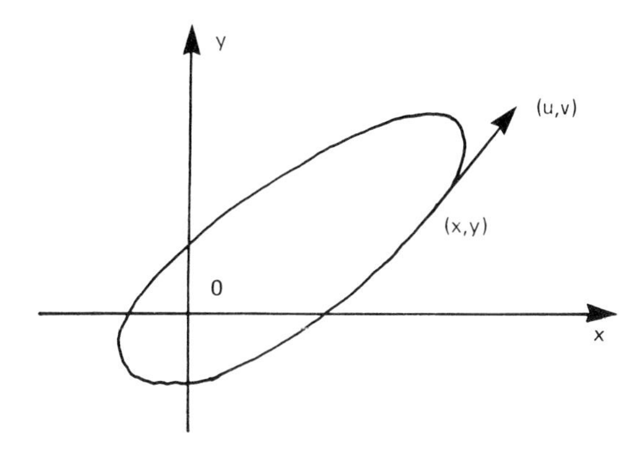

FIGURE 1

b) Spherically symmetric potential W(r), but not -1/r. Any particular trajectory is determined by its energy E and its angular momentum M. When it goes through a given position (x,y) it can have only two possible values of the momentum, as indicated in figure 2. The phase function is, therefore, two-valued, or put differently, the manifold which is determined by the phase function has two sheets over the position space. Actually it looks just like a torus.

c) Potential V(x,y) without any symmetries. Pick some arbitrary trajectory, and watch how it behaves as it goes through some arbitrary neighborhood near the position (x,y).

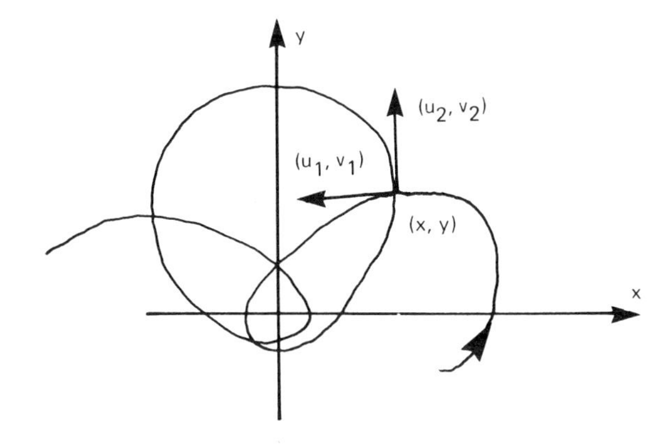

FIGURE 2

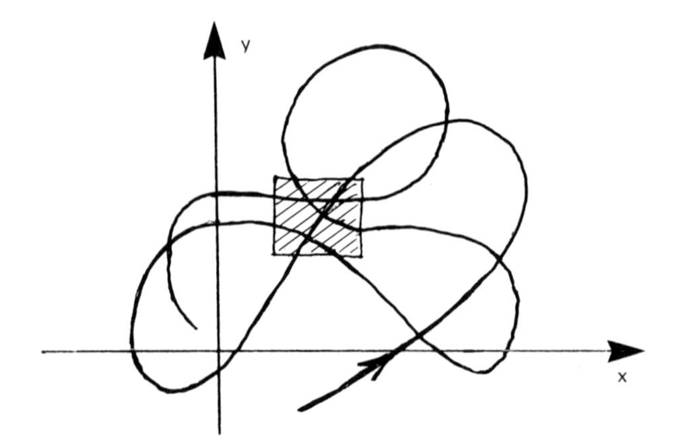

FIGURE 3

It will pass as often as you wish, but every time in a different direction, i.e. with a different momentum, as shown in figure 3.

Many students have never seen an example of the last kind. The geodesics on a surface of negative curvature and the trajectories in the anisotropic Kepler problem are exactly of this type.

The phase function in the case c) can not possibly be smooth or multivalued with some well defined multiplicity, because no function with these properties can have a gradient whose directions in any given neighborhood point every which way. If we look at $S(x,y,t)$ for either the geodesics on a surface of negative curvature or the trajectories in the AKP, we find that it gets more and more complicated as the time t increases and it does so in an exponential manner. Obviously, it would be very difficult to construct a wave function as suggested by the formula (6).

6. THE GREEN'S FUNCTION

Since the wave function is of no use, we have to use some other formalism. The main object from now on will be to study what physicists call the Green's function and some mathematicians the parametrix. It comes in two varieties, depending on the parameter which gives the information on the time dependence. If the time enters directly, we will call the Green's function sometimes also the propagator. Most often, however, we will be more interested in the Fourier analysis with respect to the time dependence. In that case the important parameter in Green's function is the frequency ω, or since we have always the relation $E = \hbar\omega$, the important parameter really becomes the energy E. But it is good to remember that both types of Green's functions are useful in establishing the relation between classical and quantum mechanics, whether or not the system is ergodic. Green's function is a solution of the inhomogeneous Schroedinger equation. Physically, it corresponds to "tickling" the quantum system at some position (x',y') at a time t', and then watching the result at some other position (x,y) and time t. We have, therefore, a quantity which depends on two sets of coordinates, $K(x,y,t;x',y',t')$. It satisfies the equation

$$i\hbar \frac{\partial K}{\partial t} - H(u,v,x,y)K = \delta(x-x')\delta(y-y')\delta(t-t') \tag{11}$$

The solution of this equation is determined, if we specify the initial conditions, $K = 0$ when t is smaller than t'. (The differential operators on the left hand side are applied to the positions (x,y) in K.)

Feynman's path integral is an explicit expression for K in terms of a summation over all conceivable paths in position space which lead from (x',y') at time t' to (x,y) at time t. Such a summation is very difficult to define in mathematically satisfactory terms for the equation (11). When we replace the factor i in front of the time derivative in (11) by -1, however, the path integral can be treated quite well. It reduces to a Wiener integral, and its theory has been investigated extensively.[10)]

If we are interested in the classical limit of K, that is the limit where $\hbar$ is considered very small compared to other quantities in the problem with the same dimension (space times momentum, or energy times time), the propagator K becomes approximated by a similar quantity $\tilde{K}$ with the same variables, x, y, t and x', y', t'. Instead of considering all conceivable paths, one takes only the classical trajectories into account, which start in (x', y') at time t' and end up in (x, y) at time t. But in doing so they have to satisfy the equations of motion (1). We will not write down the explicit expression for $\tilde{K}$, because it involves a number of details which are of no importance for us later. But it is good to keep this relation to Feynman's path integral in mind.

The information about the energy levels of the quantum system is not directly contained in K. We get this information when we Fourier analyze K with respect to time. Therefore we define the quantity

$$G(x,y;x'y';E) = \frac{1}{i\hbar} \int_0^\infty dt\, e^{iEt/\hbar} K(x,y,t;x',y'.t') \quad (12)$$

Notice that the integration goes from 0 to t. There are two points hidden in this. First, the propagator depends on t and t' only through the combination $t - t'$, because the Hamiltonian H does not depend on time. Second, the initial condition for K stipulated that K vanish when t is less than t'. Thus, the integral can always be made to converge if necessary, if the energy E in the exponent of the integrand becomes slightly complex, i.e. E acquires a small positive imaginary part.

It is easy to show that the Green's function G satisfies its own inhomogeneous stationary Schroedinger equation

$$[E - H(u,v,x,y)]G = \delta(x-x')\delta(y-y') \quad (13)$$

The energy E is now simply a parameter which varies over all the real numbers, but can even be taken to be complex with a positive imaginary part. The relation between G and the energy levels can be seen from the formula

$$G(x,y;x',y';E) = \Sigma_j \frac{\Psi_j(x,y)\cdot\Psi_j^*(x',y')}{E - E_j} \tag{14}$$

The right hand side contains a summation over all the energy levels which are numbered by the index j. The energy of the j-th level is E_j and its wave function is given by $\Psi_j(x,y)$, which is a solution of the stationary Schroedinger equation (4) for the energy E_j. If we know the Green's function G, we can find the energy levels by determining the poles of G with respect to the variable E. The corresponding wave functions are essentially the residues at those poles.

7. THE RESPONSE FUNCTION AND ITS CLASSICAL APPROXIMATION

Our goal is to determine the energy levels, but not the wave functions. That might be interesting, but it requires a lot more work. Therefore, we would like to simplify our task from the very start by dealing with an object which contains only the information we want to get. The full Green's function is really more than we can handle at this time. Essentially, we have to get rid of the position dependence in G, but keep the dependence on the energy E. Therefore we integrate it out by taking its "trace"

$$g(E) = \iint dxdy\ G(x,y;x,y;E) = \Sigma_j \frac{1}{E - E_j} \tag{15}$$

We have used the fact that the wave functions Ψ were properly normalized, i.e. the integral over all of $|\Psi_j|^2$ equals 1. The "trace" $g(E)$ is also called the response function, because it tells us how strongly the quantum system responds to an outside stimulus of a given frequency $\omega = E/\hbar$. Obviously, we need the classical approximation for g.

This is accomplished in two steps. First, the classical approximation $\tilde{G}$ of G is found. Then, we take the "trace" of $\tilde{G}$, exactly as we did for G in (15). In order to obtain $\tilde{G}$ we start with the classical approximation $\tilde{K}$ for the propagator K, and perform the Fourier transform (12) on it. With some additional work, this gives us a summation over all classical trajectories which go from (x',y') to (x,y) and move with the total energy E. In contrast to the situation where the total time t - t' for this trajectory is fixed, there may be very many such trajectories for any particular energy E. A typical situation is pictured in figure 4, where the particle goes around the origin an arbitrary number of times before it decides to stop at the endpoing (x,y). Along each one of these trajectories we have to calculate the so-called phase integral S(x,y;x',y';E)

$$S(x,y;x',y';E) = \int_{(x',y')}^{(x,y)} (u\,dx + v\,dy) \tag{16}$$

The approximate Green's function is now given by the summation

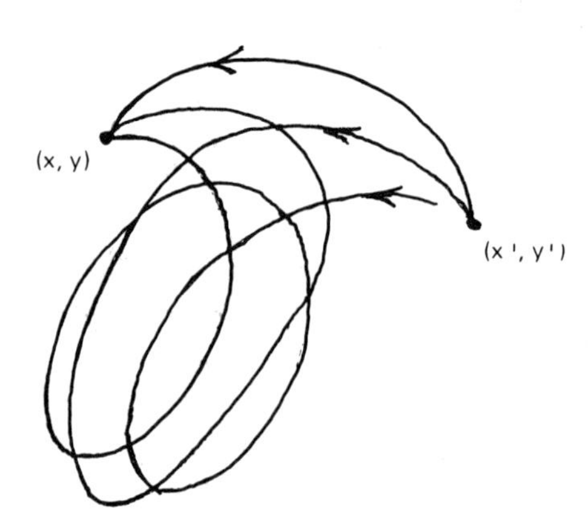

FIGURE 4

$$\tilde{G}(x,y;x',y';E) \tag{17}$$

$$= \sum_{\substack{\text{classical}\\ \text{trajectories}}} A \cdot \exp\left[\frac{i}{\hbar} S(x,y;x',y';E) - i\nu \frac{\pi}{2}\right]$$

where the amplitude factor A and the number ν of conjugate points need a little more explanation. As indicated in figure 5 we can look at all the trajectories which leave (x',y') in various directions, but all of them with the same energy E. A particular one will hit the endpoint (x,y) and its neighbors will miss going through (x,y) by various amounts. The amplitude A measures essentially the density of these neighboring trajectories around (x,y) as compared to their density at the start in (x',y'). As also shown schematically in figure 5, these trajectories may cut into each other on the way from (x',y') to (x,y), not only once but a number ν times. Every time this happens to the particular trajectory through (x,y), we call this a conjugate point of (x',y'). The quantum wave from (x',y') to (x,y) behaves as if it was reflected off a wall near a conjugate point, and like all waves it loses some of its phase when that happens. In our case the phase loss is $\pi/2$ each time, and that is exactly what the formula (17) shows.

The same integral as (15) is now calculated for the approximate Green's function $\tilde{G}$ of (17). This calculation is important

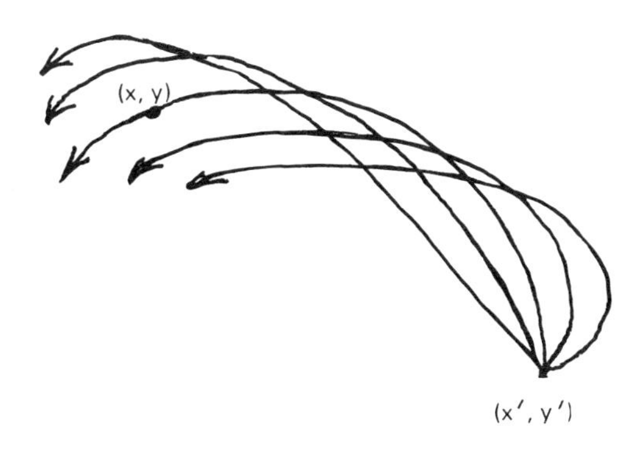

FIGURE 5

because it leads to the restriction of the summation to only the smoothly closed trajectories, the so called periodic orbits. In order to see what happens, look first at the integrand of (15). The 2 positions (x,y) and (x',y') are identified which means that the classical trajectory starts and ends up at the same place. This situation is represented in figure 6. But now this place has to be varied over all position space, if one does the integration in (15). When the position $(x,y) = (x',y')$ varies, the value of the phase integral $S(x,y;x,y;E)$ changes. Since it was one of our assumptions that the magnitude of S is much larger than $\hbar$, we can now also assume that its change as (x,y) varies must be large compared to $\hbar$. We can expect that our approximation of quantum mechanics by classical mechanics is valid only under this restriction.

This change in $S(x,y;x,y;E)$ can be calculated directly from (8) and (9), together with (10), which has a simple companion equation

$$\frac{\partial}{\partial x'} S(x,y;x',y';E) = -u'$$
$$\frac{\partial}{\partial y'} S(x,y;x',y';E) = -v' \tag{18}$$

and yields immediately

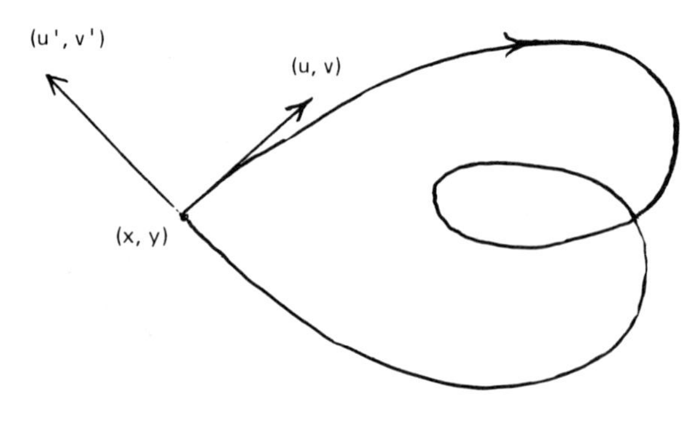

FIGURE 6

$$\frac{\partial}{\partial x} S(x,y;x,y;E) = u - u'$$
$$\frac{\partial}{\partial y} S(x,y'x,y;E) = v - v' \tag{19}$$

If the starting momentum (u,v) at (x,y) is not equal to the end momentum (u',v') again at (x,y), the phase function will vary so much that the exponential in (17) takes on all possible values indiscriminately and the integration (15) will give essentially zero. In a more careful argument one would show that the trajectories with $(u,v) \neq (u',v')$ contribute to the integral only terms of higher power in $\hbar$ than the terms where $(u,v) = (u',v')$. Therefore we will concentrate our attention exclusively on the latter trajectories, i.e. the periodic orbits.

The reasoning which lead to the restriction to periodic orbits is only the first step in a more complicated argument which will not be discussed here. It can be found in the article of the author in the Journal of Mathematical Physics in 1971 which contains references to earlier work in this area, mostly in the same journal.[5] The result is a formula which will be the basis for all our further work. The formula is written in terms of some new quantities, to wit, the action integral

$$S(E) = \oint_{\text{periodic orbit}} (udx + vdy) \tag{20}$$

which depends only on the periodic orbit and its energy E, the reduced period T_0 which measures how long (in time) it takes the particle to cover the projection of the periodic orbit into the position space, the number ν of conjugate points along the periodic orbit starting at some arbitrary place, and the stability exponent α which tells us how fast any neighboring trajectory drifts away from the periodic orbit. All this combines into

$$\tilde{g}(E) = -\frac{i}{\hbar} \sum_{\substack{\text{periodic} \\ \text{orbits}}} \frac{T_0}{2 \sinh \alpha/2} \exp\left(\frac{i}{\hbar} S(E) - i\nu \frac{\pi}{2}\right) \qquad (21)$$

Once the summation over all periodic orbits has been carried out, the only remaining variable is the energy E. We expect the approximate response function $\tilde{g}(E)$ to have some singularities, preferably poles, which we can then associate with the poles of the exact response function g(E) in (15).

In this manner we have succeeded in giving an answer, at least in principle, to Einstein's problem because the last formula can be applied to integrable as well as ergodic classical systems. The author has shown[11)] that one obtains the expected answers for the integrable systems, and it remains now to test whether this scheme works also for ergodic systems.

8. THE GEODESICS ON A SURFACE OF CONSTANT NEGATIVE CURVATURE

Surfaces of constant negative curvature (to be abbreviated by SCNC) are fairly easy to realize. Hyperbolic geometry (Lobachevsky geometry) can be modeled in several ways, and a surface of constant negative curvature is essentially a polygon whose sides are properly identified. But there is an even simpler realization as soap film which is stretched over some closed loop of wire, as shown in figure 7.

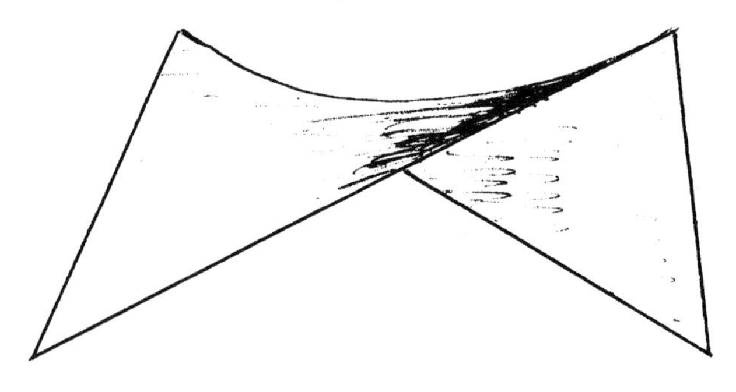

FIGURE 7

Such a soap film is not exactly what we want, however, for two reasons. Its curvature, though negative, is not necessarily constant, and an arbitrary SCNC can not be embedded or even immersed in Euclidean three dimensional space, as was first proved by Hilbert. The first objection is actually not so serious, because most of the general properties of geodesics remain the same whether the curvature is constant or not, as long as it is below a certain negative value. In particular all the ergodic qualities are independent of the constancy of the curvature. The second objection can be overcome, too, if we are willing to consider a surface with a boundary such as the one in figure 7. In that case, we have to make some assumptions about the behavior of the geodesic when it hits the boundary. The most convenient is to assume "specular reflection", i.e. the geodesic starts up again going into the surface at exactly the same location where it hit the boundary, but with a direction which is symmetric just like a billiard ball. This situation was considered by Hadamard in 1898, and he entitled his paper "On the Non-Euclidean Billiard".[12] Hadamard was the first to recognize the ergodic properties of geodesics on a SCNC. His work was continued by Birkhoff and Morse, and became an active field of research in the thirties with names like Artin, E. Hopf, and Hedlund.

For the little that I want to say about this topic it may be just as well to limit my statements to a particular model, namely Poincaré's upper half plane. Call U the domain $y > 0$ of the ordinary Euclidean (x,y) plane. It will be endowed with the metric $ds^2 = (dx^2 + dy^2)/y^2$. The geodesics are the Euclidean circles whose center lie on the x-axis. The distance between two points can be found if we first construct such a circle through the two points, call them P_1 and P_2 and call the two intersections of the circle with the real axis Q_1 and Q_2. These points can be determined by their cartesian coordinates, x and y, and they in turn can be joined together to form a complex number $z = x + iy$ in the usual manner. Thus we have the complex number z_1 and z_2 for P_1 and P_2, as well as t_1 and t_2 for Q_1 and Q_2. The latter are actually real, because Q_1 and Q_2 lie on the real axis. The distance is now given by the formula

$$S = \log\left(\frac{z_1-t_1}{z_1-t_2} : \frac{z_2-t_1}{z_2-t_2}\right) \tag{22}$$

It takes a little time to check that everything is all right with this expression, e.g. that the argument of the log is real and positive The angle between two lines at their point of intersection is just the Euclidean angle. Geometrical constructions can easily be carried out by elementary means, using only ordinary Euclidean circles and their angles.

In particular we can now construct hyperbolic polygons, i.e. polygons whose sides are geodesics, and that in turn means Euclidean circles with center on the real axis. Such a polygon can be made the "table" on which to play billiard, under the assumption that the billiard ball always moves along a geodesic. Since we know the definition of an angle in this geometry, we can apply the rule of specular reflection when the ball hits the boundary. Also we know how to measure the length along the geodesic, and therefore we can determine the time for the ball to go from one side to the other of the polygon, assuming that the ball always moves with the same constant speed w. This speed determines the total energy E through the standard formula $E = m\,w^2/2$. The momentum has the same direction as the motion of the billiard ball, and its value is mw. The action integral (16) becomes simply $m.w.L$, where L is the distance from (x',y') to (x,y) as measured in hyperbolic geometry.

There is another way of defining what the billiard ball has to do when it hits the boundary, but that requires a more careful construction of the polygon. Suppose the polygon was chosen in such a way that its sides can be identified in pairs. This imposes a number of restrictions. Corresponding sides have to be of the same length. If one tries to go around one of the corners, one keeps hitting the boundary. Every time that happens, one has to jump to the corresponding place on the corresponding boundary and continue from there. Figure 8 shows such a trajectory in an octogon where we have paired off opposite sides. In such a scheme it is fairly obvious that all the corners form one point and the "billiard table" loses all its boundary. Still, the common corner point is exceptional, unless we can fix the geometry so that the sum of the angles adds up to exactly 360 degrees.

Now we have one smooth two dimensional manifold in which the billiard ball can move without ever feeling anything that would make one place different from any other. But there are many closed loops, not necessarily geodesics, which cannot be deformed into each other. A more detailed study shows that the situation is exactly as on a sphere with two handles attach-

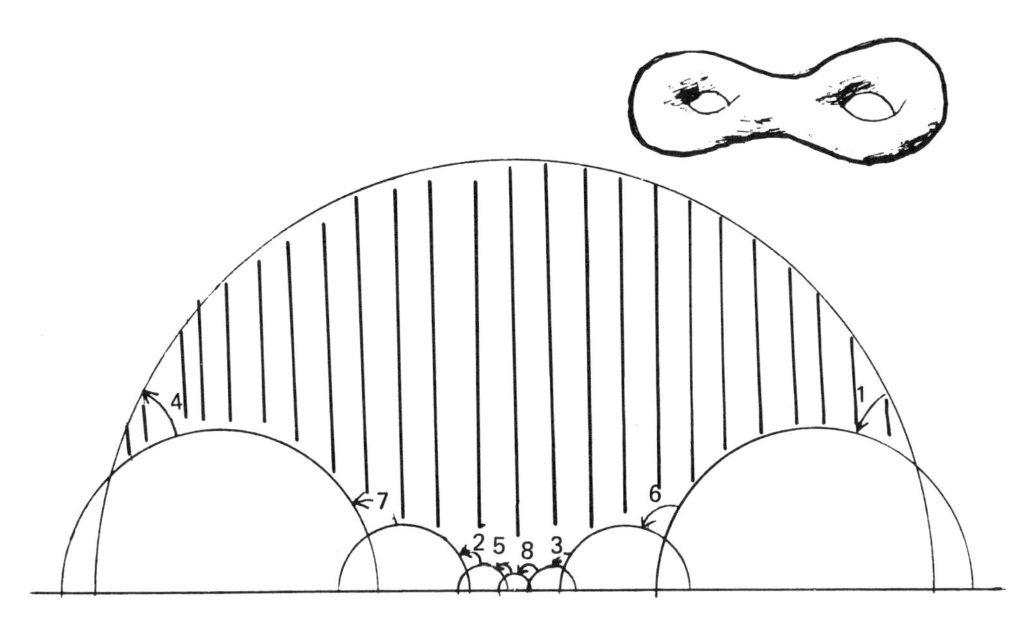

FIGURE 8

ed, or as on the double torus of figure 8, also called a surface of genus 2. When all the conditions are properly taken care of, such a manifold represents a SCNC which is smooth, compact, orientable, and complete, i.e. every geodesic can be continued indefinitely. It has a finite surface area as measured in the hyperbolic geometry. There are many different such surfaces according as the exact choice to the hyperbolic polygon. Apart from an overall scale, there are actually six real parameters which can be chosen within certain intervals in order to specify completely the SCNC. Linda Keen[13)] has shown very explicitly how to construct the most general SCNC.

9. SELBERG'S TRACE FORMULA FROM FEYNMAN'S PATH INTEGRAL

The aim of this section is to evaluate both $g(E)$ and $\tilde{g}(E)$ from (15) and (21) for the surface of constant negative curvature, any one of those which were described above. For this we need first a Hamiltonian. In the standard Riemannian metric for the Poincaré upper half plane this Hamiltonian would be simply

$$H(u,v,x,y) = y^2(u^2 + v^2)/2m \tag{23}$$

which is simply the expression of the kinetic energy in the momenta, rather than in the velocities as usual.

The stationary states Ψ for this Hamiltonian in quantum mechanics are the eigenstates of the operator which results from (23) with the help of the rule (5). One has to exercise a little care because of the factor y^2 in front which does not commute with the operator v. But it is reasonable to write Schroedinger's equation (4) in the form

$$-\frac{h^2}{2m} y^2 \left(\frac{\partial^2}{\partial x^2} + \frac{\partial^2}{\partial y^2}\right)\Psi_n = E_n\Psi_n \tag{24}$$

or with the help of the Laplacian Δ on the SCNC

$$\Delta\Psi_n + \lambda_n\Psi = 0 \quad \text{with} \quad \Delta = y^2\left(\frac{\partial^2}{\partial x^2} + \frac{\partial^2}{\partial y^2}\right) \tag{25}$$

and $\lambda_n = 2m\,E_n/\hbar^2$. In other words, getting the spectrum of the Hamiltonian (23) is the same as finding the eigenvalues of the Laplacian. Finding the propagator and the Greens function carries right on through, but in (15) one has to insert the volume element $1/y^2$ under the integral sign.

The trouble arises when we want to use the Ansatz (6) which was derived ultimately from Feynman's path integral. The function S(x,y,t) in the exponential is to be computed along a geodesic by integrating equation (8). But there is no doubt that Feyman's path integral has to be modified when the system moves through a curved space. One finds this already when working in ordinary polar coordinates where the two angular variables vary essentially over a sphere. A considerable amount of ink has been spilled over this issue, and the outcome is not clear. In our case it has to be kept in mind that we are primarily interested in the long-time behavior of the trajectories. The recipe to be adopted in that situation is fairly clear. One has to add to the right hand side of (8) the term $\hbar^2 C/8m$, where C is the Gaussian curvature at the position (x,y). In the case of a sphere $C = 1/R^2$, and in our case the Gaussian curvature is simply $-1/R^2$. In our calculations we will always set $R = 1$. Notice that this additional term acts exactly like an additional potential energy in the Hamiltonian. Also it is small, if we consider Planck's constant $\hbar$ as small compared to the other quantities. Nevertheless the presence of this extra term makes a crucial difference both on the sphere and on the SCNC. The correction from the curvature of the underlying space will make its appearance when we try to do the time integration in (12) for the approximate propagator K. This integral is done by the stationary phase method which means that we look at the exponent in the integrand which is

$$\frac{ms^2}{2t} - \frac{1}{4}\frac{\hbar^2}{2mR^2}t + Et \qquad (26)$$

apart from a common factor $i/\hbar$, and we determine its minimum as a function of t. The variable s is the geodesic distance from (x,y) to (x',y'). This minimum is found to be

$$s\sqrt{2m\left(E - \frac{\hbar^2}{8mR^2}\right)} \qquad (27)$$

and enters into (17) instead of the integral (16). The latter would be given by the simpler expression $s(2mE)^{1/2}$ which one gets when the curvature C vanishes as in Euclidean space in the absence of a potential. The amplitude A of the new expression (17) for the approximate Greens function $\tilde{G}$ is also based on the exponent (27) rather than the simpler $s(2mE)^{1/2}$.

In order to calculate the resonance function $\tilde{g}(E)$ in (21), replace the distance s between the initial point (x,y) and the final point (x',y') by the length of the closed geodesic L. The reduced period T_0 is the derivative of the new action integral (27) with respect to E, where L has to be replaced with the length of the projection L_0 of the geodesic into position space. All closed geodesics in a space of negative curvature are unstable, and their stability exponent α equals the length L. Moreover there are never any conjugate points so that $\nu = 0$.

Now we are ready to write out the complete expression for both $g(E)$ based on (15) and for $\tilde{g}(E)$ based on (21). The main variable is the energy E, but (27) makes it clear that we might be better off if we use the variable

$$z = \frac{1}{2} - \frac{i}{\hbar}\sqrt{2mE - \frac{\hbar^2}{4}} \qquad (28)$$

E varies along the real axis, with a small imaginary part in case of doubt, as we emphasized in the definition of the Fourier transform (12). Also a free particle has mostly positive energies, even on a curved surface, so that the root is well defined on the right hand side of (28), with a positive imaginary part. That makes the variable z well defined on a neighborhood of the critical line with $\mathrm{Re}(z) > 1/2$. By simple arithmetic we get

$$\tilde{g}(E) = \frac{-m}{\hbar^2 (z-1/2)} \Sigma \frac{L_0}{1-\exp(-L)} e^{-zL} \tag{29}$$

from (21). On the right hand side of (15) we express E in terms of z as given by (28), and introduce the eigenvalues λ_n of the Laplacian instead of the eigenvalues of the Hamiltonian E_n as in (25). Thus

$$g(E) = -\frac{2m}{\hbar^2} \Sigma_n \frac{1}{z(z-1)+\lambda_n} \tag{30}$$

The question now is whether (29) is in any way related to (30) in some kind of approximate manner. Our argument has been entirely heuristic, and we have no idea of the errors which were introduced when we replaced the Greens function in (12) by (17), and then further approximated the trace (15) by the sum over all periodic orbits (21).

But it now turns out that Selberg's trace formula can be written as

$$(2z-1)\Sigma_{n=0}^{\infty}\left(\frac{1}{z(z-1)+\lambda_n} - \frac{1}{z+n}\right) \tag{31}$$

$$= \Sigma_{\text{per. orbits}} \frac{L_0}{1 - \exp(-L)} e^{-zL}$$

Its derivation is quite different from our line of reasoning, and uses the large degree of symmetry which is inherent in our hyperbolic polygons as smooth, closed, orientable surfaces of constant negative curvature. Our argument did not take advantage of this symmetry, and is therefore basically valid (to the extent that it is valid at all) even for surfaces whose curvature is not constant. Therefore it comes as a pleasant surprise to find that Selberg's trace formula (31) says simply that $g(E) = \tilde{g}(E)$, with one small correction.

According to a well known formula of Hermann Weyl the number of eigenvalues $N(\xi)$ of a Laplacian below the value ξ in two dimensions is proportional to ξ asymptotically for large ξ, with a constant of proportionality equal to the area divided by 4π. The total area of a closed, smooth, orientable surface of constant curvature and genus g is given by $4\pi(g-1)R^2$ where R is its radius of curvature, in our case $R = 1$. Thus the number of eigenvalues $\lambda_n < \xi$ is asymptotically equal to $(g-1)\xi$. Therefore, the series (30) has its terms decreasing like $1/n$, and does not converge as it stands. The extra terms on the left hand side of (31) insure the convergence. They do not contain any new information, it almost seems, but that is not correct. They do arise even in our heuristic derivation except we did not talk about them. In taking the trace (15) using the classical approximation (17), we neglected the term which comes from the trajectories of zero length, which start at a point (x,y) and end there without ever having gone anywhere. They contribute exactly the extra terms on the left hand side of (31), and that can also be seen in Selberg's derivation.

Another comment is necessary on the convergence of (29) and the right hand side of (31). It can be shown relatively easily that the number of closed geodesics of length L smaller than ξ grows exponentially as $\exp(\xi)$. We can therefore, replace asymptotically the summation over the periodic orbits by an integration over ξ with an integrand $\exp(\xi)\ \exp(-z\xi)$ where ξ goes from 0 to infinity. This integral clearly converges only when the real part of z is larger than 1. (29) and the right hand side of (31) are defined as functions of the complex variable z only for $\mathrm{Re}(z) > 1$. But the left hand side of (31) provides an analytic continuation.

10. A LITTLE PHYSICS ABOUT DONOR IMPURITIES IN A SEMICONDUCTOR

The movement of a billiard ball on a surface of constant negative curvature can be understood mathematically thanks to its high degree of symmetry which leads to Selberg's trace formula. But as a physical system it is not very satisfactory, because it bears so little relation to anything that can be actually observed. A physicist or a chemist would like to see an example of ergodic classical motion which is realized in the laboratory if not in nature. That would permit measurements to be made, rather than only calculations. The electron in the neighborhood of a donor impurity is such an example, and one which can also be treated mathematically in rather complete fashion. It will be discussed in the rest of this paper in some detail, but first of all its physical situation will be explained for those readers who have a background in solid state physics.

The (by now) classical semiconductors silicon and germanium form beautiful cubic crystals where the atoms are arranged in the same lattice structure as the carbon atoms in the diamond crystal. Each atom is surrounded by four neighbors and the electron clouds around each atom are distorted in such a manner as to form bridges between neighboring atoms. When an electric field is applied, these electron clouds deform rather easily in a way which tends to cancel out the effects of the applied electric field. The ratio between the deformation of electron clouds and the electric field is called the dielectric constant κ of the crystal, and it is very large, 11.4 in silicon and 15.4 in germanium. This fact is important for the success of our theory.

A donor impurity is a foreign atom which has replaced one of the Si or Ge atoms in the lattice. Its effective nuclear charge is too big by one unit of positive electric charge, and in order to maintain electric neutrality, it brings with it an extra electron. Thus the donor impurity "donates" an electron to the Si or Ge crystal. The question is to know what happens to that extra electron.

Without the donor impurity both silicon and germanium are insulators, just like their famous relative, diamond. In our present day language, we would describe this simple fact in the following words. The electrons of the crystal are in energy levels which form bands, that is to say, these energy levels fill up certain intervals on the energy axis which are separated by intervals without any energy levels. Each energy level accomodates exactly one electron, and one knows the total number of electrons to be accomodated, so that one starts from the bottom until everybody has found its place. At that point, it turns out that we have just filled up a band, and if we had one more electron, it would have to start in a new band at an energy which is separated from the others by a sizable gap of about 1 electron-volt. An electron has to be promoted across that gap before it can move around in response to an applied electric field. This can happen only if we provide the crystal with the right kind of impurities, or if we heat up the crystal.

Once an electron finds itself in the energy band above the gap. called the conduction band because it permits electric conductivity to arise, its response to the electric field depends on the detailed structure of the conduction band. One does not have to know everything about this band, but only about its states of lowest energy, its so called minima. Here we encounter the first complication: there are six minima of equal energy in Si, and four in Ge. Thus the electron has a choice which the physicists call a degeneracy. Moreover, one has to know the shape of these minima, although they all look alike. When talking about the shape it is implied that there is a parameter which can be varied, and for each value of which there is a well defined energy. This parameter is variously called a wave vector, or crystal momentum, or simply momentum because it plays exactly that role, although in a somewhat devious manner. The energy as a function of the momentum in the neighborhood of a minimum is given by a quadratic function of the momentum, but there is no guarantee that this quadratic function is simply proportional to the square of the momentum as it is in ordinary mechanics.

In both Si and Ge the quadratic function has one small and two equal large eigenvalues, and if we orient the axes accordingly, we can write

$$\frac{p_x^2}{2m_1} + \frac{p_y^2}{2m_2} + \frac{p_z^2}{2m_3} \tag{32}$$

in the neighborhood of each minimum. The coefficients in the quadratic function have been written so as to reflect the expression for the kinetic energy in ordinary mechanics, where the mass appears in the denominator. The values of these masses can be found experimentally. They are for

$$\begin{aligned} &\text{Si:} \quad m_1 = .916\ m_e \ \text{and}\ m_2 = m_3 = .1905\ m_e \\ &\text{Ge:} \quad m_1 = 1.588\ m_e \ \text{and}\ m_2 = m_3 = .0815\ m_e \end{aligned} \tag{33}$$

where we have used the mass m_e of the free electron as the scale.

We are now ready to discuss the fortunes of the extra electron which is "donated" to the crystal by the donor impurity. First of all, it feels the attraction of the extra nuclear charge, but that attraction is weakened by the high dielectric constant of the crystal. Therefore the electron moves in an orbit around the impurity atom which is much larger than such an orbit would be in free space. This orbit covers many lattice sites to the point where the whole lattice structure gets washed out completely as if there was a continuous medium rather than a discontinuous, discrete lattice. In this medium the electron has to negotiate between the effective kinetic energy as given by (32) and the potential energy as given by the ordinary Coulomb field, that is (electric charge)^2/distance,but now weakened by the dielectric constant. The result is a set of energy levels inside the energy gap, i.e. below the minima of the conduction band, which are our main goal in this investigation. Because of the degeneracy of the minima, each of these energy levels is split into as many sublevels as there are minima. This splitting can be well understood independently of the main level structure which we are studying, and therefore we ignore it from here on.

A last complication will be mentioned so that one can make the comparison with experiment. A donor impurity can be an atom of lithium, phosphorus, arsenic antimony, or bismuth. The energy levels of the extra electron are shifted by an amount which depends on how much the electron "sees" of the inside of the impurity atom. The resulting shift is well understood again, independently of the main features of the electron states which come from the outside. This so called chemical shift again will be ignored.

The Hamiltonian for the whole 3-dimensional problem can now be written as

$$H(u,v,w;x,y,z) = \frac{u^2}{2\mu} + \frac{v^2+w^2}{2\nu} - \frac{1}{\sqrt{x^2+y^2+z^2}} \tag{34}$$

in terms of the position (x,y,z) and the momenta (u,v,w) where $\mu = (m_1/m_2)^{1/2}$, $\nu = (m_2/m_1)^{1/2}$. The energy E can be used as a scaling parameter, so that all calculations need only be made for the special value $H = -\frac{1}{2}$. For bound states, i.e. $E < 0$, the all important action integral becomes

$$S = \left(\frac{m_0 e^4}{-2\varkappa^2 E}\right)^{1/2} \int (u\,dx + v\,dy + w\,dz) \tag{35}$$

where the integral is independent of E, and we have $m_0 = (m_1 m_2)^{\frac{1}{2}}$, e is the electronic charge, and $\varkappa$ is the dielectric (= 11.4 for Si and 15.4 for Ge).

11. SCHROEDINGER'S EQUATION FOR THE DONOR IMPURITY

There is really no problem in formulating the eigenvalue problem after the simplifications of the preceeding section. One finds

$$-\frac{\hbar^2}{2m_1}\frac{\partial^2\psi}{\partial x^2} - \frac{\hbar^2}{2m_2}\frac{\partial^2\psi}{\partial y^2} - \frac{\hbar^2}{2m_3}\frac{\partial^2\psi}{\partial z^2} - \frac{e^2}{\varkappa r}\psi = E\psi \tag{36}$$

an equation which was first discussed by Kohn and Luttinger[14] in 1955. They adopted the method of the quantum chemists of making a reasonable Ansatz for the wave function $\psi(x,y,z)$ and then minimizing the expectation value of the energy with respect to the parameters in the wave function. Thus they wrote

$$\psi = C \exp\left(-\sqrt{\frac{x^2}{a^2} + \frac{y^2}{b^2} + \frac{z^2}{c^2}}\right) \tag{37}$$

for the groundstate, which is essentially the wave function for the hydrogen atom, but where the scales for the x, y, and z axes, i.e. the parameters a, b, and c, are chosen such that the integral

$$\int d^3x \left\{\frac{\hbar^2}{2m_1}\left(\frac{\partial\psi}{dx}\right)^2 + \frac{\hbar^2}{2m_2}\left(\frac{\partial\psi}{\partial y}\right)^2 + \frac{\hbar^2}{2m_3}\left(\frac{\partial\psi}{\partial z}\right)^2 - \frac{e^2}{\varkappa r}\psi^2\right\} \Big/ \int d^3x\,\psi^2 \tag{38}$$

becomes as small as possible. The resulting approximate value of the lowest eigenvalue is probably quite good.

If one wants to get some of the excited states, the Ansatz (37) has to be generalized by introducing other functions such as polynomials in x,y, and z multiplying the function (37) with the minimal parameters. One then considers linear combinations of such functions and determines their coefficients again essentially by finding the minimum for (38). Such a calculation was carried out by Faulkner[15)] using 9 independent functions, and yields a set of approximate excited states. But this method gets poorer the more one moves away from the ground state. In constrast,if the approximation through classical mechanics works at all, then it should become better the higher the quantum number, i.e. the higher the exicted state.

12. THE POINCARÉ MAP FOR THE ANISOTROPIC KEPLER PROBLEM

The Hamiltonian (34) for the anisotropic Kepler Problem (AKP) was written in 3 dimensions, but we notice right away the symmetry between the second and third dimension, y and z. This leads to a constant of motion, an angular moment around the x-axis, M, which is conserved along any trajectory. If $M = 0$, the trajectory lies in a plane through the x-axis, and the problem becomes essentially two dimensional. If $M \neq 0$, one can still reduce the equations of motion to those of a system of two degrees of freedom, but these equations are more complicated and all the analysis of the following sections is not valid for them any longer. Although our results are restricted to what looks like a purely two-dimensional version of the AKP, the three-dimensional origin will occasionally surface again, such as when it comes to counting the number of conjugate points.

The first task is to gain a complete overview over all possible trajectories. This is done with the help of the Poincaré map as follows. A particular trajectory is defined by its initial conditions, and we choose for this purpose the time when the trajectory crosses the x-axis, i.e. when $y = 0$. (Since $M = 0$ and the trajectory lies in a plane, we take the plane $z = 0$.) If we know the values of u and x at that time, we can calculate the value of v from the condition that $H(u,v,x,0) - - 1/2$, and we get

$$v = \pm \sqrt{2\nu}\left[\frac{1}{|x|} - \frac{1}{2}\left(1 + \frac{u^2}{\mu}\right)\right]^{1/2} \tag{39}$$

The double sign indicates the possibility of either going upwards or downwards from the x-axis. In either case the quantity under the root has to be positive,

$$|x|\left(1 + \frac{u^2}{\mu}\right) < 2 \tag{40}$$

The domain for the variables u and x satisfying this condition is shown in figure 9. Let us now assume that our trajectory started at time t_0 with the values u_0 and x_0 for u and x on the x-axis, going upwards, i.e. with $y = 0$ and $v_0 > 0$. At some later time $t_1 > t_0$ the trajectory will cross the x-axis

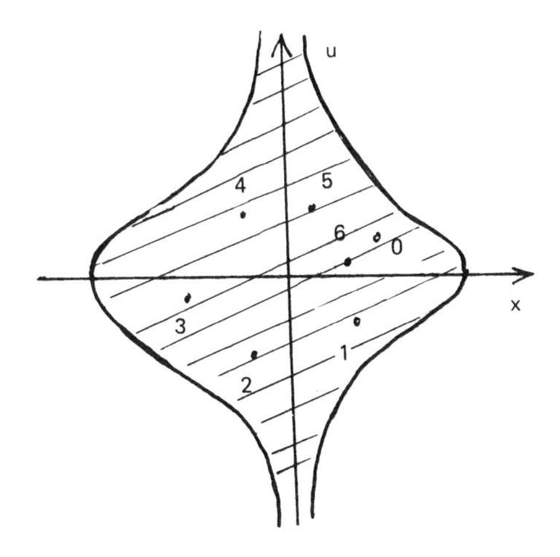

FIGURE 9

again for the first time after t_0, and the values of u and x at that moment will be u_1 and x_1 with $v_1 < 0$. Then comes a time t_2 with values u_2 and x_2 and $v_2 > 0$, etc. The resulting sequence of couples $(u_0,x_0),(u_1,x_1),(u_2,x_2),\ldots$ can be plotted in the domain (40) or the figure 9 as a sequence of points which are numbered 0, 1, 2,... The particular sequence which is shown corresponds to a trajectory like the one in figure 10. Contrary to our intuition from the ordinary Kepler problem, such a behavior can be found in the AKP for a certain set of initial conditions. Notice that the trajectory stays inside a circle of radius 2 in the (x,y) plane.

Finding the point (u_1,x_1) from (u_0,x_0) can be viewed as a map of the domain (40) into itself which can be found in detail by integrating the equations of motion (1) with the Hamiltonian (34) and $z = 0$ as well as the initial conditions (u_0,x_0) and $y = 0$. This is the Poincaré map which can be shown quite generally to conserve the element of area in the (u,x) plane. In order to discuss this map further, it seemed advantageous to give the domain a slightly different shape by a simple area conserving transformation which produces a rectangle, namely

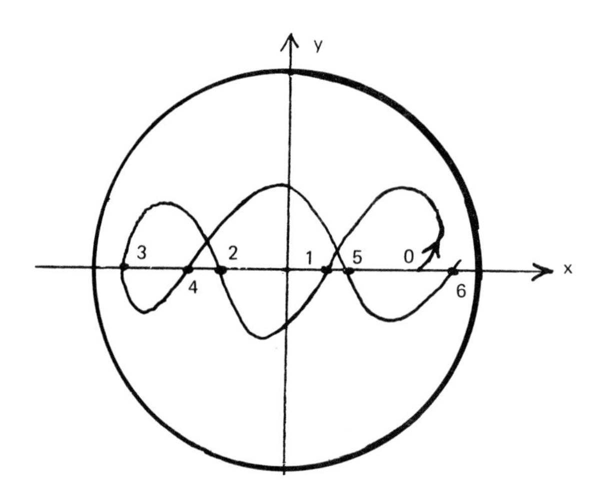

FIGURE 10

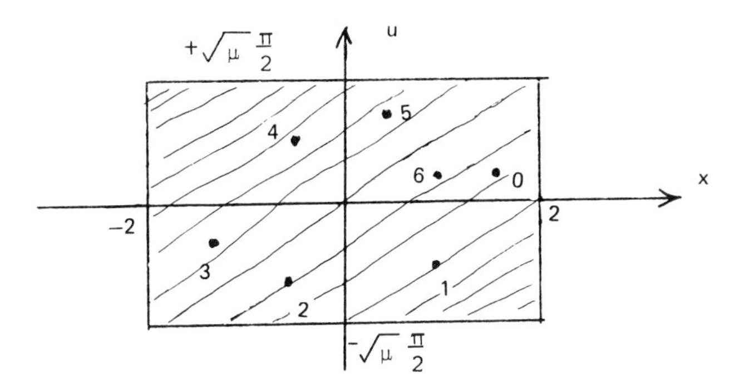

FIGURE 11

$$U = \sqrt{\mu}\ \text{arctg}\ \frac{u}{\sqrt{\mu}}, \quad X = x(1 + \frac{u^2}{\mu}) \qquad (41)$$

In the new coordinates the domain of the initial conditions is given by

$$-\sqrt{\mu}\ \frac{\pi}{2} < U < \sqrt{\mu}\ \frac{\pi}{2}, \quad -2 < X < +2 \qquad (42)$$

which is shown in figure 11, together with the sequence of points corresponding to the ones in figure 9. Clearly the map (40) is continuous except possibly at the boundaries, and the Poincaré map can be defined just as well for the rectangle (42) as for the more complicated shape (40). Therefore, we will henceforth always speak about the map $(U_{j+1}, X_{j+1}) = P(U_j, X_j)$, for integer j.

13. THE MAP INTO THE BINARY SEQUENCES

The Poincaré map contains all the essential information about the AKP, but we need a more detailed picture, at least a more intuitively appealing one, which tells in some qualitative manner how different trajectories behave. For instance, we may want to know only whether a particular trajectory goes back and forth between the half planes $x > 0$ and $x < 0$, or whether it stays in one of those two half planes for a number of intersections with the x-axis, before it goes on to the other half plane. This kind of information is given in the form of a binary sequence $a = (\ldots, a_{-2}, a_{-1}, a_0, a_1, a_2, \ldots)$, where $a_i = \text{sign}(x_i)$. This binary sequence is infinitely long in both directions, and the negative indices correspond to the intersections of the trajectory in the time before t_0. There is a whole literature about such binary sequences, and the spaces of all these is provided with a topology. Once that has been accomplished, mathematicians like to call such binary sequences Bernoulli sequences because of the obvious connection with probability theory. The topology which is needed in the AKP is different, and the term Bernoulli sequence should be avoided, although in some of his earlier publications the author used this term unfortunately, and caused a certain amount of confusion thereby.

The topology which is natural in the AKP follows if we map the binary sequences into two real numbers, ξ and η, by the simple formulas

$$\xi = \Sigma_{j\leq 0}\, a_j \left(\frac{1}{2}\right)^{-j-1} \qquad \eta = \Sigma_{j>0}\, a_j \left(\frac{1}{2}\right)^{j} \tag{43}$$

The number ξ can be said to describe the past of the trajectory, while the number η describes the future. The two numbers are obviously inside a square

$$-1 \leq \xi \leq +1 \qquad -1 \leq \eta \leq +1 \tag{44}$$

If one knows the initial conditions (U_0, X_0) inside the rectangle (42), it is easy to find the binary sequence a and from there the numbers ξ and η from the formula (43). If necessary, the equations of motion have to be integrated numerically, but as long as we need ξ and η only to certain accuracy, we know how many intersections we have to find in the

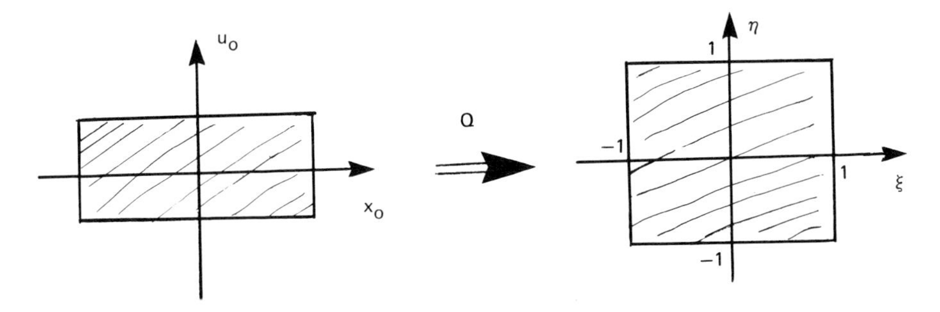

FIGURE 12

past and in the future. Thus, a mapping Q is defined from the rectangle (42) into the square (44), as shown very schematically in figure 12

The properties of this map are the crucial instrument which permits us to solve our problem of enumerating all the periodic orbits in the AKP.

14. A THEOREM AND A STRONG CONJECTURE

The map Q from the initial conditions into the square (44) was first proposed by the author and studied numerically in rather extensive detail by integrating the equations of motion numerically The essential properties then emerged quite clearly, so that one could try to prove them analytically. Such an effort was undertaken independently and by different methods by R. Devaney and the author. They both succeeded in showing one half of what they wanted to prove, and failed in the second half for which there was nevertheless very substantial evidence on the basis of the numerical calculations. There have been many claims by various specialists in the general area of dynamical systems that the second, unproven half of the claims could not possibly be correct. The numerical evidence is so strong, however, that the claims are justified as a practical description of what happens, and they should be accepted at least in that sense.

THEOREM. Every point in the square (44) is the image of at least one point in the rectangle (42) through the map Q as given by (43).

The proofs of this theorem are quite lengthy and technically involved, and will not be discussed. But a certain number of comments will help to show the significance of the statement above. It says in effect that whatever binary sequence one might propose, one can find initial conditions (u_0, x_0) such that the corresponding trajectory has intersections with the x-axis which follow exactly the given sequence. The only exceptions to this claim are the two "uniform" sequences, (...+++++...) and (...-----...) whose image lies on the boundary of the square (44). The ordinary Kepler ellipse would be given by the alternating sequence (...+-+-+-+-...) which gets mapped into the interior point (-1/3,+1/3), and there is indeed a very simple periodic orbit in the AKP with the binary sequence. There are two situations which require special attention. Assume that the binary sequence for the past is given in some arbitrary, but fixed way, and that the future is determined by either one of two sequences of binary numbers as follows:

1) $(\ldots, a_{-2}, a_{-1} \cdot a_0, +1, -1, -1, -1, -1, \ldots)$
2) $(\ldots, a_{-2}, a_{-1}, a_0, -1, +1, +1, +1, +1, \ldots)$

In both cases the binary sequences end up being uniform into the future, but with opposite signs. Both these sequences get mapped into the same point of the square (44) with $\eta = 0$. In the topology of the Bernoulli sequences these two point are not even near each other, while in our topology they happen to be identical. If our topology makes any sense, then there should be only one trajectory corresponding to these two sequences. How can that be?

The answer comes from studying the trajectories which have a long but finite sequence of identical binaries, as if there was finally a sign change in the above two sequences as one continues into the future. One finds that such a trajectory passes very close to the origin of the (x,y) plane just before the long string of identical binaries starts. In doing so the trajectory approaches the origin along the y-axis just barely to the right in the case 1), and just barely to the left in case 2). Then the trajectory swings sharply around the origin and moves down the x-axis in many short wiggles in the case 1), or up the x-axis with many consecutive crossings at a very short distance from one another in the case 2). The two situations are quite analogous as shown in figure 13. It is then obvious what happens in the limit when the string of identical binaries becomes ever longer. The trajectory approaches the origin from the y-axis and makes a collision. It does not matter whether the trajectory came rather from the right as suggested by the sequence 1) or from the left as in 2). In the limit the two trajectories are identical in the past. It is then quite natural to assign to them the same two numbers ξ and $\eta = 0$. They are collision trajectories. One can study the approach to collision in detail, and find that any one of

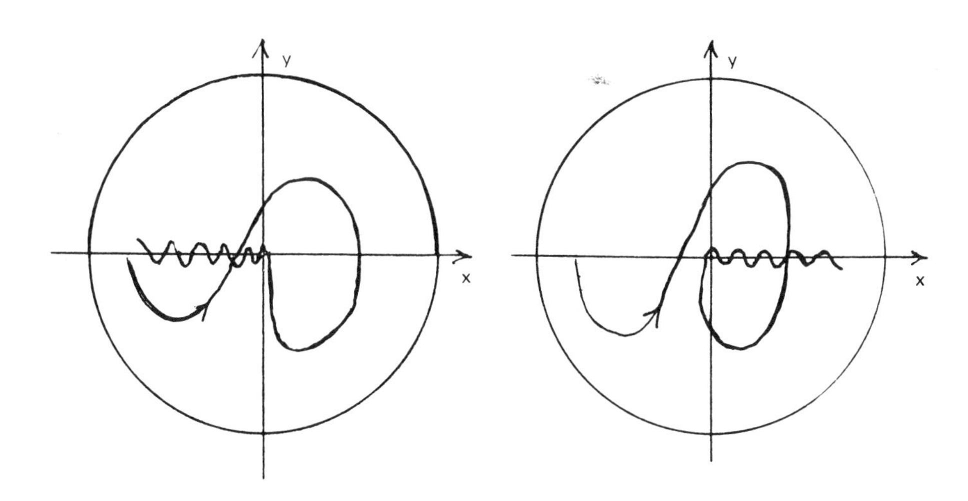

FIGURE 13

the intersections in the past like (u_{-1}, x_{-1}) varies continuously, as η goes to zero continuously, while ξ approaches some given non zero value. Quite generally a value of either ξ or η which equals a rational number whose denominator is a power of 2, indicates a collision. A trajectory can start in an inverse collision (come out of the origin along the y-axis) and end up in a collision after undergoing some arbitrary sequence of crossings with the x-axis. The collision trajectories form a one-parameter family which can be calculated rather effectively, and serves, therefore, as a way to pin down other trajectories which have a similar sequence of binaries outside the collision.

CONJECTURE. Every point in the square (44) is the image of exactly one point in the rectangle (42) under the mapping Q given by (43).

This is the statement which neither R. Devaney nor the author has been able to prove, and whose correctness has been doubted by many specialists in the field. We will accept it as empirically confirmed, just like an experiment in physics. It says that the binary sequence as given by the two numbers ξ and η specifies uniquely the initial conditions. The theorem at the beginning of this section has already assured us that there is always such a trajectory. In particular, a periodic orbit is uniquely specified by the periodic sequence of its intersections with the x-axis. Therefore, we can enumerate all the periodic orbits by enumerating all the periodic binary sequences.

In a way we have solved our main problem by this simple corollary of the conjecture. A periodic binary sequence never leads to collision, because there will never be an arbitrarily long sequence of identical binaries if the periodic binary sequence has a finite period. A number of periodic orbits are presented in figures 14 through 18 from calculations for the

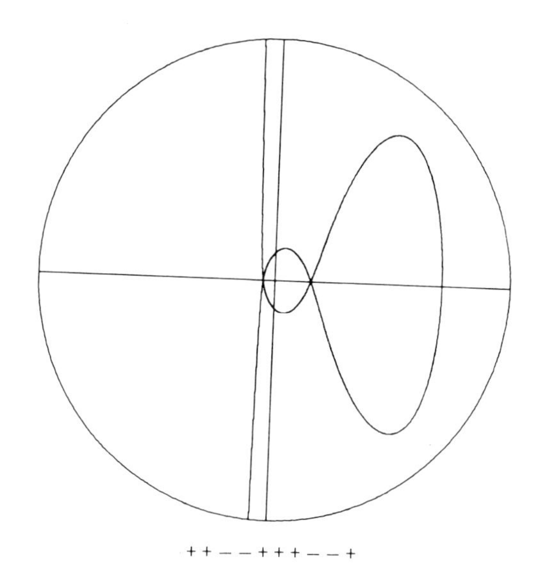

+ + – – + + + – – +

Mass Ratio	NC ID	Stability
5,00000	10 33	7,58666
X Initial	Eta	Period
1,417669899921	0.612903225806	24,05197
U Initial	Xi	Action
0.000000000000	0.225806451613	24,05229

FIGURE 14

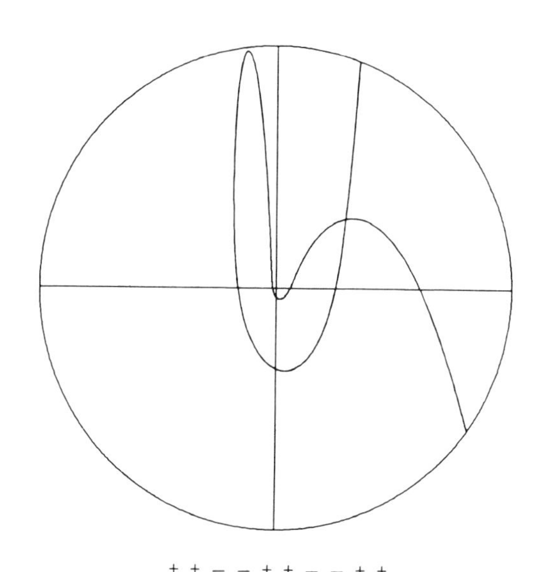

+ + – – + + – – + +

Mass Ratio	NC ID	Stability
5,00000	10 21	7,69322
X Initial	Eta	Period
1.632728028495	0.601173020528	23,41778
U Initial	Xi	Action
-0.337298904897	0.601173020528	23,41793

FIGURE 15

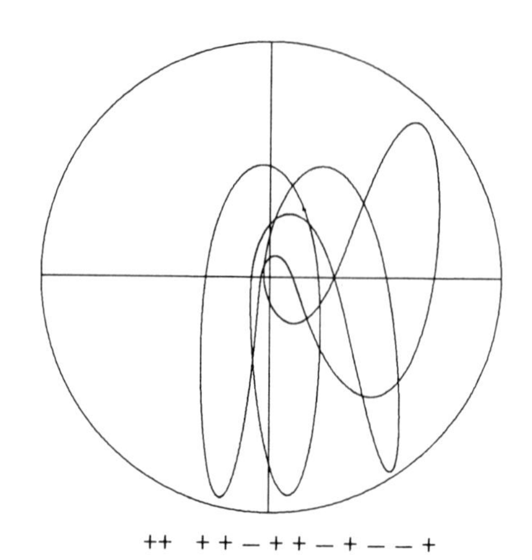

++ + + − + + − + − − +

Mass Ratio	NC	ID	Stability
5,00000	10	26	7.00001
X Initial	Eta		Period
1;478647023910	0. 706744868035		24.91656
U Initial	Xi		Action
0.123693620128	0.178885630499		24.91538

FIGURE 16

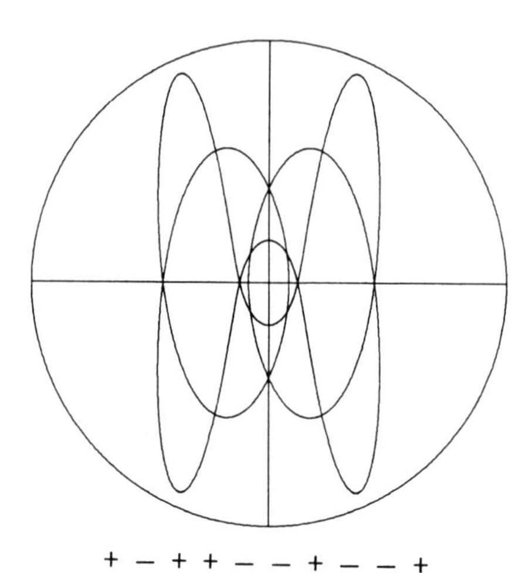

+ − + + − − + − − +

Mass Ratio	NC	ID	Stability
5,00000	10	38	6.28369
X Initial	Eta		Period
0.972582170434	0.393939393939		26.09105
U Initial	Xi		Action
−0.187861966538	0.151515151515		26.09079

FIGURE 17

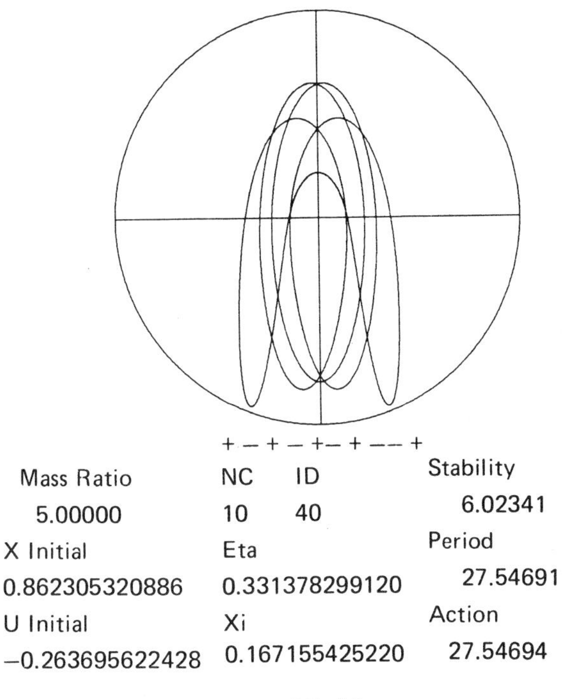

FIGURE 18

mass ratio 5. They show a number of features which can be recognized from the corresponding binary sequence without integrating the equations of motion. This aspect of the uniqueness conjecture will be discussed shortly in order to give the reader an impression of what this conjecture entails. For more detail there is a paper by the author with many numerical results.

Consider a sequence of binaries where $a_{-i} = a_i$. If we set the origin of the time coordinate so that $t = 0$ when the trajectory crosses the x-axis at (x_0, u_0), then we have complete symmetry under time reversal, i.e. the transformation which changes t into $-t$. Under such a transformation, however, $u(t)$ goes into $-u(-t)$. But under our assumption about the binary sequence together with the uniqueness, we must get the same trajectory when we look from $t = 0$ into the future or into the past. Therefore, we have $u_0 = u(0) = -u(0) = -u_0$, i.e. $u_0 = 0$. The trajectory intersects the x-axis at a right angle, as shown in figure 14, or stated differently, the trajectory is symmetric with respect to the x-axis.

By a similar argument, the trajectory is seen to be symmetric with respect to the y-axis, if we have the relation $a_{1-i} = -a_i$. The most striking of these symmetries which can be recognized simply by looking at the binary sequence, arises when we have the relation $a_{1-i} = a_i$, because it then follows from the uniqueness that $(u_0, x_0) = (-u_1, x_1)$. The particle returns to the same place of intersection with the x-axis with its velocity exactly reversed. It retraces its own path in position space. This can only happen, if it has come to a momentary rest between the times t_0 and t_1. Such a momentary rest implies that all the energy has been converted into potential energy, and the particle is at the boundary of position space, $x^2 + y^2 = 4$. Such an orbit is shown in figure 15.

15. FURTHER COMMENTS ON INITIAL CONDITIONS AND BINARY SEQUENCES

A trajectory closes itself smoothly only after an even number of intersections with the x-axis. Therefore the period of its associated binary sequence is even. If we want to enumerate all possible periodic binary sequences, we shall put down first all those of length 2, then those of length 4, then 6, etc. For a length 2N there are a total of 2^{2N} sequences, but the first and the last of these, so to speak, +++++ and -----, do not count. Correspondingly, we can now assert that there are $2^{2N} - 2$ periodic orbits with 2N x-axis crossings in the AKP. In this way of counting, we have distinguished two orbits when they were started at different x-crossings. Also, we have not taken into account the fact that some orbits have special symmetries like the ones which were discussed at the end of the last section. Or they can be composed of a shorter periodic orbit which is repeated more than once, as in (+-+-+-) which is three times the orbit (+-). Therefore the effective number of periodic orbits with a given number of x-axis crossings is less than $2^{2N} - 2$ by some factor like 2N. Still, their number increases tremendously with N. That is typical for ergodic systems, contrary to the intuitive notion that an ergodic system has only few periodic orbits. In contrast, the number of periodic orbits in an integrable system increases only like a polynomial of the length of the orbit, its degree equal to the number of degrees of freedom.

The mapping Q from the rectangle (42) into the square (44) is continuous in the natural topologies of these domains. But this map is not differentiable. This latter fact can be established by discussing in detail the trajectories near a collision. It can be formulated in a slightly unprecise, but basically correct manner as follows: consider a small shift in the initial conditions (U_0,X_0) by an amount δ, and the corresponding small shift ε in the (ξ,η) plane. Then we have the relation $\varepsilon \sim \sqrt{\delta}$ with some constant of proportionality which depends on the values of U_0 and X_0. This type of continuity, called Hoelder continuity, makes it very difficult to find the periodic orbits by numerical integration of the equations of motion.

In order to find a particular periodic orbit as given by its binary sequence, we have first to guess its initial conditions (U_0,X_0) and then find the corresponding binary sequence, i.e. the numbers ξ and η, by establishing the consecutive intersections with the x-axis. Since the numbers ξ and η are not quite correct, we have to add a small shift to (U_0,X_0) and do the whole thing over again in the hope of getting closer to the required values of ξ and η. But if we had to make a shift of .01 in the values of U_0 and X_0, we probably ended up with an error of .1 in ξ and η in a direction which we could not predict. Thus we may not have gotten closer to the solution.

The Poincaré map P was defined as going from (U_0,X_0) to (U_1,X_1) as in figure 11. If we now map (U_0,X_0) into

(ξ_0,η_0) by Q, and similarly (U_1,X_1) into (ξ_1,η_1), then we can just as well consider the map directly from (ξ_0,η_0) to (ξ_1,η_1) as defined by the sequence of maps $Q^{-1}PQ$. This is well defined because of the uniqueness conjecture, and can be visualized very simply. Instead of cutting the infinite sequence $(\ldots,a_{-1},a_0,a_1,a_2,\ldots)$ in two between a_0 and a_1, we now cut it between a_1 and a_2, in order to calculate ξ_1 and η_1 from (43). It is somewhat easier to describe the inverse process, i.e. going from (ξ_1,η_1) to (ξ_0,η_0). We squeeze the square (44) down to half its height, and broaden it to twice its width. Then we cut it vertically in half, and put the right half on top of the left half, just like the baker rolling his dough. The procedure is shown in figure 19. It is universally called the baker's transformation.

The Poincaré map P conserves the area as a consequence of the equations of motion (1). This property distinguishes a Hamiltonian system from other dynamical systems which are studied nowadays by mathematicians and physicists. It expresses the absence of friction, that is not so much the conservation of energy as the absence of conversion into heat. It prevents the formation of attractors, strange or otherwise, to which the trajectories in phase space converge.

The bakers transformation $Q^{-1}PQ$ quite obviously conserves also an area, that is the area in the (ξ,η)-plane. It is essential to understand that these two areas are not related to

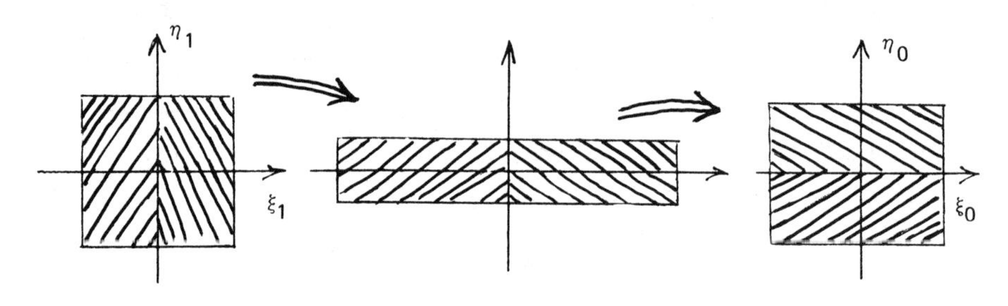

FIGURE 19

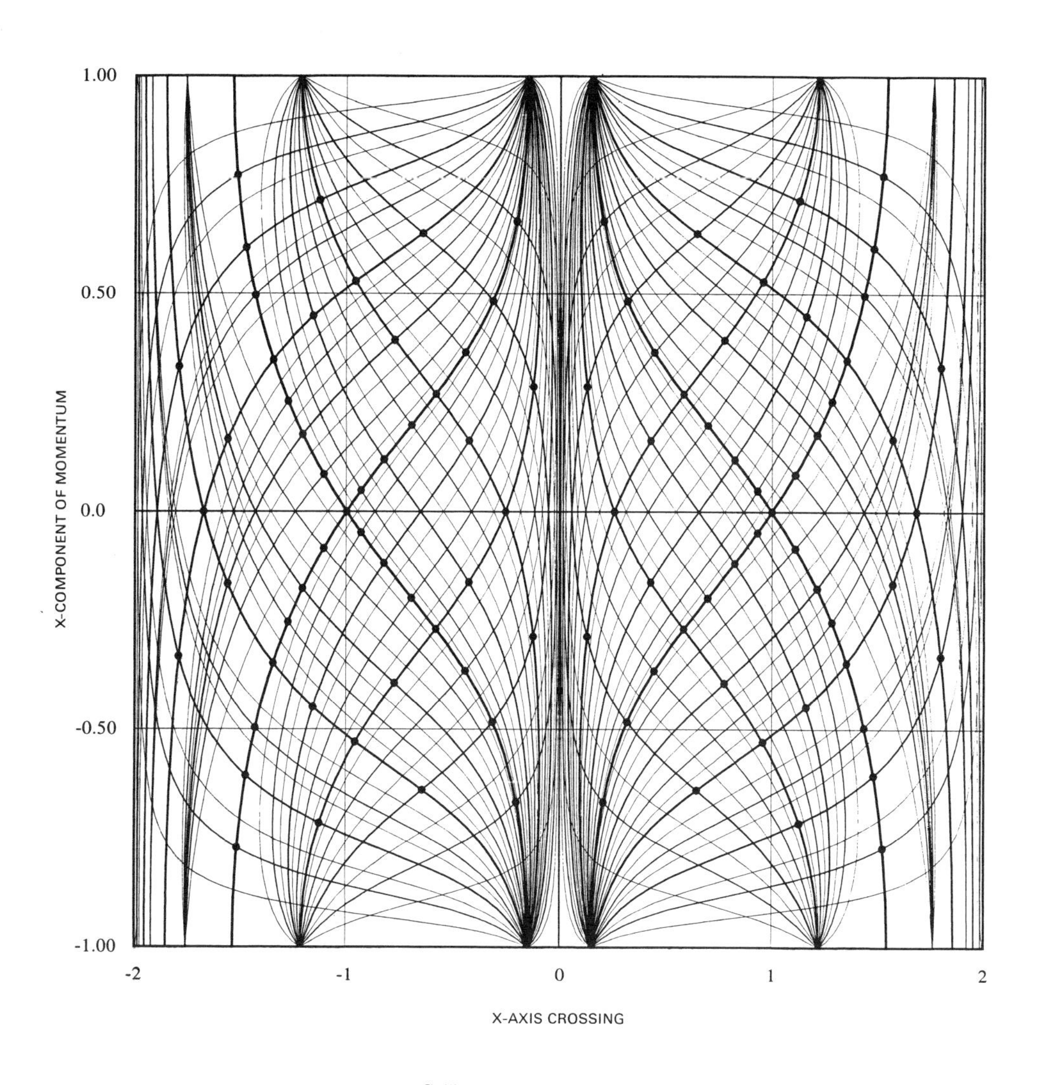

Collision Trajectories: Ratio = 5.00

FIGURE 20

each other. One might have thought at first that it is possible to take any small neighborhood V in the (U,X) plane which is then mapped by Q into a small neighborhood W in the (ξ,η) plane. From our estimate of the continuity of P it follows that W shrinks to zero when V does. Therefore, there could have been a limit for the ratio area(W)/area(V) as the neighborhood V is contracted to a point. Such a thing does not occur, however, and this can be seen directly from looking at the curves of constant ξ and the curves of constant η in the (U,X) plane. This is demonstrated in figure 20, where the consecutive values of ξ and η are chosen equidistantly. If the conserved areas in the (ξ,η) plane were in a simple ratio with the conserved areas in the (U,X) plane, even if that ratio varies with the point (U,X), there would not be any detectable local variations in the areas of the little distorted squares in figure 20. Thus, there are two conserved densities in the AKP, but one cannot be expressed in terms of the other. This surprising coexistence of two such unrelated conserved areas is a general feature of dynamical systems, and has been shown to occur under fairly general conditions. But as always in this business, there are very few examples which show this feature in an explicit manner. Even in the case of the geodesics on a surface of constant negative curvature, there is no explicit description of a conserved area other than the area in phase space of the Poincaré map.

16. THE ACTION INTEGRAL AS A FUNCTION OF THE BINARY SEQUENCES

The binary sequences gives us the tool which we need in order to carry out the enumeration over all the periodic orbits in the calculation of the response function (21). But at this point we are still in a difficult situation. For any given periodic binary sequence we have to find the initial conditions (U_0,X_0), and then do the integration of the equations of motion so as to compute the value of the action integral and the stability exponent. The question comes up whether one cannot make a shortcut, and proceed directly from the periodic binary sequence to the action integral and the stability exponent.

Let us consider the periodic orbits with a fixed number 2N of intersections with the x-axis. There are 2^{2N} binary sequences of length 2N, and we can extend each one from $-\infty$ to $+\infty$ by repeating them over and over again. To each belongs a periodic orbit together with a value for the action integral S and the stability exponent α. We will include even the two uniform sequences (+++++) and (-----), and assign to them, somewhat arbitrarily the values $S = 0$ and $\alpha = 0$. If we choose the variables u,v,w,x,y,z in the normalization of the Hamiltonian (34), the action integral S can be split into two factors as shown in (35), where the first factor depends on the energy E and the various physical parameters, while the second factor, to be called Φ from now on, depends only on the particular periodic orbit. Thus, Φ is a function only of the binary sequence which describes the periodic orbit, and the only remaining parameter in the Hamiltonian (34) which is the mass ratio $\mu/\nu = m_1/m_2$. Similarly, the stability exponent α depends only on the binary sequence and the mass ratio. If we take the mass ratio to be a parameter in our investigation, then we are lead to examine the possible functions Φ and α of the variable $a = (a_1, a_2, a_3, \ldots, a_{2N-1}, a_{2N})$, where each a_i takes on only the two values +1 or -1.

It is not difficult to write down the most general function of the 2N binary variables a_i. Since the square of a_i is alway to 1, a polynomial in the a_i can be at most of first degree in any one of them. Thus we have a constant term, then 2N terms in each a_i, then N(2N - 1) terms in $a_i a_j$ with $i \neq j$, etc., altogether 2^{2N} terms, each one with an arbitrary coefficient. That corresponds exactly to the possible 2^{2N} values which an arbitrary function of the 2N binary variables can assume. This representation has several advantages, because the choice of the coefficients in the polynomial is intimately connected with the symmetries of the function to be represented.

The most obvious of these symmetries comes from the fact that the value of Φ or of α is the same for all binary sequences which result from one another by cyclic permutation, such as $(a_1,a_2,\ldots a_{2N-1},a_{2N}),(a_{2N},a_1,a_2,\ldots,a_{2N-1})$, etc. These all yield the same periodic orbit, except the first intersection with the x-axis is different, but the value of Φ or α is independent of the numbering of the x-axis crossings. Thus the coefficients of the first order terms in the polynomial have to be all identical, while the second order terms $a_i a_j$ can have coefficients which depend only on the difference $i - j$, etc. When any orbit is reflected on the y-axis, all the values of the corresponding binary sequence change sign, but neither the value of Φ nor of α changes. Therefore the polynomial representing Φ or α can have only terms of even order. In general one can group all the terms in the polynomial together which belong to the same symmetry, including the symmetries which were discussed in connection with our main conjecture. The author has described the details of this discussion elsewhere.[16] One finds, of course, that there are exactly as many coefficients to be determined as there are periodic orbits of different shape. As N varies from 1 to 8 one finds 2, 4, 8, 18, 44, 122, 362, 1162 such different shapes. They can easily be enumerated in the form of binary sequences, but not in the form of explicit initial conditions. Clearly, we gain nothing as long as we have to use the polynomial in the binary variables in its full generality. even if we take into account the special features of the AKP. Therefore we are now ready to sacrifice some of the more complicated dependences of Φ and α on the binary variables, in return for a simpler overall expression. How far we are willing to go is a matter of taste and the ability to use the result in practical calculations.

At this point it helps at least a physicist to notice the analogy with the theory of magnetism. Each binary variable can be interpreted as the direction of an atomic magnetic moment, either up or down, in a chain of 2N atoms. The energy of a particular arrangement of these spins or magnetic moments (these names are used interchangeably) is generally assumed to depend only on two physical causes. The most obvious is an external magnetic field which would give rise to a linear term in the polynomial. As we have seen already, such a term is absent in our case. The other physical cause for the interaction of the spins is the so called exchange which manifests itself as a second order term of the type $a_i a_j$ in the energy of particular arrangement. This is also sometimes called a two-body interaction because it depends only on the states of the atoms with the numbers i and j. Such an exchange interaction is usually assumed to depend only on the difference $i - j$, presumably because all the atoms are identical in nature and are positioned at equal intervals so that the difference $i - j$ is proportional to their distance. The coefficient for a term $a_i a_j$ decreases therefore as the absolute value of the difference $i - j$ increases. It is further customary to think of the chain of atoms as being disposed on a circle in order to avoid any complications which might arise from the presence of boundaries, or equivalently, the chain is repeated over and over again, and the interaction is allowed to be effective between any two atoms in this infinite chain, although it decreases, of course, with distance. Thus, there are formally terms like $a_i a_{j-2N}$, $a_i a_{j+2N}$, $a_i a_{j+4N}$, etc. where $a_{j-2N} = a_j = a_{j+2N} = a_{j+4N} =$ etc.

17. AN EFFECTIVE APPROXIMATION FOR THE ACTION INTEGRAL

After the discussion in the preceeding section, it is now tempting to imitate the theory of magnetism and to approximate the action integral as a function of the binary sequence by no more than two body terms. We can see exactly the closeness of this approximation if we first get the complete representation as a polynomial in the binaries a_i, and then examine the coefficients of the higher order terms. Such a complete representation for a given value of N requires that we know the action integral for all the periodic orbits with 2N x-axis crossings. That is a major calculation which was carried out for the mass-ratio 5 and N = 1,2,3,4, and 5. For other mass ratios, the calculations were done only up to N = 4. It seems very difficult to devise a general algorithm which is not too wasteful on the many easy orbits and is still able to pin down the few difficult ones. But the calculations up to N = 5 are very instructive.

Once the values of the action integrals are known for all the periodic orbits of a given N, the coefficients in the corresponding polynomial are readily obtained. For this purpose we notice that the function a_i, $a_i a_j$, $a_i a_j a_k$, etc. for different combinations of i, j, k, etc. can be considered as orthogonal functions over some space, in this case over the space of all binary sequences of length 2N. Since we have computed the values of the function to be represented, we get the coefficients in the expansion by the standard procedure, taking into account the various symmetries and the resulting simplifications. This method gives us a valuable check on any approximation we might use, because we have the completeness relation between the sum of the squares of function values and the sum of the squares of the coefficients. Again the details can be found in the paper of the author referred to earlier.

For the action integral there is a rather large constant term which increases linearly with N in fairly good approximation. The second order terms are much smaller and decrease as a function of the difference $i - j$ in such a manner that they decrease as $i - j$ goes from 1 to N, but are symmetric with respect to N, exactly as one would find it in a chain of spins on a circle. All the higher order terms are so small that the sum of their squares is negligible with respect to the squares of the second order terms, and a fortiori the constant term. All this holds for the action integral, but not for the stability exponent. The only simple statement for the latter concerns the constant term which is again the largest and increases roughly linearly with N.

At this point the second order terms in the action integral are still entirely the result of a numerical calculation. In order to get an even closer comparison with the chain of magnetic moments, we will now try to approximate the second order coefficients with an expression which assumes that the interaction between a pair of spins decays exponentially with the distance between them. The decay rate will be chosen to fit the second order coefficients from the earlier calculation as closely as possible. This is done by fitting exactly the minimum action integral (which is zero), the maximum (which increases exactly proportionally to N), and the average value. The effective formula for Φ then becomes

$$\Phi \simeq 2N\tau \cosh^2\left(\frac{\gamma}{2}\right) - \frac{\tau}{2} \sinh \gamma \; \Sigma_{i=1}^{2N} \Sigma_{j=\infty}^{+\infty} a_i a_j \exp(-\gamma|j-i|) \tag{45}$$

which has only two parameters. The overall scale τ can be obtained at once from the simple periodic orbit of length 2 (with binary sequence +- and $\Phi = 2\tau$), and the decay rate γ

which imitates the behavior of the second order coefficients. The expression (45) is definitely not exact, but it must give the essential features of the dependence correctly at least in some asymptotic sense for large N.

It is unfortunate that there is no such magical formula for the stability exponent. The best we seem to be able to do, is the very crude approximation where we set

$$\alpha \cong 2N\beta \tag{46}$$

The parameters τ, γ, and β have different values for different mass ratios. One can argue that it is more important to find a good approximation for the action integral Φ than for the stability exponent. The singularities in the response function $\tilde{g}(E)$ as approximated by (21) come primarily from the constructive interference of the terms in the sum over all periodic orbits, i.e. from the exponents of a large number of terms differing by multiples of $2\pi i$. The real part of these exponents have very little effect on this constructive interference, although they are important in determining the convergence of the series.

A last comment before actually calculating the sum over all periodic orbits is concerned with their correct counting. We have already remarked that the same periodic orbit shows up as 2N binary sequences, all of them of length 2N, but differing from one another by cyclic permutations. If we sum over all binary sequences, then we will in general count each periodic orbit 2N times. Thus, we expect to divide each term by 2N, and this will be certainly correct for all those periodic orbits which have no further symmetry. But there may be difficulties in other orbits. E.g. if the particular orbit under study consists of repeating a shorter orbit of length 2M so that N = K.M, then there will be only 2M different binary sequences of length 2N belonging to the periodic orbit. If we divide all the terms in the summation over the binary sequences by 2N, then we seem to have undercounted this particular one by a factor K. But at this point we have to take into account the factor T_0 in front of the exponential in (21). In our particular case T_0 is not the full period T because the particle goes K times around a shorter periodic orbit. Instead we have $T_0 = T/K$. Therefore, the factor T_0 exactly picks up what seemed at first an error of undercounting. One can check in all cases of symmetric orbits that the summation over the binary sequences is correct as long as T_0 in (21) is replaced by T/2N.

18. THE SUMMATION OVER ALL PERIODIC ORBITS

The formula (21) for the approximate response function $\tilde{g}(E)$ is now replaced by a summation over all binary sequences of even length 2N and N goes from 1 to ∞, but simultaneously we replace the reduced period T_0 by the full period over 2N, as was explained at the end of the last section. This replacement allows us a further simplification. The action integral S(E) and the full period T are related by the well known formula

$$T = \frac{dS}{dE} \tag{47}$$

Since the stability exponent in the AKP does not depend on the energy E, one can integrate the formula (21) on both sides with respect to E from -∞ to some upper limit E. At the same time, we can use the explicit dependence on E as given in (35), and get the expression

$$\int_{-\infty}^{E} \tilde{g}(E)\,dE = \sum_{\substack{\text{bin. seq.} \\ \text{length } 2N}} \frac{1}{2N} \frac{1}{2 \sinh \frac{\alpha}{2}} \exp(-s\Phi) \tag{48}$$

where we have again turned the E-axis in the complex plane by 90° through the use of the variable $s = -i(-m_0 e^4/2\kappa^2 E\hbar^2)^{1/2}$.

We will expect the sum to converge for Re(s) = 0. The three-dimensional origin of the two-dimensional AKP raises its head to require another modification. The details of the following argument are explained in a forthcoming paper by the author,[17] but it is important to indicate at least the nature of this modification because it is essential in the comparison of our approximate results with the exact solutions of Schrodinger's equation.

If we consider the three-dimensional Green's function of the AKP for states with angular momentum, zero, we notice that it looks formally like the two-dimensional Green's function, except that y is now the distance from the x-axis. Therefore, the coordinate y should not be allowed to take on negative values. The two-dimensional Green's function tells us more than we need, because y can be both positive and negative. The two-dimensional Green's function can be split into two parts. one which is even with respect to a change in the sign of y, and another which is odd. Only the first part is of interest to us. In pursuing this idea, we finally come to the conclusion that the response function contains the summation over a second set of periodic orbits in the plane AKP. These orbits start with (U_0,X_0) and $v_0 > 0$, but the intersection (U_{2N+1},X_{2N+1}) coincides with (U_0,X_0), i.e. $U_0 = U_{2N+1}$ and $X_0 = X_{2N+1}$, while $v_{2N+1} < 0$ necessarily. Such an orbit is, of course, periodic, i.e. closes smoothly, after 4N + 2 intersections with the x-axis, but in the expression for the approximate Green's function we have to insert the action integral, the stability exponent, the number of conjugate points, and the reduced period from (U_0,X_0) to (U_{2N+1},X_{2N+1}) only. In particular there are now effectively 2(2N + 1) = 4N + 2 conjugate points which lead to a factor (-1). In a formal way of speaking, we can associate a periodic orbit to each binary sequence of length 2N + 1. All its characteristic attributes, like its action integral and its stability exponent, are simply half of the value for the ordinary periodic orbit whose binary sequence consists of 4N + 2 binaries with $a_{i+2N+1} = a_i$. Thus, the length of the binary sequences will now vary over all integers M from 1 to ∞. The resulting response function $\tilde{g}(E)$ corresponds to the three-dimensional AKP with vanishing angular momentum around the x-axis.

The formula (48) gets changed into

$$\int_{-\infty}^{E} \tilde{g}(E)\, dE = -\Sigma_{\text{bin. seq.}} \frac{(-1)^M}{M} \frac{1}{2 \sinh \frac{\alpha}{2}} \exp(-s\Phi) \qquad (49)$$

where the summation now goes over all binary sequences of finite length M, and the quantities Φ and α are interpreted as explained in the preceeding paragraph. Notice that this expression is correct if we don't use the approximate expressions for Φ and α which were derived in the preceeding section. It is indeed rather remarkable that the approximate response function $\tilde{g}(E)$ can be reduced to this simple form for the AKP, with no other approximation than the general arguments which went into (21). It was this very approximation which turned out to be exact in the case of the geodesics of a surface of constant negative curvature.

In order to find the singularities in (49) explicitly, we now use the further approximation for Φ and α of the preceeding section, i.e. (45) and (46). In addition, we replace $2\,\mathrm{Sinh}(\alpha/2)$ by $\exp(+\alpha/2)$ because α becomes very large as M increases, and the singularities are determined by the terms with large M. In this way we get the final expression

$$\int_{-\infty}^{E} \tilde{g}(E)\, dE \simeq -\Sigma_{\text{bin. seq.}} \frac{(-1)^M}{M} \exp(-s\Phi - M\frac{\beta}{2}) \qquad (50)$$

where Φ is now the function (45). If we disregard for the moment the fact that the variable s is complex, and in particular purely imaginary in our domain of interest, and if we forget the factor $(-1)^M/M$, then we recognize in (50) the expression for the grand canonical ensemble of a linear chain of spins with an exponentially decaying exchange interaction. The variable s corresponds to the inverse temperature, and the stability exponent $\beta/2$ plays the role of the chemical potential. In a certain sense, all we are doing is to look for the singularities in the complex temperature plane of this thermodynamic model. Such an investigation must be somehow related to finding phase transitions.

19. THE CALCULATION OF THE SINGULARITIES

Since we have now made connection with a standard problem in statistical mechanics, there is hope that we will be able to evaluate the expression (50). The main ingredients are the binary sequences and the functional form of Φ. It is a quadratic function of the binaries, whose coefficients are invariant under a translation of the indices and decay exponentially with the difference in the indices. These features allow us to apply a certain number of tricks which were mainly invented by Mark Kac,[18] for this particular case, although they had been used in somewhat different circumstances and in different form before. They are known by the name of transfer operator methods. In our case such an operator negotiates the transition from the binary (alias spin) with index i to the binary with index $i + 1$. The sum over all binary sequences of a given length M appears then as the trace of the M-th power of that transfer operator.

Again the details are worked in the forthcoming paper of the author, but it may be instructive to follow the general procedure in order to get an idea of the origin of the approximate energy levels. Let us therefore assume that we have found an operator A such that the sum over all binary sequences of length M can be expressed as trace(A^M). In our case this operator A is an integral operator which acts on real functions in the interval from $-\infty$ to $+\infty$, and has a good kernel whose absolute value squared can be integrated and converges well. A depends on the parameters which are left in (50), that is, s, γ, and β. The variable s always occurs in the combination $z = s\tau \sinh \gamma$ which will be used from now on.

The trace of the M-th power of A is difficult to evaluate, unless we succeed in diagonalizing the operator A. It turns out that this can be done with some computational labor, if we choose a set of orthonormalized functions in the interval $(-\infty,+\infty)$. Such a set is available very conveniently through the eigenfunctions of the harmonic oscillator, the Hermite polynomials. The operator A can be written as a matrix whose elements turn out to be elementary functions of the three parameters z, γ, β. These matrix elements can be calculated starting with the lowest order by using simple recursion formulas. One can show that these matrix elements decay rather rapidly as soon as their order exceeds the absolute value of the parameter z. Thus, we get away with truncating the matrix A provided we limit ourselves to values of z within some upper bound. If the computer lets us handle 50 by 50 matrices, then we can effectively obtain the eigenvalues of the operator A for the absolute value of z up to about 12.

Let us now take a fixed value of the mass ratio, say 4.80, which in turn fixes the value of γ to .622 and of β to 0.75. The energy dependence is hidden in z through the variable s which was defined in (48). The eigenvalues of A are numbered by an index n which runs through the positive integers starting at 0. We will show explicitly their dependence on z by writing $\lambda_n(z)$. Since we have

$$\text{trace}(A^M) = \Sigma_{n=0}^{\infty} \lambda_n^M \tag{51}$$

the expression (51) becomes immediately

$$\int_{-\infty}^{E} \tilde{g}(E)\,dE = -\Sigma_M \Sigma_n \frac{(-1)^M}{M} \lambda_n^M = \Sigma_n \log(1 + \lambda_n) \tag{52}$$

or in somewhat fancier language

$$\exp\left(\int_{-\infty}^{E} \tilde{g}(E)dE \right) = Z(s) = \Pi_{n=0}^{\infty} (1 + \lambda_n(z)) \tag{53}$$

Notice that the quantity $\tilde{g}(E)$ which is our main goal, becomes the logarithmic derivative of $Z(s)$, and we can write after some trivial reorganization

$$\frac{m_0 e^4}{\kappa^2 \hbar^2 s^3} \tilde{g}(E) = - \frac{Z'(s)}{Z(z)} = \Sigma_{n=0}^{\infty} \frac{-\tau \sinh \gamma}{1 + \lambda_n(z)} \cdot \frac{d\lambda_n}{dz} \tag{54}$$

where we consider each eigenvalue $\lambda_n(z)$ as an analytic complex valued function of z. This brings us to the end of our story. The approximate response function $\tilde{g}(E)$ has a pole with residue 1, just as we wanted it, whenever an eigenvalue $\lambda_n(z)$ of the operator A becomes equal to -1. In practice, we let z run down the imaginary axis and watch what happens to the various eigenvalues $\lambda_n(z)$. Most of them will start out with a very small absolute value, but move gradually toward the unit circle. Then they will "mill around" the unit circle in rather unpredictable manner, and move close to the point -1. Whenever that happens, we are near a pole, and we can find its exact location by little effort. The poles do not lie exactly on the imaginary z-axis, or equivalently, on the imaginary s-axis. But they are close enough that one can associate a well defined energy level with each of them. The small imaginary part in the value of the approximate energy level seems to be the price which we have to pay for wanting to approximate the energy levels of a system with classically ergodic behavior.

In working out the details one can see that the operator A transforms even functions into even ones, and odd functions into odd ones. Therefore the matrix elements never connect an even and an odd Hermite polynomial, and we get two entirely distinct spectra. After some additional analysis one finds

TABLE 1. Energy levels for a donor impurity in sillicon. Designation of the levels in columns 1 and 5, earlier results in columns 2 and 6, results of the present work in columns 3 and 7, Faulkner's results in columns 4 and 8. Units are millielectronvolts.

Even parity				Odd parity			
1s	36.81	29.06	31.27	2p	9.20	10.48	11.51
2s		8.35	8.83	3p		5.10	5.48
3d	4.09	4.63	4.75	4p		3.12	3.33
3s		3.64	3.75	4f	2.30	2.30	2.33
4d		2.82	2.85	5p		2.15	2.23
4s		2.11	2.11	5f		1.64	1.62
5d		1.87	1.87	6p		1.49	1.52
5g	1.47	1.53	1.52	6f		1.32	1.20
5s		1.43	1.38	6h	1.02	1.09	1.10

that the even eigenvalues $\lambda_n(z)$ are associated with energy levels which are symmetric with respect to the coordinate x, while the odd eigenvalues $\lambda_n(z)$ are associated with the energy levels which are antisymmetric with respect to the coordinate x. We call this symmetry in the energy levels their parity, and we list the levels in two separate columns according to their parity. Table 1 gives the levels which were obtained in this fashion along with the levels which were calculated by Faulkner from a rather limited set of basis functions with the variational principle. In a general way one can say that Faulkner's results are good for the low lying levels, while our results should get better the higher the level. Therefore, it is not surprising that there should be such good agreement for most of the energy levels.

20. SUMMARY AND CONCLUSION

This has been an informal presentation of a rather large program of research where the results are only partial at this time. One of the main goals has been to show that there is a fairly systematic approach to understanding the relations between classical and quantum mechanics, even when the classical system is ergodic. The two examples of the geodesics on a surface of constant negative curvature, and of an electron around a donor impurity in a semiconductor could be carried to completion. In the first case, this was done on the strength of a famous mathematical result which had only to be interpreted, while in the second case it was possible to combine analytical and numerical evidence to make a connection with a well known problem in statistical mechanics. In either case, it would be foolish to claim that an alternative method has been established to find approximate eigenvalues of Schrodinger's equation, because the summation over all periodic orbits is much more difficult to carry out than the standard variational procedure for finding the energy levels.

The methods and arguments in this report are, therefore, to be taken more in the vein of an existence proof than a practical recipe for the possibility of going from a classically ergodic dynamical system to its quantum analog. Some people have doubted the feasibility of such a transition, and it is evident, indeed, ever since Einstein's paper in 1917, that the usual method works only when the classical trajectories lie on invariant tori in phase space. Even with the methods in this paper there remain many questions: can one see in the spectrum of the quantum system that the classical system is ergodic? what does ergodic behavior mean in a quantum system? are there any simplifications when the number of degrees of freedom increases? Many more examples have to be studied before these questions can be answered.

REFERENCES

1) A. Einstein, Verh. Dtsch. Phys. Ges. 19, 82 (1971).
2) A. Pais, Rev. Mod. Phys. 51, 861-914 (1979).

3) J. Hadamard, J. Math. Pure Appl. 4, 27 (1898); M. Morse, Am. J. Math. 43, 33 (1921); E. Artin, Abhandl, Math. Sem. Hamburg 3, 170 (1924).

4) A. Selberg, J. Indian, Math. Soc. 20, 47 (1956).

5) M.C. Gutzwiller, J. Math. Phys. 12, 343 (1971).

6) M.C. Gutzwiller, J. Math. Phys. 14, 139 (1973).

7) M.C. Gutzwiller, J. Math. Phys. 18, 806 (1977).

8) R.C. Devaney, J. Differential Equations 29 253 (1978); Inventiones Math. 45, 221 (1978), and in The Structure of Attractors in Dynamical Systems, ed. by N.C. Markley, Lecture Notes in Mathematics Vol. 668, Springer, N.Y. 1978.

9) B. Randol, Trans. Am. Math. Soc. 236, 209 (1978).

10) Cf. M. Kac, Probability and related topics in physical sciences, Interscience publishers, London and New York 1958, or Integration in function spaces and some of its Applications, Lezioni Fermiane, Pisa, 1980. L.S. Schulman, Techniques and Applications of path Integration, Wiley, New York, 1981

11) M.C. Gutzwiller, J. Math. Phys. 11, 1791 (1970).

12) J. Hadamard, "Sur le billiard non-Euclidien", Soc. Sci. Bordeaux, Proces Verbaux, 1898, 147 (1898).

13) L. Keen, Acta Math. 115, 1 (1966); Annals of Math. 84 404 (1966).

14) W. Kohn and J.M. Luttinger, Phys. Rev. 96, 1488 (1954).

15) R.A. Faulkner, Phys. Rev. 184, 713 (1969).

16) M.C. Gutzwiller, in Classical Mechanics and Dynamics Systems, ed. Z. Nitecki, Marcel Dekker, 1981.

17) M.C. Gutzwiller, to be published. A preliminary report is given in Phys. Rev. letters 45, 150 (1980).

18) M. Kac, Phys. Fluids 2, 8 (1959).

DIOPHANTINE APPROXIMATION OF COMPLEX NUMBERS

Asmus L. Schmidt

Matematik Institut
University of Copenhagen
Copenhagen, Denmark

This article contains a general survey of the theory of regular and dually regular chains as a convenient tool to treat several number theoretic problems for complex numbers, the real analogs of which are usually treated by means of regular continued fractions.

The theory was initiated by the author in 1967 (see [6]), where a number of new approximation results were obtained for the fields $\mathbb{Q}(i\sqrt{m})$, m = 1,2,3,7, by the rather crude method using Farey triangles and Farey quadrangles. As a matter of fact Farey hexagons play a similar role in the field $\mathbb{Q}(i\sqrt{11})$ (see [12]), as do 4-dimensional Farey simplices in the case of quaternions (see [7], [8]).

However, in 1975 (see [9]) I could present the much better theory of regular and dually regular chains in the Gaussian case, which among other things led to a complete determination of the discrete part of the Gaussian Markoff and Hurwitz spectra, the result being strikingly similar to Markoff's and Hurwitz' classical theory. Important further information has been added in the later papers [10], [11], and very recently an explicit ergodic theory has been developed (see [13]).

It should be mentioned that the discrete parts of the Markoff and Hurwitz spectra could be determined also for $m = 11$ (see [12]) and for $m = 3$ (see [14]). However, the algorithms used in these cases are admittedly poorer than in the Gaussian case.

Consequently I have chosen to concentrate myself entirely to the Gaussian case, where I believe the theory is now equally rich as that of regular continued fractions as presented for instance in the classical book of O. Perron (see [3]). For the historical aspects of the subject one should consult [9].

Finally I would like to mention that a young American mathematician, Norman Richert, has treated the entire subject very thoroughly in his thesis (see [4]). In particular he has written several BASIC computer programs associated with different aspects of the theory.

1. BASIC DEFINITIONS

For any matrix $M = \begin{pmatrix} a & b \\ c & d \end{pmatrix} \in GL(2,\mathbb{Z}[i])$ we let as usual $M: \hat{\mathbb{C}} \to \hat{\mathbb{C}} = \mathbb{C} \cup \{\infty\}$ also be the map $z \to Mz = (az+b)/(cz+d) \in PGL(2,\mathbb{Z}[i])$. We denote by $\|Mz\| = |cz+d|^{-4}$ the Jacobian of the map M at the point z.

Two numbers $\xi, \eta \in \mathbb{C}\setminus\mathbb{Q}(i)$ are called <u>properly</u> (resp. <u>improperly</u>) <u>equivalent</u> if there exists a matrix $M \in GL(2,\mathbb{Z}[i])$ with $\eta = M\xi$, and with $\det M = \pm 1$ (resp. $\pm i$).

The matrices $V_1, V_2, V_3, E_1, E_2, E_3, C, S \in GL(2,\mathbb{Z}[i])$ are basic and defined as follows

$$V_1 = \begin{pmatrix} 1 & i \\ 0 & 1 \end{pmatrix}, \quad V_2 = \begin{pmatrix} 1 & 0 \\ -i & 1 \end{pmatrix}, \quad V_3 = \begin{pmatrix} 1-i & i \\ -i & 1+i \end{pmatrix},$$

$$E_1 = \begin{pmatrix} 1 & 0 \\ 1-i & i \end{pmatrix}, \quad E_2 = \begin{pmatrix} 1 & -1+i \\ 0 & i \end{pmatrix}, \quad E_3 = \begin{pmatrix} i & 0 \\ 0 & 1 \end{pmatrix},$$

$$C = \begin{pmatrix} 1 & -1+i \\ 1-i & i \end{pmatrix}, \quad S = \begin{pmatrix} 0 & -1 \\ 1 & -1 \end{pmatrix}$$

The sets $\mathcal{T}, \mathcal{T}^* \subset \hat{\mathbb{C}}$ are defined by

$$\mathcal{T} = \{z = x + iy \mid y \geq 0\} \cup \{\infty\}$$

$$\mathcal{T}^* = \{z = x + iy \mid 0 \leq x \leq 1,\ y \geq (x-x^2)^{1/2}\} \cup \{\infty\}$$

Also

$$\mathcal{V}_j = V_j(\mathcal{T}),\ \mathcal{E}_j = E_j(\mathcal{T}^*),\ \mathcal{C} = C(\mathcal{T}^*),\ \mathcal{V}_j^* = V_j(\mathcal{T}^*),\ \mathcal{C}^* = C(\mathcal{T}) \tag{1.1}$$

and we notice that

$$\mathcal{T} = \mathcal{V}_1 \cup \mathcal{V}_2 \cup \mathcal{V}_3 \cup \mathcal{E}_1 \cup \mathcal{E}_2 \cup \mathcal{E}_3 \cup \mathcal{C}, \tag{1.2}$$

$$\mathcal{T}^* = \mathcal{V}_1^* \cup \mathcal{V}_2^* \cup \mathcal{V}_3^* \cup \mathcal{C}^*$$

where the unions are disjoint except for boundary points (see Fig. 1 and Fig. 1*).

We notice the following relations:

$$S^3 = I \tag{1.3}$$

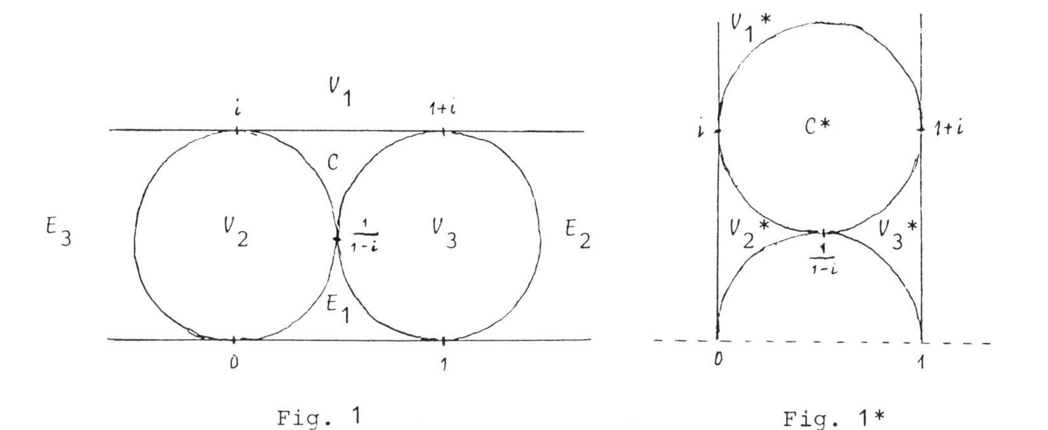

Fig. 1 Fig. 1*

$$SV_jS^{-1} = V_{j+1}, \quad SE_jS^{-1} = E_{j+1}, \quad SCS^{-1} = C \tag{1.4}$$

$$\det V_j = 1, \quad \det E_j = i, \quad \det C = -i \tag{1.5}$$

$$V_j^{-1} = \bar{V}_j, \quad E_j^{-1} = \bar{E}_j, \quad C^{-1} = -\bar{C} \tag{1.6}$$

$$\mathcal{V}_{j+1} = S(\mathcal{V}_j), \quad \mathcal{E}_{j+1} = S(\mathcal{E}_j), \quad \mathcal{C} = S(\mathcal{C}) \tag{1.7}$$

$$\mathcal{V}^*_{j+1} = S(\mathcal{V}^*_j), \quad \mathcal{C}^* = S(\mathcal{C}^*) \tag{1.8}$$

$$\mathcal{V}_{2-j} = R(\mathcal{V}_j), \quad \mathcal{E}_{2-j} = R(\mathcal{E}_j), \quad \mathcal{C} = R(\mathcal{C}) \tag{1.9}$$

$$\mathcal{V}^*_{2-j} = R(\mathcal{V}^*_j), \quad \mathcal{C}^* = R(\mathcal{C}^*) \tag{1.10}$$

In particular the relations (1.7)-(1.10) show that the subdivisions (1.1), (1.2) of $\mathcal{J}$, $\mathcal{J}^*$ possess an S_3-symmetry, where $S_3 = \langle S,R \rangle$, $S: z \to -1/(z-1)$, $R: z \to 1-\bar{z}$. The index j is always considered modulo 3.

Let

$$\mathcal{T}_{V_j} = \{V_j^m \mid m \in \mathbb{N}\}, \quad 1 \le j \le 3$$

$$\mathcal{T}_V = \bigcup_1^3 \mathcal{T}_{V_j}, \quad \mathcal{T}_E = \{E_1, E_2, E_3\}$$

$$\mathcal{T} = \mathcal{T}_V \cup T_E \cup \{C\}$$

A <u>regular product</u> is a finite product $T_0T_1\dots T_n$ such that

(i) $T_\nu \in \mathcal{T}$ for $0 \le \nu \le n$,

(ii) $T_\nu, T_{\nu+1} \in \mathcal{T}_{V_j}$ is not allowed for any $1 \le j \le 3$, $0 \le \nu < n$,

(iii) $T_\nu \in \mathcal{T}\backslash\mathcal{T}_E$ if $\det T_0T_1\dots T_{\nu-1} = \pm i$.

A <u>dually regular product</u> is a finite product $T_0T_1\ldots T_n$ satisfying (i), (ii) and

(iii*) $T_\nu \in \mathcal{T}\setminus\mathcal{T}_E$ if $\det T_0T_1\ldots T_{\nu-1} = \pm 1$.

For any regular product $M_n = T_0T_1\ldots T_n$ the Farey-set $F(T_0T_1\ldots T_n)$ is defined as

$$F_n = F(T_0T_1\ldots T_n) = \begin{cases} M_n(\mathcal{J}) & \text{if} \quad \det M_n = \pm 1 \\ M_n(\mathcal{J}^*) & \text{if} \quad \det M_n = \pm i \end{cases}$$

For any dually regular product $M_n = T_0T_1\ldots T_n$ the <u>dual Farey set</u> $F^*(T_0T_1\ldots T_n)$ is defined as

$$F_n^* = F^*(T_0T_1\ldots T_n) = \begin{cases} M_n(\mathcal{J}^*) & \text{if} \quad \det M_n = \pm 1 \\ M_n(\mathcal{J}) & \text{if} \quad \det M_n = \pm i \end{cases}$$

The sets $M_n(\mathcal{J})$ are called <u>circular</u>, and the set $M_n(\mathcal{J}^*)$ are called <u>triangular</u>. In particular

$$\mathcal{V}_j = F(V_j), \quad \mathcal{E}_j = F(E_j), \quad \mathcal{C} = F(C), \quad \mathcal{J} = F(I) \qquad (1.11)$$

$$\mathcal{V}_j^* = F^*(V_j), \quad \mathcal{C}^* = F^*(C), \quad \mathcal{J}^* = F^*(I) \qquad (1.12)$$

A <u>regular chain</u> is an infinite product $T_0T_1\ldots T_n\ldots$ such that $T_0T_1\ldots T_n$ is a regular product for each $n \in \mathbb{N}_0$. A <u>dually regular chain</u> is defined similarly.

For a (dually) regular chain $T_0T_1\ldots T_n\ldots$ the sequence of (dual) Farey sets F_n (resp. F_n^*) is strictly decreasing for $n \in \mathbb{N}_0$, and we denote by $\xi_0 = [T_0T_1\ldots T_n\ldots]$ the unique $\xi_0 \in \bigcap_0^\infty F_n$ (resp. $\bigcap_0^\infty F_n^*$). ξ_0 is then said to be <u>represented</u> by the (dually) regular chain $T_0T_1\ldots T_n\ldots$, and we also write $T_0T_1\ldots T_n\ldots = \mathrm{ch}\,\xi_0$ (resp. $\mathrm{ch}^*\xi_0$). It follows that $\xi_0 \in \mathcal{J}\setminus\mathbb{Q}(i)$ (resp. $\mathcal{J}^*\setminus\mathbb{Q}(i)$). Let

$$M_n = T_0 T_1 \dots T_n = \begin{pmatrix} p_1^{(n)} & p_2^{(n)} \\ q_1^{(n)} & q_2^{(n)} \end{pmatrix}, \quad \begin{matrix} p_3^{(n)} = p_1^{(n)} + p_2^{(n)} \\ q_3^{(n)} = q_1^{(n)} + q_2^{(n)} \end{matrix}$$

Then

$$p_1^{(n)}/q_1^{(n)} = M_n(\infty), \quad p_2^{(n)}/q_2^{(n)} = M_n(0), \quad p_3^{(n)}/q_3^{(n)} = M_n(1)$$

and therefore

$$\lim_{n\to\infty} p_j^{(n)}/q_j^{(n)} = \xi_0 \quad \text{for} \quad 1 \le j \le 3$$

The fractions $p_j^{(n)}/q_j^{(n)}$, $n \in \mathbb{N}_0$, $1 \le j \le 3$, are called the <u>convergents</u> for $\xi_0 = [T_0 T_1 \dots T_n \dots]$.

If $T_0 T_1 \dots T_n \dots$ is a (dually) regular chain, then so is $T_n T_{n+1} \dots$ for each $n \in \mathbb{N}_0$, and actually of the same or dual type as $T_0 T_1 \dots T_n \dots$ according as $\det T_0 T_1 \dots T_{n-1}$ is ± 1 or $\pm i$; $\xi_n = [T_n T_{n+1} \dots]$ is called the n'th <u>complete quotient</u>.

A <u>regular double chain</u> is a two-way infinite product $\Pi_{-\infty}^{\infty} T_n$ such that $T_{n+1} T_{n+2} \dots$ and $T_n T_{n-1} \dots$ for each $n \in \mathbb{Z}$ is a regular chain and a dually regular chain or vice versa.

2. PROPERTIES OF REGULAR AND DUALLY REGULAR CHAINS

I. Representation of complex numbers

Every $\xi_0 \in \mathcal{J} \setminus \mathbb{Q}(i)$ has either one or two representations by a regular chain $T_0 T_1 \dots T_n \dots$, the representation being unique precisely when $\xi_0 \in \partial \mathcal{J} = \mathbb{R}$ or if ξ_0 is not properly equivalent to a real number. Similarly every $\xi_0 \in \mathcal{J}^* \setminus \mathbb{Q}(i)$ has either one or two representations by a dually regular chain $T_0 T_1 \dots T_n \dots$, the representation being unique precisely when $\xi_0 \in \partial \mathcal{J}^*$ or ξ_0 is not improperly equivalent to a real number.

In fact, given $\xi_0 \in \mathcal{J}\backslash\mathbb{Q}(i)$ (resp. $\mathcal{J}^*\backslash\mathbb{Q}(i)$), a (dually) regular chain representing ξ_0 is determined by the requirement $\xi_0 \in F(T_0 \ldots T_n)$ (resp. $F^*(T_0 \ldots T_n)$) for all $n \in \mathbb{N}_0$, the existence and "uniqueness" stemming from (1.1), (1.2), (1.11), (1.12).

It is sometimes useful to extend the definition of a (dually) regular chain by allowing T_0 also to be V_1^{-m}, $m \in \mathbb{N}$. Then all $\xi_0 \in \mathbb{C}\backslash\mathbb{Q}(i)$ have (essentially unique) representations by regular chains, and all $\xi_0 \in \mathbb{C}\backslash\mathbb{Q}(i)$ with $0 \leq \mathrm{Re}\ \xi_0 \leq 1$ have (essentially unique) representations by dually regular chains.

II. Equivalence

Assuming that $\xi_0, \eta_0 \in \mathbb{C}\backslash\mathbb{Q}(i)$ are both not properly equivalent to real numbers, then ξ_0 is properly equivalent to η_0 exactly when the unique regular chains $T_0 T_1 \ldots T_n \ldots$ and $U_0 U_1 \ldots U_m \ldots$ for ξ_0 and η_0 end essentially in the same way, i.e. there exist n_0, m_0, j_0 such that

$$\det T_0 \ldots T_{n_0} = \pm \det U_0 \ldots U_{m_0} \quad \text{and} \quad T_{n_0+\nu} = S^{j_0} U_{m_0+\nu} S^{-j_0}, \qquad \nu \in \mathbb{N}$$

Similar results hold for proper equivalence between two numbers represented by dually regular chains, and for improper equivalence between two numbers, one of which is represented by a regular chain the other by a dually regular chain.

III. Approximation

Let $\xi_0 \in \mathbb{C}\backslash\mathbb{Q}(i)$, and let p/q, $p,q \in \mathbb{Z}[i]$, $q \neq 0$, be an irreducible fraction. The approximation of p/q to ξ_0 is measured by

$$c(p/q) = (|q|\,|q\xi_0 - p|)^{-1}$$

and the <u>approximation</u> <u>constant</u> $C(\xi_0)$ is defined as

$$C(\xi_0) = \limsup_{p/q} c(p/q)$$

It is well known (see [1], [2], or [6]) that $c(p/q) > \sqrt{3}$ for infinitely many p/q for each $\xi_0 \in \mathbb{C}\setminus\mathbb{Q}(i)$. In particular $C(\xi_0) \geq \sqrt{3}$ for all $\xi_0 \in \mathbb{C}\setminus\mathbb{Q}(i)$.

Suppose now that ξ_0 is represented by a (dually) regular chain $T_0T_1 \ldots T_n \ldots$. Then it is an important fact that each p/q with $c(p/q) \geq 1 + 1/\sqrt{2}$ is a convergent for $\xi_0 = [T_0T_1 \ldots T_n \ldots]$. Since $\sqrt{3} > 1 + 1/\sqrt{2}$, the approximation constant $C(\xi_0)$ is determined by the convergents alone, hence

$$C(\xi_0) = \limsup_{n,j} c_j^{(n)}$$

where

$$c_j^{(n)} = c(p_j^{(n)}/q_j^{(n)}) = (|q_j^{(n)}||q_j^{(n)}\xi_0 - p_j^{(n)}|)^{-1}$$

Now it can be shown that

$$c_1^{(n)} = |\xi_{n+1} + q_2^{(n)}/q_1^{(n)}|$$

$$c_2^{(n)} = |\xi_{n+1}^{-1} + q_1^{(n)}/q_2^{(n)}|$$

$$c_3^{(n)} = |(\xi_{n+1} - 1)^{-1} + q_1^{(n)}/q_3^{(n)}|$$

or generally

$$c_j^{(n)} = |S^{1-j}([T_{n+1}T_{n+2}\ldots]) - S^{1-j}\,\overline{T_n \ldots T_0(\infty)}|$$

so that

$$C(\xi_0) = \limsup_{n,j} |S^j([T_{n+1}T_{n+2}\ldots]) - S^j\overline{T_n\ldots T_0}(\infty)|, \quad (2.1)$$

where the bar means complex conjugation.

Two numbers which are properly or improperly equivalent have the same approximation constant.

IV. Quadratic irrationals

Let $\xi_0 \in \mathbb{C}\setminus\mathbb{Q}(i)$ (resp. with $0 \leq \operatorname{Re} \xi_0 \leq 1$). Then ξ_0 is quadratic over $\mathbb{Q}(i)$ if and only if ξ_0 is represented by a periodic (dually) regular chain $T_0T_1\ldots T_n\ldots$, i.e. constants $h \in \mathbb{N}_0$, $k \in \mathbb{N}$ exist such that $T_n = T_{n+k}$ for all $n \geq h$. The (dually) regular chain is then written $T_0\ldots T_{h-1}\overline{T_h\ldots T_{h+k-1}}$. Especially it is purely periodic, when we may choose $h = 0$.

A quadratic irrational ξ_0 with algebraic conjugate ξ_0' is called reduced (resp. dually reduced) if $\xi_0 \in \mathcal{J}$ and $\overline{\xi_0'} \in \mathcal{J}^*$ (resp. $\xi_0 \in \mathcal{J}^*$ and $\overline{\xi_0'} \in \mathcal{J}$). It is an important additional fact, that a quadratic irrational is represeted by a purely periodic (dually) regular chain if and only if it is (dually) reduced. Also if $\xi_0 = [\overline{T_0\ldots T_{k-1}}]$, then $\overline{\xi_0'} = [\overline{T_{k-1}\ldots T_0}]$.

V. The regular chain for $\sqrt{D}$

Let $D = a + ib$, $a,b \in \mathbb{N}$, be a non-square Gaussian integer in the first quadrant. Then it follows from the results in IV that except for $D = 1 + i$, $1 + 2i$, $2 + i$ (cf. Table 1) the regular chain for $\sqrt{D}$ has the form

$$\sqrt{D} = [V_1^{m_0}E_2\overline{T_2\ldots T_{k+1}}], \quad m_0 \in \mathbb{N}_0 \qquad (2.2)$$

where k is the shortest period with $\det T_2\ldots T_{k+1} = \pm 1$.

TABLE 1

a	b	preperiod	period
1	1	V_3	$V_3\overset{*}{E}_2V_1\overset{*}{C}S^{-1}$
1	2	V_3	$E_2V_1CS^{-1}$
1	3	V_1E_2	$V_3CV_2E_3V_2^{\,2}S$
1	4	V_1E_2	$CE_3CV_3E_1V_3^{\,2}S^{-1}$
1	5	V_1E_2	$CV_1V_2^{\,2}V_1V_2E_3CV_3E_1V_3^{\,2}S^{-1}$
2	1	V_3	$E_2V_3V_1CS^{-1}$
2	2	E_2	$CE_2V_1^{\,2}$
2	3	E_2	$V_2V_1CE_3V_2^{\,2}S$
2	4	V_1E_2	$CV_2E_3V_2^{\,2}S$
2	5	V_1E_2	$CV_2^{\,2}V_1^{\,2}E_2V_1^{\,2}$
3	1	E_2	$CV_2E_3V_1V_2^{\,2}S$
3	2	E_2	$CV_1E_1V_3^{\,2}S^{-1}$
3	3	E_2	$\overset{*}{C}V_3^{\,2}E_2\overset{*}{C}E_2V_1^{\,2}$
3	4	perfect	square
3	5	V_1E_2	$\overset{*}{C}V_2^{\,3}E_1V_2\overset{*}{C}V_1E_2V_1^{\,3}$
4	1	E_2	$\overset{*}{C}V_2^{\,3}\overset{*}{E}_1V_3^{\,3}S^{-1}$
4	2	E_2	$CV_2E_1V_3^{\,3}S^{-1}$
4	3	E_2	$CV_3E_1CE_3V_2^{\,3}S$
4	4	E_2	$CE_1V_3^{\,3}S^{-1}$
4	5	V_1E_2	$CE_1V_2^{\,2}CV_1E_2V_1^{\,3}$
5	1	E_2	$CV_2E_1V_2^{\,2}V_3^{\,3}S^{-1}$
5	2	E_2	$CE_1CV_3V_1V_3E_2V_3V_1^{\,3}$
5	3	E_2	$CE_1\overset{*}{C}E_3CV_3^{\,3}E_2CE_1CE_3V_2^{\,3}S$
5	4	E_2	$V_1V_2CE_2V_1^{\,3}$
5	5	V_1E_2	$V_1V_3CV_2E_3V_2^{\,3}S$

The periods in (2.2) are of two types:

(a) $k = 3\ell$ and $T_{n+\ell} = S^j T_n S^{-j}$ for $n \geq 2$ and a fixed $j = \pm 1$. In the table only the first third of the period is written and supplemented by S^j.

(b) The remaining case. In the table the full period ending with V_1^m, $m \geq 2$, is written.

A number of formulas for $\mathrm{ch}\sqrt{D}$ exist, of which I indicate only a sample

$$\mathrm{ch}\sqrt{(a+bi)^2+1} = V_1^{b-1}E_2\,\overline{V_1^{a-2}CV_3^{2b-2}E_1V_3^{a+1}S^{-1}}$$

$$\mathrm{ch}\sqrt{(a+bi)^2-1} = V_1^{b-1}E_2\,\overline{V_1^{a-2}CV_2^{2b-1}E_3V_2^{a}S}$$

$$\mathrm{ch}\sqrt{(a+bi)^2+i} = V_1^{b}E_2\,\overline{V_1^{a-2*}CV_2^{2a-1}E_1V_2^{2b-1*}CV_1^{2b-1}E_2V_1^{a+1}}$$

$$\mathrm{ch}\sqrt{(a+bi)^2-i} = V_1^{b-1}E_2\,\overline{V_1^{a-2*}CV_3^{2a-2}E_2V_3^{2b-2*}CV_1^{2b-2}E_2V_1^{a}}$$

These formulas are valid, when all exponents are in $\mathbb{N}_0$.

Some of the regular chains above possess a certain skew symmetry with respect to matrices C or E_j in the period (indicated by an asterisk). Matrices in the period lying symmetric to such matrices are then either equal or deviates systematically by a $(j-1\ j+1)$ permutation of their subscripts.

VI. The complex Pellian and non-Pellian equations

For D a nonsquare Gaussian integer the equations

$$x^2 - Dy^2 = \pm 1 \tag{2.3}$$

$$x^2 - Dy^2 = \pm i \tag{2.4}$$

are called the <u>Pellian</u> and <u>non-Pellian equations</u>, respectively. For D in the first quadrant a fundamental solution of (2.3) is obtained from (2.2) as $(p_{j+1}^{(\ell)}, q_{j+1}^{(\ell)})$ if the period is of type (a) and as $(p_1^{(k+1)}, q_1^{(k+1)})$ if the period is of type (b). Also (2.4) has solutions if and only if the period for ch$\sqrt{D}$ possesses the skew symmetry described in V. For the examples mentioned in V this happens for $D = 1 + i,\ 3 + 3i,\ 3 + 5i,\ 4 + i,\ 5 + 3i$ and $(a+bi)^2 \pm i$.

VII. Hurwitzian chains

A (dually) regular chain is called Hurwitzian if it is of the form

$$T_0 T_1 \cdots T_h \; T_{h+1} \cdots T_{h+k} \cdots$$

where for fixed m $(1 \leq m \leq k)$

$$T_{h+m+nk} = \begin{matrix} C, E_{j_m} \\ f_m(n) \\ V_{j_m} \end{matrix} \qquad \text{for all } n$$

the choice depending only on m, and where $f_m \in \mathbb{Q}[x]$ maps $\mathbb{N}_0 \to \mathbb{N}_0$.

The main result is that if ξ_0 has a Hurwitzian (dually) regular chain then so has $\eta_0 = M\xi_0$ for any $M = \begin{pmatrix} a & b \\ c & d \end{pmatrix}$ with $a,b,c,d \in \mathbb{Z}[i]$ and $\det M \neq 0$.

Examples of Hurwitzian chains are:

$$\text{ch}(J_0(2/(a-bi))/J_1(2/(a-bi)))$$
$$= V_1^{-b-1} E_2 \overline{V_{1-n}^{ab+a-2} CV_{-n}^{bn+2n-1} E_{2-n}}\Big|_{n=0}^{\infty}, \quad a \geq 2,\ b \geq 1$$

$$\text{ch}(\coth(1/(a-bi)))$$

$$= V_1^{-b} E_2 V_1^{-1} \overline{V_{n+1}^{2an+a-1} CV_{n+2}^{2bn+2n-2} E_{n+1}}\Big|_{n=0}^{\infty}, \quad a \geq 2,\ b \geq 1$$

ch(exp(1/(a-bi)))

$$= V_3^{b-1}\,\overline{E_{n+2}V_n^{2an+a-1}V_{n+2}CV_{n+1}^{2bn+3b-2}}\Big|_{n=0}^{\infty},\quad a \geq 1,\ b \geq 1$$

ch(exp(2/(2a+1-(2b+1)i)))

$$= V_3^{b}\,\overline{E_{2-n}V_{-n}^{(4a+2)n+a-1}CV_{2-n}V_{1-n}^{(4b+2)n+3b}E_{-n}}$$

$$\overline{V_{1-n}^{(4a+2)n+3a+1}V_{-n}CV_{2-n}^{(4b+2)n+5b+1}}\Big|_{n=0}^{\infty},\quad a \geq 1,\ b \geq 0$$

ch(exp(2/(2a+1-2bi)))

$$= V_3^{b-1}\,\overline{E_2V_3^{(6a+3)n+a-1}CV_1^{24bn+12b-2}E_3V_1^{(24a+12)n+12a+5}C}$$

$$\overline{V_2^{6bn+5b-2}E_1V_2^{(6a+3)n+5a+2}V_1CV_3^{6bn+7b-2}}\Big|_{n=0}^{\infty},\quad a \geq 1,\ b \geq 1$$

ch(exp(2/(2a-(2b+1)i)))

$$= V_3^{b}\,\overline{E_{n+2}V_n^{6an+a-1}CV_n^{12an+6a-1}E_{n+1}V_n^{(12b+6)n+6b+1}CE_nC}$$

$$\overline{V_n^{(6b+3)n+5b}E_{n+2}V_n^{6an+5a-1}V_{n+2}CV_{n+1}^{(6b+3)n+7b+2}}\Big|_{n=0}^{\infty},\quad a \geq 1,\ b \geq 0$$

VIII. Ergodic theory

Let X be the disjoint union $\mathfrak{J} \cup \mathfrak{J}^*$. Except for a subset of Lebesque measure zero, the regular and dually regular chains provide a unique representation of X. Hence the shift transformation T is uniquely defined a.e. on X by

$$z = [T_0T_1\cdots T_n\cdots] \rightarrow Tz = [T_1\cdots T_n\cdots]$$

where $T_0T_1 \ldots T_n \ldots$ is the unique (dually) regular chain for $z \in X$, and with z and Tz in the same or opposite of the two sets $\mathcal{J}$ and $\mathcal{J}^*$ according as $\det T_0 = \pm 1$ or $\pm i$.

A measure μ on X is T-invariant if $\mu(Y) = \mu(T^{-1}Y)$ for all Borel subsets Y of X. Assuming that μ has a density function f with respect to Lebesque measure, then μ is T-invariant if and only if

$$f(z) = \Sigma_{M \in Adm(z)} f(Mz) \|Mz\| \quad \text{for a.a } z \in X \tag{2.5}$$

where

$$Adm(z) = \begin{cases} \mathcal{T} \setminus \mathcal{T}_E \setminus \mathcal{T}_{V_j} & \text{if } z \in \mathcal{V}_j \\ \mathcal{T} \setminus \mathcal{T}_E & \text{if } z \in \mathcal{E}_1 \cup \mathcal{E}_2 \cup \mathcal{E}_3 \cup \mathcal{C} \\ \mathcal{T} \setminus \mathcal{T}_{V_j} & \text{if } z \in \mathcal{V}_j^* \\ \mathcal{T} & \text{if } z \in \mathcal{C}^* \end{cases}$$

Defining $\tilde{f}: X \to \mathbb{R}_+ \cup \{0\}$ by

$$\tilde{f}(z) = \begin{cases} f(z) & \text{if } z \in \mathcal{E}_1 \cup \mathcal{E}_2 \cup \mathcal{E}_3 \cup \mathcal{C} \text{ or } \mathcal{C}^* \\ \Sigma_{m=0}^{\infty} f(V_j^m z) \|V_j^m z\| & \text{if } z \in \mathcal{V}_j \text{ or } \mathcal{V}_j^* \end{cases}$$

with inversion formula

$$f(z) = \begin{cases} \tilde{f}(z) & \text{if } z \in \mathcal{E}_1 \cup \mathcal{E}_2 \cup \mathcal{E}_3 \cup \mathcal{C} \text{ or } \mathcal{C}^* \\ \tilde{f}(z) - \tilde{f}(V_j z) \|V_j z\| & \text{if } z \in \mathcal{V}_j \text{ or } \mathcal{V}_j^* \end{cases} \tag{2.6}$$

the condition (2.5) for f is replaced by the following smooth functional equation for $\tilde{f}$:

$$\tilde{f}(z) = \begin{cases} \Sigma_{j=1}^{3} \tilde{f}(V_j z)\|V_j z\| + \tilde{f}(Cz)\|Cz\|, & z \in \mathfrak{J} \\ \Sigma_{j=1}^{3} \tilde{f}(V_j z)\|V_j z\| + \Sigma_{j=1}^{3} \tilde{f}(E_j z)\|E_j z\| + \tilde{f}(Cz)\|Cz\| & z \in \mathfrak{J}^* \end{cases} \quad (2.7)$$

The main results are as follows:

(1) Equation (2.7) has the positive, continuous, S_3-symmetric solution

$$\tilde{f}(z) = \begin{cases} g(z), & z \in \mathfrak{J}^* \\ \frac{1}{\pi}(h(z)+h(Sz)\|Sz\|+h(S^{-1}z)\|S^{-1}z\|), & z \in \mathfrak{J} \end{cases} \quad (2.8)$$

where

$$g(z) = \frac{1}{y^2}, \quad z \in \mathfrak{J}^*$$

$$h(z) = \frac{1}{xy} - \frac{1}{x^2} \arctan \frac{x}{y}, \quad z \in \mathfrak{J}$$

By (2.6) and (2.8) f (and hence μ) is given explicitly.

(2) The transformation T is ergodic on the measure space (X,μ)

(3) μ is uniquely determined except for a multiplicative constant. With the normalization made

$$\frac{1}{2}\mu(X) = \mu(\mathfrak{J}) = \mu(\mathfrak{J}^*) = 24/\sqrt{15} \arccos \frac{1}{4} - 2\pi$$

$$\frac{1}{2}\tilde{\mu}(X) = \tilde{\mu}(\mathfrak{J}) = \tilde{\mu}(\mathfrak{J}^*) = \pi$$

(4) Let $z \in X$ have the (dually) regular chain

$$T_0(z)T_1(z)\ldots T_n(z)\ldots$$

with (dual) convergents $p_j^{(n)}(z)/q_j^{(n)}(z)$. Define

$$e(V_j^m) = m, \quad e(E_j) = e(C) = 1$$

$$\Phi(x) = \pi - 2\frac{\arccos x}{\sqrt{1-x^2}}$$

$$L = \frac{1}{\mu(X)} \int_X \log|c_0(z)z + d_0(z)|d\mu$$

where

$$T_0(z) = \begin{pmatrix} a_0(z) & b_0(z) \\ c_0(z) & d_0(z) \end{pmatrix}$$

Then the following holds for a.a.z $\in$ X:

$$p(T_n(z) = V_j^m) = \mu(\mathfrak{J})^{-1}(\Phi(\tfrac{1}{2m}) - 2\Phi(\tfrac{1}{2(m+1)}) + \Phi(\tfrac{1}{2(m+2)})) \tag{2.9}$$

where p is the frequency function.

$$\lim_{n\to\infty} \frac{1}{n} \Sigma_{\nu=0}^{n-1} e(T_\nu(x)) = \pi/\mu(\mathfrak{J}) \tag{2.10}$$

$$\lim_{n\to\infty} |q_j^{(n)}(z)|^{1/n} = \exp(L) \quad \text{for} \quad 1 \leq j \leq 3 \tag{2.11}$$

The precise evaluation of the constant L remains an open problem. A detailed account of the ergodic theory is given in [13].

3. THE GAUSSIAN MARKOFF AND HURWITZ SPECTRA

For a quadratic form $f = (\alpha,\beta,\gamma)$ given by

$$f(x,y) = \alpha x^2 + \beta xy + \gamma y^2, \qquad a,\beta,\gamma \in \mathbb{C}$$

with discriminant $\delta = \delta(f) = \beta^2 - 4\alpha\gamma$, the <u>minimum</u> μ is defined as

$$\mu = \mu(f) = \inf_{(x,y)\in\mathbb{Z}[i]^2\setminus\{(0,0)\}} |f(x,y)|$$

The <u>Gaussian Markoff spectrum</u> is the set

$$\mathfrak{M} = \{\lambda(f) = \sqrt{|\delta(f)|}/\mu(f) \mid \delta(f) \neq 0\}$$

Similarly the <u>Gaussian Hurwitz spectrum</u> is the set

$$\mathcal{H} = \{C(\xi) \mid \xi \in \mathbb{C}\setminus\mathbb{Q}(i)\}$$

Now by (2.1)

$$\mathcal{H} = \{\limsup_{n,j} |S^j([T_{n+1}T_{n+2}\cdots]) - S^j\overline{T_n\cdots T_0(\infty)}|\} \qquad (3.1)$$

where $T_0T_1\cdots T_n\cdots$ is any regular chain, and we want to establish that similarly

$$\mathfrak{M} = \{\sup_{n\in\mathbb{Z},j} |S^j([T_{n+1}T_{n+2}\cdots]) - S^j(\overline{[T_nT_{n-1}\cdots]})|\}, \qquad (3.2)$$

where $\Pi_{-\infty}^{\infty} T_n$ is any regular double chain.

It should be noticed that these expressions are completely analogous to those for the usual (real) Markoff and Hurwitz spectra in terms of regular continued fractions.

A form f_0 with discriminant $\delta_0 \neq 0$ and with roots $\xi_0, \eta_0 \in \mathbb{C}\setminus\mathbb{Q}(i)$ is called <u>reduced</u> (resp. <u>dually reduced</u>) if (with suitable notation) $\xi_0 \in \mathcal{J}$, $\overline{\eta_0} \in \mathcal{J}^*$ (resp. $\xi_0 \in \mathcal{J}^*$, $\overline{\eta_0} \in \mathcal{J}$). Let f_0 be reduced with ch $\xi_0 = T_0T_1,\ldots,$ ch* $\overline{\eta_0} = T_{-1}T_{-2}\cdots$. Then we associate with f_0 a <u>double chain</u> of forms

$$\ldots f_{-2}, f_{-1}, f_0, f_1, f_2, \ldots$$

defined successively by $f_{n+1} = f_n \circ T_n$ given by

$$f_{n+1}(x,y) = f_n(x',y'), \binom{x'}{y'} = T_n\binom{x}{y}, \quad n \in \mathbb{Z}$$

It follows that $\prod_{-\infty}^{\infty} T_n$ is a regular double chain, and that the roots ξ_n, η_n of f_n (with suitable notation) satisfy $\xi_n = [T_n T_{n+1} \cdots]$, $\bar{\eta}_n = [T_{n-1} T_{n-2} \cdots]$ for all $n \in \mathbb{Z}$, so that $f_n = (\alpha_n, \beta_n, \gamma_n)$ is reduced or dually reduced for each $n \in \mathbb{Z}$. Also

$$\mu(f_0) = \inf_{n\in\mathbb{Z}}\{|f_n(1,0)|, |f_n(0,1)|, |f_n(1,1)|\}$$

$$= \inf_{n\in\mathbb{Z}}\{|\alpha_n|, |\gamma_n|, |\alpha_n+\beta_n+\gamma_n|\}$$

where

$$\sqrt{|\delta(f_0)|} = \sqrt{|\delta(f_n)|} = |\alpha_n||\xi_n-\eta_n|, |\gamma_n|$$

$$= |\alpha_n||\xi_n\eta_n|, |\alpha_n+\beta_n+\gamma_n| = |\alpha_n|(1-\xi_n)(1-\eta_n)|$$

Now every form f with discriminant $\delta \neq 0$ is equivalent to a reduced form f_0, and hence

$$\lambda(f) = \lambda(f_0) = \sup_{n\in\mathbb{Z}}\{|\xi_n-\eta_n|, |\xi_n^{-1}-\eta_n^{-1}|, |(\xi_n-1)^{-1}-(\eta_n-1)^{-1}|\}$$

$$= \sup_{n\in\mathbb{Z},j} |S^j([T_n T_{n+1} \cdots]) - S^j\overline{[T_{n-1} T_{n-2} \cdots]}|$$

This gives formula (3.2) for $\mathfrak{M}$.

If, in particular, f_0 is a reduced form with coefficients in $\mathbb{Z}[i]$, then ξ_0 is a reduced quadratic irrational, while $\bar{\eta}_0 = \bar{\xi}'_0$ is dually reduced. Consequently

$$\text{ch } \xi_0 = \overline{T_0 \cdots T_{k-1}}, \quad \text{ch*} \bar{\eta}_0 = \text{ch*} \bar{\xi}'_0 = \overline{T_{k-1} \cdots T_0}$$

so that the regular double chain $\prod_{-\infty}^{\infty} T_n = \overline{T_0 \cdots T_{k-1}}$ is purely periodic (in both directions). In this case

$$\mu(f_0) = \min_{0\le n<k} \{|f_n(1,0)|, |f_n(0,1)|, |f_n(1,1)|\}$$

so that $C(\xi_0) = \lambda(f_0)$ is easily computed.

If, conversely, $\xi_0 = \overline{[T_0 \cdots T_{k-1}]}$ is a reduced quadratic irrational, and $T_0 \cdots T_{k-1} = M = \begin{pmatrix} a & b \\ c & d \end{pmatrix}$, then $f_0 = f_M = (c, d-a, -b)$ is a reduced form with coefficients in $\mathbb{Z}[i]$ and with ξ_0 as a root.

A few examples (all of importance later on) will illustrate this:

EXAMPLE 1. $f_0 = (1,-1,1)$ is reduced with ch $\xi_0 = \vec{C}$. Hence $f_n = (-i)^n f_0$ for $n \in \mathbb{Z}$. Thus $\mu(f_0) = 1$, and since $\delta(f_0) = -3$, one finds

$$C(\xi_0) = \lambda(f_0) = \sqrt{3} = \sqrt{4-1/1^2}$$

EXAMPLE 2. $f_0 = (1+i,-2,1)$ is reduced with ch $\xi_0 = \overline{E_1 C}$. Hence $f_{n+2} = f_n$ for $n \in \mathbb{Z}$, and $f_1 = (-1+i,2,-1)$. Thus $\mu(f_0) = 1$, and since $\delta(f_0) = -4i$, one finds

$$C(\xi_0) = \lambda(f_0) = 2$$

EXAMPLE 3. $f_0 = (5,-5+8i,3-4i)$ is reduced with ch$\xi_0 = \overline{E_1 C E_1 V_1 C V_1}$. Hence $f_{n+6} = f_n$ for $n \in \mathbb{Z}$, and $f_1 = (7i,6-7i,-3+4i)$, $f_2 = (7,-7+6i,4-3i)$, $f_3 = (5i,8-5i,-4+3i)$, $f_4 = (5i,-2-5i,1+6i)$, $f_5 = (5,-5+2i,6+i)$. Thus $\mu(f_0) = 5$, and since $\delta(f_0) = -99$, one finds

$$C(\xi_0) = \lambda(f_0) = \sqrt{99}/5 = \sqrt{4-1/5^2} = 1.98997\ldots$$

EXAMPLE 4. $f_0 = (-3i,3i,1-2i)$ is reduced with ch $\xi_0 = \overline{V_2 E_2 V_2 C E_1 C V_3 E_3 V_3 C E_2 C V_1 E_1 V_1 C E_3 C} = \overline{V_2 E_2 V_2 C E_1 C S}$. Hence $f_{n+6} = f_n \circ S^{-1}$ for $n \in \mathbb{Z}$, and $f_1 = (2-i,-4+i,1-2i)$, $f_2 = (2-i,-3+2i,2+i)$, $f_3 = (2+i,-1-2i,2+i)$,

$f_4 = (1-4i, 2+5i, -1-2i)$, $f_5 = (4+i, -3-4i, 1+2i)$. Thus $\mu(f_0) = \sqrt{5}$, and since $\delta(f_0) = 15 + 12i$, one finds

$$C(\xi_0) = \lambda(f_0) = \sqrt{\frac{3}{5}\sqrt{41}} = 1.96007\ldots$$

The main result on the Gaussian Markoff and Hurwitz spectra is as follows:

$$\mathfrak{M} \cap]0,2[= \mathcal{H} \cap]0,2[= \{\sqrt{\tfrac{3}{5}\sqrt{41}}\} \cup \{\sqrt{4 - \frac{1}{\Lambda^2}} \mid \Lambda = 1, 5, 29, 65, \ldots\},$$

where the possible Λ occur in solutions $(\Lambda_1, \Lambda_2; M_1, M_2)$ of the diophantine equations

$$\Lambda_1 + \Lambda_2 = 2M_1M_2, \quad 2\Lambda_1\Lambda_2 = M_1^2 + M_2^2 \qquad (3.3)$$

or equivalently of the single diophantine equation

$$M_1^2 + M_2^2 + 2\Lambda^2 = 4M_1M_2\Lambda$$

The solutions of (3.3) conssitute a tree of neighboring solutions

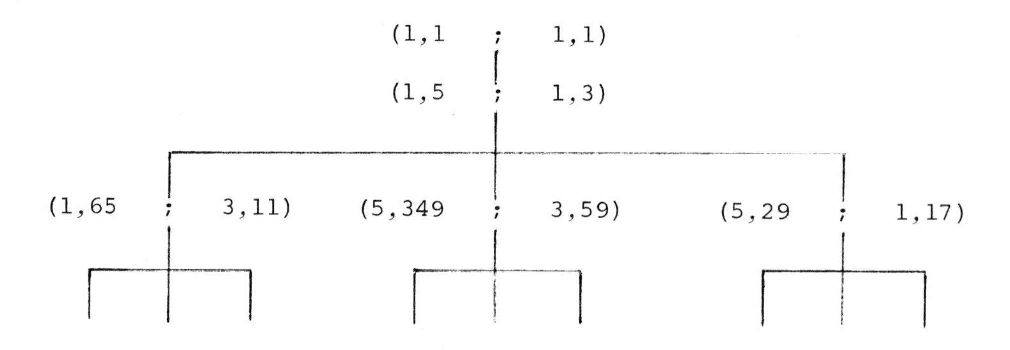

A corresponding tree of Markoff symbols (see [11])

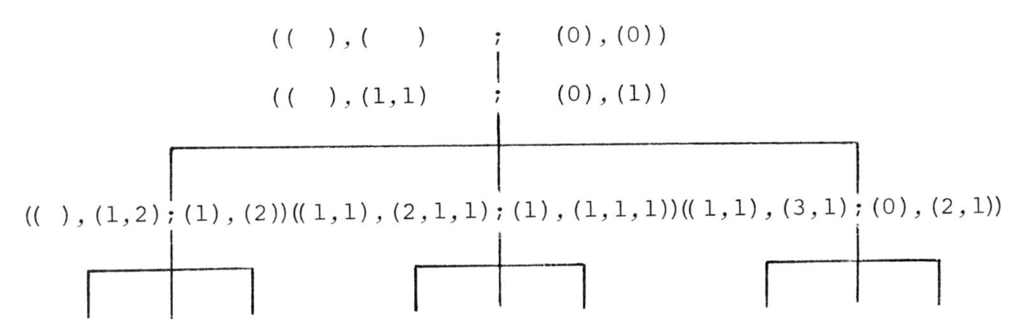

associates with each Λ and M a definite Markoff symbol $R = (r_0, r_1, \ldots, r_k)$ and thereby a certain period $\pi(R) = \rho-1, \rho', \rho'', \ldots, \rho'', \rho', \rho$ or ρ, the length of which is even for Λ and odd for M. Now let

$$U_n = E_1 C^{2n+1} \text{ for } n \in \mathbb{N}_0,$$

$$U'_n = E_1 V_1 C^{2n-1} V_1 \text{ for } n \in \mathbb{N}, \quad U'_0 = U_0 = E_1 C$$

If $\pi(A) = \alpha_1, \alpha_2, \ldots, \alpha_{2q-1}, \alpha_{2q}$ has even length, then we define

$$G_A = U_{\alpha_1} U'_{\alpha_2} U_{\alpha_3} U'_{\alpha_4} \ldots U_{\alpha_{2q-1}} U'_{\alpha_{2q}}$$

If $\pi(B) = \beta_1, \beta_2, \ldots, \beta_{2q-1}$ has odd length, then we define

$$H_B = U_{\beta_1} U'_{\beta_2} \ldots U_{\beta_{2q-1}} U'_{\beta_1} U_{\beta_2} U'_{\beta_3} \ldots U_{\beta_{2q-2}} U'_{\beta_{2q-1}}$$

Then $\lambda(f) = \sqrt{4-\Lambda^{-2}} < 2$ if and only if f is equivalent to a multiple of the integral form f_{G_A}, where A is the Markoff symbol of even period length corresponding to Λ. However $\lambda(f_{H_B}) = 2$ for all Markoff symbols B of odd period length.

Also $\lambda(f) = \sqrt{\frac{3}{5}\sqrt{41}}$ if and only if f is equivalent to a multiple of $(-3i, 3i, 1-2i)$ or its complex conjugate.

Similarly $C(\xi) = \sqrt{4-\Lambda^{-2}} < 2$ if and only if ξ is equivalent to a root ξ_0 with regular chain $\overline{U_{\alpha_1} U'_{\alpha_2} \ldots U_{\alpha_{2q-1}} U'_{\alpha_{2q}}}$

of the form f_{G_A}. And $C(\xi) = \sqrt{\frac{3}{5}\sqrt{41}}$ if and only if ξ is equivalent to a root of $(-3i,3i,1-2i)$ or its complex conjugate.

Table 2 containing the 11 smallest values of Λ together with the corresponding Markoff symbols A, periods $\pi(A)$, and matrices G_A is given on Page 375.

A similar Table 3 containing the 11 smallest values of M is given on Page 376.

Due to the fact that only V_1, E_1, C appear in the (dually) regular chain for the roots of the forms f_{G_A} and f_{H_B}, it follows that these roots are invariant under the map R, hence lie on the line $\text{Re } z = \frac{1}{2}$.

For any form $f = (\alpha,\beta,\gamma)$ with roots on this symmetry line one may define

$$\mu_1(f) = \inf_{x,y \in \mathbb{Z}[i], \text{Re } x/y = \frac{1}{2}} |f(x,y)|; \quad \lambda_1(f) = \sqrt{|\delta(f)|/\mu_1(f)}$$

and accordingly the "symmetric Markoff spectrum"

$$\mathfrak{M}_1 = \{\lambda_1(f) \mid \delta(f) \neq 0, \text{ both roots of } f \text{ on } \text{Re } z = \tfrac{1}{2}\}$$

The main result on $\mathfrak{M}_1$ is now that

$$\mathfrak{M}_1 \cap]0,2[= \{\sqrt{4-\Lambda^{-2}}\} \cup \{\sqrt{4-2M^{-2}}\}$$

where the possible Λ, M occur in solutions $(\Lambda_1,\Lambda_2;M_1,M_2)$ of (3.3). Furthermore $\lambda_1(f) = \sqrt{4-\Lambda^{-2}}$ (resp. $\sqrt{4-2M^{-2}}$) if and only if f is equivalent to a multiple of some f_{G_A} (resp. f_{H_B}), where A,B are Markoff symbols of Λ,M.

For a detailed account on the Gaussian Markoff and Hurwitz spectra one should consult [4], [9], [10], [11], [15], [16].

TABLE 2

Λ	A	$\pi(A)$	G_A
1	( )	undefined	C^2
5	(1,1)	0,1	$E_1CE_1V_1CV_1$
29	(3,1)	0,0,0,1	$(E_1C)^3E_1V_1CV_1$
65	(1,2)	1,2	$E_1C^3E_1V_1C^3V_1$
169	(5,1)	0,0,0,0,0,1	$(E_1C)^5E_1V_1CV_1$
349	(2,1,1)	0,1,1.1	$E_1CE_1V_1CV_1E_1C^3E_1V_1CV_1$
901	(1,3)	2,3	$E_1C^5E_1V_1C^5V_1$
985	(7,1)	0,0,0,0,0,0,0,1	$(E_1C)^7E_1V_1CV_1$
4549	(3,2)	1,1,1,2	$E_1C^3E_1V_1CV_1E_1C^3E_1V_1C^3V_1$
5741	(9,1)	0,0,0,0,0,0,0,0,0,1	$(E_1C)^9E_1V_1CV_1$
11521	(1,1,2,1)	0,0,1,0,0,1,0,1	$(E_1C)^2E_1C^3(E_1C)^2E_1V_1CV_1E_1CE_1V_1CV_1$

TABLE 3

M	B	π(B)	H_B
1	(0)	0	$(E_1C)^2$
3	(1)	1	$E_1C^3E_1V_1CV_1$
11	(2)	2	$E_1C^5E_1V_1CV_1$
17	(2,1)	0,0,1	$(E_1C)^2E_1C^3(E_1C)^2E_1V_1CV_1$
41	(3)	3	$E_1C^7E_1V_1C^5V_1$
59	(1,1,1)	0,1,1	$E_1CE_1V_1CV_1E_1C^3E_1CE_1C^3E_1V_1CV_1$
99	(4,1)	0,0,0,0,1	$(E_1C)^4E_1C^3(E_1C)^4E_1V_1CV_1$
153	(4)	4	$E_1C^9E_1V_1C^7V_1$
339	(1,2,1)	0,0,1,0,1	$(E_1C)^2E_1C^3E_1CE_1C^3(E_1C)^2E_1C)^2E_1V_1CV_1E_1CE_1V_1CV_1$
571	(5)	5	$E_1C^{11}E_1V_1C^9V_1$
577	(6,1)	0,0,0,0,0,0,1	$(E_1C)^6E_1C^3(E_1C)^6E_1V_1CV_1$

REFERENCES

[1] Ford, L.R., On the closeness of approach of complex rational fractions to a complex irrational number. Trans. Amer. Math. Soc. 27 (1925), 146-154.

[2] Perron, O., Uber die Approximation einer komplexen Zahl durch Zahlen des Körpers K(i). I. Math. Ann. 103 (1930), 533-544; II. Math. Ann. 105 (1931), 160-164.

[3] Perron, O., Die Lehre von den Kettenbrüchen I. Teubner, Stuttgart, 1954.

[4] Richert, N., Diophantine approximation of complex numbers. Unpublished thesis, Claremont Graduate School, California, USA, 1980.

[5] Richert, N., A canonical form for planar Farey sets. Proc. Amer. Math. Soc. 83(1981), 259-262.

[6] Schmidt, A.L., Farey triangles and Farey quadrangles in the complex plane. Math. Scand. 21 (1967), 241-295.

[7] Schmidt, A.L., Farey simplices in the space of quaternions. Math. Scand. 24 (1969), 31-65.

[8] Schmidt, A.L., On the approximation of quaternions. Math. Scand. 34 (1974), 184-186.

[9] Schmidt, A.L., Diophantine approximation of complex numbers. Acta. Math. 134 (1975), 1-85.

[10] Schmidt, A.L., On C-minimal forms. Math. Ann. 215 (1975), 203-214.

[11] Schmidt, A.L., Minimum of quadratic forms with respect to Fuchsian groups I.J. reine angew, Math. 286/287 (1976), 341-368.

[12] Schmidt, A.L., Diophantine approximation in the field $\mathbb{Q}(i\sqrt{11})$. J. Number Theory 10 (1978), 152-176.

[13] Schmidt, A.L., Ergodic theory for complex continued fractions. Monatsh. Math. 93(1982), 39-62.

[14] Schmidt, A.L., Diophantine approximation in the Eisensteinian field. J. Number Theory (to appear).

[15] Vulakh, L. Ya., On the Markov spectrum for the imaginary quadratic fields $\mathbb{Q}(i\sqrt{D})$, where $D \not\equiv 3 \pmod 4$. Vestnik Moskov, Univ. Ser. I Mat. Meh. 6 (1971), 32-41.

[16] Vulakh, L. Ya., On the Markov spectrum for the Gaussian field. Izv. Vyss. Ucebn, Zaved. Mathematika 2 (1973), 26-40.

TRAJECTORIES ON REIMANN SURFACES

Mark Sheingorn*

Institute for Advanced Study
Princeton, New Jersey

The purpose of this paper is to describe the mathematical theory of the behavior of trajectories (geodesics) on Riemann surfaces. Except for a few results in §4, the paper is entirely expository.

As the Chudnovsky seminar (Columbia, Spring 1981) has shown, there is considerable interest in these sorts of questions beyond the mathematical community. This paper is intended to be comprehens-ible to such people. Thus only the main ideas of the proofs will be provided, along with references allowing the reader to pursue further detail. One last acknolwedgement: as will be seen, the topic under discussion is vast. The make up of this article, then, is subject to the author's own experience (automorphic forms via number theory) and taste and is certainly not exhaustive. No claim of superiority is made for this particular slice of the pie.

Research partially supported by the NSF
*Current Affiliation: Department of Mathematics, Baruch College, City University of New York, New York, New York

§1. PRELIMINARIES AND NOTATION

Although we shall be dealing with Riemann surfaces, our technique will be to treat them by using their representations by Fuchsian groups. We begin by describing this representation. (Lehner [L] is a good, brief introduction to these matters).

Let $H^+ = \{z | z = x + iy, y \geq 0\} \cup \{\infty\}$. (By H we mean $H^+ - \{\mathbb{R} \cup \{\infty\}\}$.) H^+ is called the upper half plane. It's topology is induced by the euclidean one, together with the neighborhoods of ∞ arising from the stereographic projection. The mapping $z \to az+b/cz+d$, $\begin{pmatrix} a & b \\ c & d \end{pmatrix} \in SL(2,\mathbb{R})$[1)], is called a linear fractional transformation. These are conformal homeomorphisms of H onto itself. They are the only such transformations of H. A subgroup Γ of $SL(2,\mathbb{R})$ is called discontinuous if, for some $z_0 \in H$ $\{A(z_0) | A \in \Gamma\}$ has no accumulation point in H. (It can be checked that this definition does not depend on z_0.) One reason this condition is natural is that without it there would be no hope of producing meromorphic automorphic functions on Γ, i.e. functions satisfying $f(z) = f(Az)$, $z \in H$, $A \in \Gamma$. Such discontinuous subgroups of $SL(2,\mathbb{R})$ are called Fuchsian groups.

So let Γ be a Fuchsian group. A set F which contains exactly one point from each orbit under Γ is called a fundamental set. I.e. (i) $\forall\, z \in H$, $\exists z_0 \in F$ and $A_0 \in \Gamma \cdot\ni\cdot A_0(z_0) = z$ and (ii) $z_0, z_1 \in F \Rightarrow z_0 \cap \{A(z_1) | A \in \Gamma\} = \emptyset$. Now this set F may be rather wild and in order to get a reasonable set, we must introduce the hyperbolic metric for H.

H can be given a metric, compatible with its euclidean topology, called the Poincaré metric, defined by the area-element $dxdy/y^2$ and line-element $|dz|/y$. We list two of its properties.

[1)] Notes follow the text

(1.1). The hyperbolic lines (geodesics) are half-circles perpendicular to the real axis (including vertical lines).

(1.2). Elements of $SL(2,\mathbb{R})$ preserve distance and area, i.e. are isometries.

Now using a device known as the Ford circles one can construct a <u>fundamental region</u> R for Γ which has the following properties:

(1.3). R is open and hyperbolically convex.

(1.4). $\partial R \cap H$ consists of hyperbolic arcs called <u>sides</u> of R. These arcs are congruent (in pairs) by elements of Γ.

(1.5). a fundamental set for Γ consists of $R \cup$ {a portion of the sides}.

(1.6). The transformations which pair the sides generate Γ. (Even the relations of Γ may be recovered from R.)

(1.7). If Γ is finitely generated, then R has a finite number of sides (and conversely).

(1.8). ∂R meets the Real axis $\cup\{\infty\}$ in points called vertices (whose sides are paired by $A \in \Gamma$ which thus fixes the vertex) or intervals (called <u>free sides</u>) whose emanating sides are paired. See figure 1.

Now the connection with Riemann surfaces is simply this: if we glue R together as indicated by the pairing transformations of Γ, we get a 2-manifold. It has an analytic structure which it inherits from the plane and the fact that the A's $\in \Gamma$ are themselves conformal. Thus it is a Riemann surface.

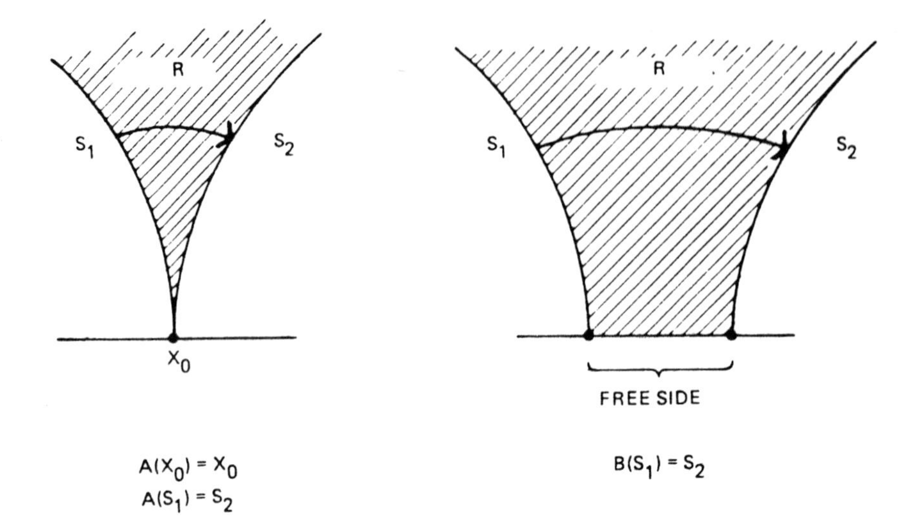

FIGURE 1

(1.9). $\cup_{A \in \Gamma} A(R) = H^+$, so that the (hyperbolically congruent) "tiles" $A(R)$ tessellate H^+. (Recall $A(R) \cap B(R) = \emptyset$; $A,B \in \Gamma$; $A \neq B$.) Further, with a finite number of simple exceptions (e.g., the complex plane), every Riemann surface arises in this way. One can study Riemann surfaces by studying Fuchsian groups.

We close this section by describing how the trajectories arise in this context. Let the Riemann surface be S. It can be represented by a Fuchsian group Γ so that the glued fundamental region (usually written H/Γ) is S. Now as we have said, H has a hyperbolic metric. Because $SL(2,\mathbb{R})$ acts isometrically on H, this metric may be projected onto S. The pull back to H of a trajectory on S is a trajectory on H, i.e. a hyperbolic line*[2]. We will study trajectories on S via their pull backs to H.

§2. AN EXAMPLE

Every number theorists' favorite Fuchsian group: $SL(2,\mathbb{Z})$ (the modular group), a.k.a. $\Gamma(1)$.

This group has a fundamental region defined as $R = \{z \mid |z| > 1;\ -1/2 < \operatorname{Re} z < 1/2\}$:

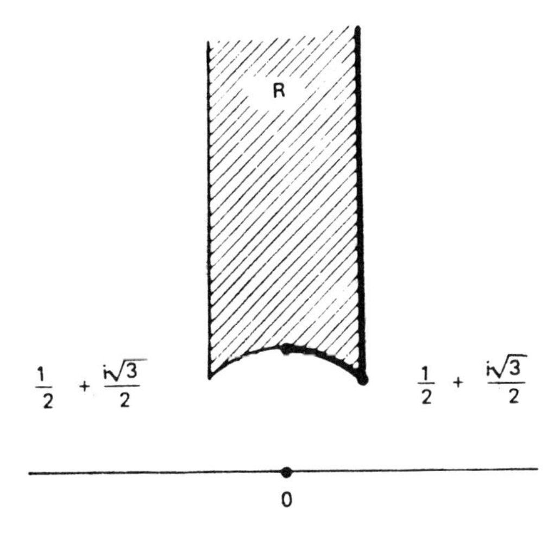

FIGURE 2

(A fundamental set is given by the heavily shaded portion of the boundary $\cup$ R.)

$\Gamma(1)$ is generated by $T = \begin{pmatrix}1 & 1\\ 0 & 1\end{pmatrix}$ and $S = \begin{pmatrix}0 & 1\\ -1 & 0\end{pmatrix}$, that is $z \overset{T}{\to} z + 1$ and $z \overset{S}{\to} -1/z$. T^{-1} takes the heavily shaded vertical side into the light one and S takes the heavily shaded circular side into the light one. Thus $H/\Gamma(1)$ is a once punctured (at ∞) sphere--not a very exciting (or generic) Riemann surfact.

Next we sill describe the tesselation of H by $\Gamma(1)$. "At $i\infty$", the powers of T (positive and negative) give (euclidean) translations of R (figure 3). The element $\begin{pmatrix}p & \cdot\\ q & \cdot\end{pmatrix}$[*3] takes ∞ to p/q. Thus the configuration of figure 3 is mapped to "a flower" at p/q: (figure 4) the height of this flower is about $1/q^2$, regardless of p. The collection of these flowers (including the one at ∞) fill out H. Each "petal" of the flower is a fundamental region for $\Gamma(1)$. The two sides of the petal anchored at p/q are paired by an element of $\Gamma(1)$ <u>fixing p/q</u> (just as $z \to z + 1$ fixes ∞). In fact, powers of this same element (positive and negative) fill out the whole flower. The other side is broken in two and each piece is paired with the other by a element fixing the break point, just as $z \to -1/z$ fixes i.

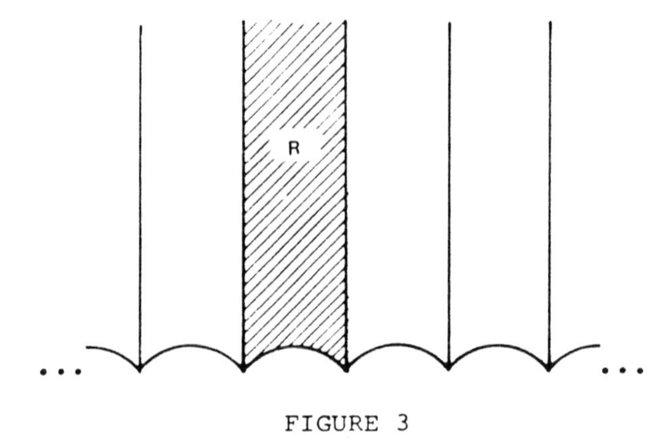

FIGURE 3

We now turn to the trajectories on $H/\Gamma(1)$. As we have said, this is tantamount to choosing a hyperbolic line in H. For simplicity at this stage, we assume the hyperbolic line (call it ℓ) is vertical i.e. the trajectory ends at ∞, the puncture. As the trajectory heads down to the real axis it passes through various flowers and petals thereof (see figure 5). Each one of these petals will have associated with it an element of $\Gamma(1)$ which maps it to the original fundamental region R of figure 2. Of course the segment of ℓ going through the petal maps to a segment going through R. Thus the entirety of ℓ can be mapped, by a sequence of transformations in $\Gamma(1)$, onto a set of segments passing through R^{*4}.

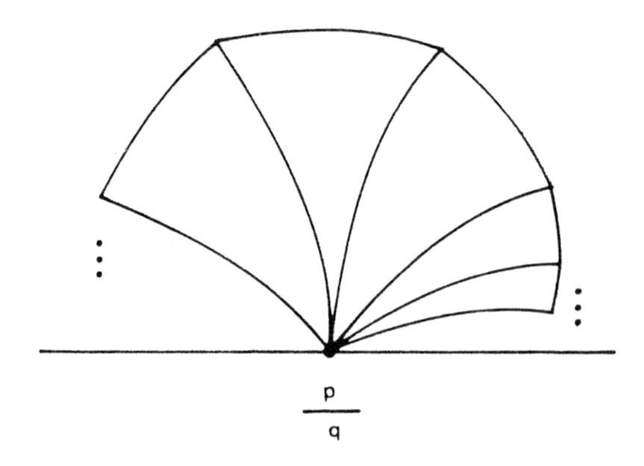

FIGURE 4

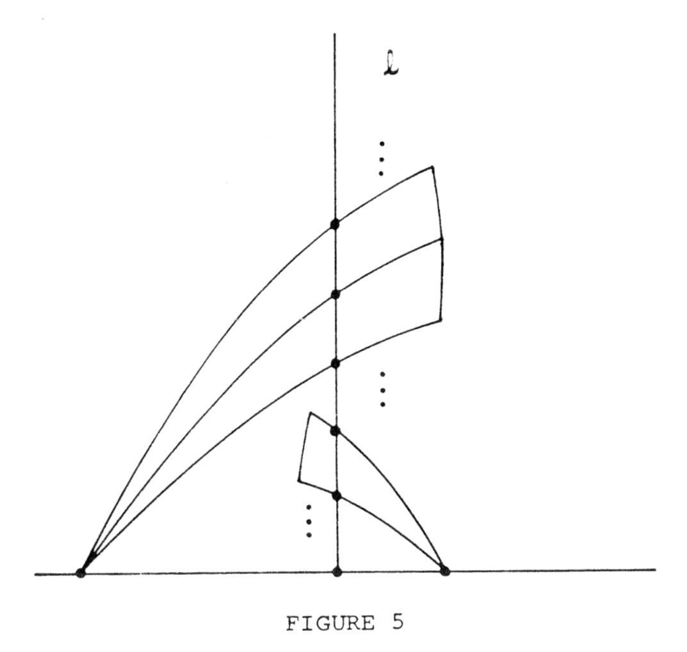

FIGURE 5

See figure 6. (If one now imagines R being glued together to form $H/\Gamma(1)$, the segments "become" the trajectory.)

Now the next step is to introduce a mechanism which makes the explicit computation of endpoints of the segments of ℓ and the transformations A and B possible. Since we are dealing here with $SL(2,\mathbb{Z})$, it is not surprising that the mechanism is number-theoretic--the continued fraction algorithm. That continued fractions could be applied in this context was noticed first, apparently, by Ford [F]*[5]. Artin [A], and Myrberg [My1], [My2] are the more usual citations. First we must set our notation.

Let ζ be any real number. Choose an integer a_0 such that $0 \leq \zeta - a_0 < 1$. At this point, it is easy to see that

$$\zeta - a_0 = \cfrac{1}{a_1 + \cfrac{1}{a_2 + \cfrac{1}{a_3 + \dots}}} \qquad a_i \geq 1;\ i = 1,2,3,\dots$$

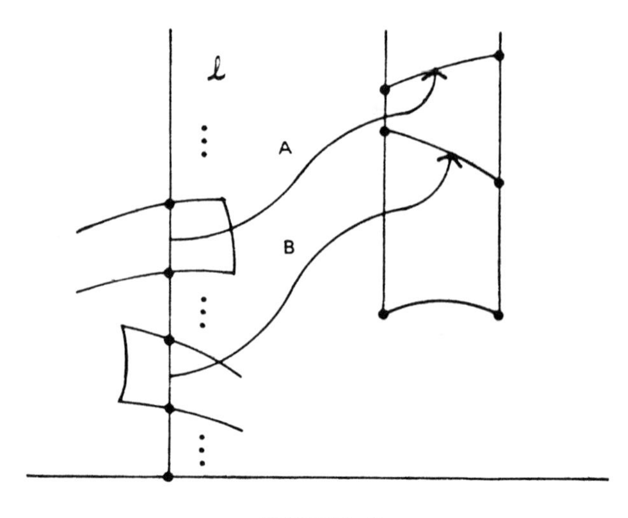

FIGURE 6

where $\zeta - a_0$ uniquely determines the integers $a_i \geq 1$.*6 So

$$\zeta = a_0 + \frac{1}{a_1 + \dots} \tag{2.1}$$

This is called the continued fraction expansion of ζ. It is finite if and only if ζ is rational. (2.1) is usually written $\zeta = [a_0, a_1, a_2, \dots]$. The a_i's are called <u>partial denominators</u> of ζ. The fraction $p_n/q_n = [a_0, a_1, \dots, a_n)$ is called (the nth) <u>convergent</u> of ζ. Finally it is a fact that

$$p_n q_{n+1} - q_n p_{n+1} = (-1)^{n+1} \tag{2.2}$$

Details for all of the above can be found in Khinchin [K].

We are now ready to specify the ingredients of figure 6. In the following, a stream-lined, idealized version is presented. Yet it is accurate in its essentials. Call ζ the "base" of ℓ. Then the nth*7 segment on ℓ is $[\zeta + i/q_n^2, \zeta + i/q_{n+1}^2]$. The element of $\Gamma(1)$ mapping this to R is given by $A_n = \begin{pmatrix} \pm q_{n+1} & \pm p_{n+1} \\ q_n & -p_n \end{pmatrix}$, where the $\pm$ is chosen (consistant with (2.2)) as n is odd or even. See figure 7.

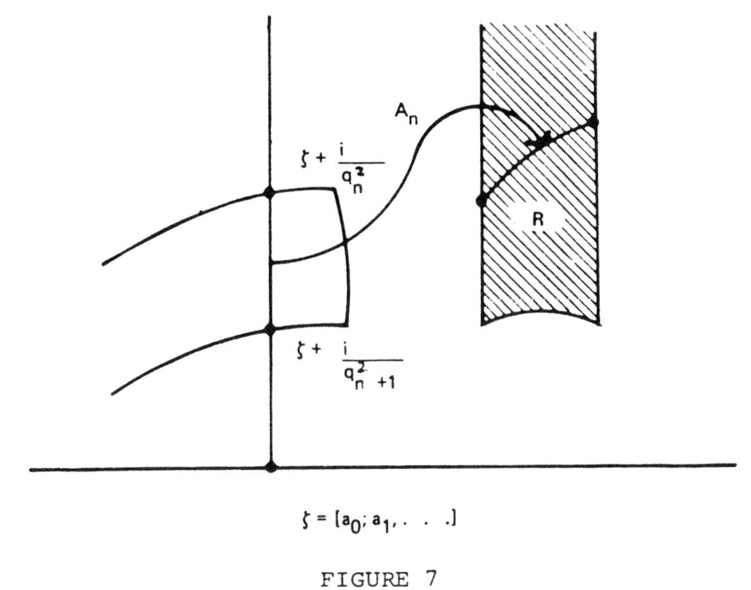

FIGURE 7

Thus we see that, through figure 7, the continued fraction expansion of ζ completely describes the course of the trajectory (in R) that ℓ represents. We close with a simple sort of inference that this insight allows.

The continued fraction algorithm also satisfies

$$q_{n+1} = a_{n+1}q_n + q_{n-1} \tag{2.3}$$

and

$$0 \leq \left|\frac{q_{n+1}\zeta - p_{n+1}}{q_n\zeta - p_n}\right| \leq 1 \tag{2.4}$$

Now (2.4) says $0 \leq |A_n(\zeta)| \leq 1$. Also, $a_{n+1} \leq |A_n(\infty)| = |q_{n+1}/q_n| \leq a_{n+1} + 1$, by (2.3) and the fact that q_n is monotone increasing. By (1.2), $A_n(\ell)$ is a hyperbolic line. We have just shown that diameter of the semicircle $A_n(\ell)$ is about a_{n+1}. Conclusion: if ζ has bounded partial denominators, the trajectory eventually leaves a neighborhood of ∞ in R. In particular, this trajectory could not be dense in R. Since there are uncountably many such ζ, there are uncountably many such trajectories.*[8] The above discussion is set out in more detail in Artin [A].

§3. GLOBAL PROPERTIES OF THE SET OF TRAJECTORIES ON A RIEMANN SURFACE

The preceding section gives an indication of the status of this study in the 20's. We might summarize it by saying that for a given surface a mechanism for describing the trajectories was created which permitted: (a) construction of trajectories having certain interesting properties and (b) characterization (in terms of the mechanism) of certain interesting classes of trajectories. See for example, Morse [Mo].

In the 30's a dramatic shift occurred. Instead of focussing on one trajectory and determining its path, the emphasis, clearly influenced by the burgeoning ergodic theory. was on showing that almost all trajectories had various kinds of regularity in the limit. This work produced a theory of enormous power and beauty*[9] But the computational aspects of the theory are obscure: can one tell whether a given trajectory is a "good" one, i.e. has the limiting behavior? How quickly does the limiting behavior manifest itself?

The names associated with these developments are Hedlund and Hopf [Ho]. Seidel [S3] should also be mentioned. Our account relies heavily on the excellent survey given in Hedlund [He].

We begin with the concepts of phase space Ω and the geodesic flow. Ω is the set fundamental $R \times [0,2\pi)$, and, thinking of $[0,2\pi)$ as having the euclidean topology and Lebesque measure, endowing this set with the product topology and metric. Ω is visualized as a point of S together with a direction. Consider a point $P \in \Omega$. P gives a point $p \in R$ and a direction θ. Choose the trajectory at p in the direction θ. For $-\infty < t < \infty$ proceed along this trajectory a (hyperbolic) distance t. This brings us to a new point p_t and a new angle with respect to the local x-axis, θ_t. See figure 8.

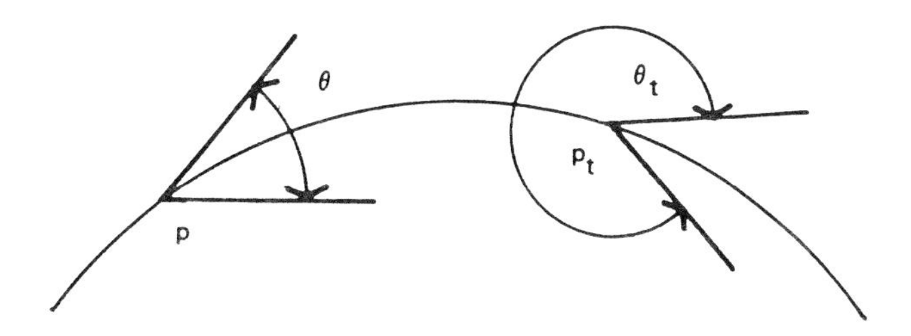

FIGURE 8

That is, this brings us to a different point of Ω, P_t. The mapping $T_t : P \to P_t$, $\forall P \in \Omega$ has three properties:

(i) $T_0(P) = P$; $T_t(T_s(P)) = T_{t+s}(P)$

(ii) $T_t(P)$ is continuous in t and P

(iii) If A is μ-measurable then $\mu(A) = \mu(T_t(A))$. Here μ is the measure defined on Ω.)

Thus T is a <u>flow</u>. It is called the <u>geodesic flow</u>. Now for fixed $P \in \Omega$, the set $\{P_t | -\infty < t < \infty\}$ is called a <u>motion</u> (or trajectory) in Ω.

We will now describe some representative sorts of regularity for the geodesic flow.

(A) REGIONAL TRANSITIVITY. Given D and D^*, two open sets in Ω, there exists a t such that $D_t \cap D^* \neq \emptyset$. (Here $D_t = \{P_t | P \in D\}$.)

(B) METRIC TRANSITIVITY. Given M and M^*, two measurable sets of positive measure in Ω, there exists a t such that $M_t \cap M^* \neq \emptyset$.

(C) MIXTURE. Given M, M^* and M^{**}, three sets of positive measure in Ω, $\lim_{t\to\pm\infty} \mu(M_t \cap M^*)/\mu(M_t \cap M^{**}) = \mu(M^*)/\mu(M^{**})$. Clearly (C) $\Rightarrow$ (B) $\Rightarrow$ (A).

Now (A) and (B) can be restated in terms of trajectories:

(A'). There exists a dense trajectory. (This means that the points of the motion are dense in Ω.)

(B'). For almost all trajectories, the mean time of sojourn in a measurable set M is equal to $\mu(M)/\mu(\Omega)$.

REMARKS. The mean time of sojourn is defined as $\lim_{\beta-\alpha\to\infty} L_{\alpha,\beta}(P,M)/(\beta-\alpha)$ where $L_{\alpha,\beta}(P,M)$ is the <u>linear</u> measure of $M \cap \{P_t | \alpha \leq t \leq \beta\}$. We assume in this definition that $\mu(\Omega)$ is finite.

Mixture can also be stated in "local" terms, but to do this one needs to introduce a second flow--the horocylic flow-and this would take us too far afield.

Now the major results are these.

1. If Γ is such that $S = H/\Gamma$ has finite area, then the geodesic flow is a mixture.

2. If Γ has no free sides, then the geodesic flow on S is regionally transitive.

3. If Γ has free sides, and thus H/Γ has infinite area,*[10] then almost all trajectories eventually leave and compact subset of S forever.

In their fullest generality, all these results are due to Hedlund (see [He] for the history of the problem). However one class of intermediate results deserves special mention. Hopf [Ho] showed that the geodesic flow is metrically transitive when H/Γ has finite area. He did this with a very novel method.*[11] Namely, using the fact that the failure of metric transitivity is equivalent to the existence of a Γ-invariant set in Ω which is neither of zero nor full measure, he studied this condition in terms of the existence of bounded, harmonic, Γ-invariant (or <u>automorphic</u>), functions. This study leads to the aforementioned result elegantly. But it is rather removed from an analysis of the behavior of individual trajectories.*[12]

Hedlund's proofs are in the tradition of the 20's as described at the outset of this section. Still, much of the analysis takes place on the real line--the terminal point of the trajectory. Indeed much of the technique does not depend on which trajectory one is using, as long as it has a given terminal point. As such, it does not shed much light on individual trajectories. There are no computations involving specific trajectories.

The preceeding paragraph is meant as observation and not disparagement. Nothing can obscure the obvious fact that the work described in this section was a triumph of its time. It is nothing less than some of the best work of some of the best mathematicians of the day.

§4. CONTINUED FRACTIONS FOR GENERAL FUCHSIAN GROUPS

In this section we give a definition of the continued fraction expansion of a trajectory in H/Γ. The expansion is given in terms of the group Γ. It bears unmistakable similarity to the expansions of Myrberg and Morse, though it is not the same and was arrived at by analogy to $SL(2,\mathbb{Z})$. We shall describe how the situation reduces to the standard continued fractions for $SL(2,\mathbb{Z})$ and then state some applications in that case. Finally, we shall give some applications of the general case and suggest the direction of our future study. At the outset let me state that this mechanism*[13] is meant to allow the study of individual trajectories--their properties and approximations and the construction of trajectories with special properties.

We begin by introducing our continued fraction expansion (mod Γ) of the trajectory ℓ. Fix a fundamental region R_0 in H. A trajectory on $S = H/\Gamma$ determines a unique hyperbolic

line ℓ through R_0. Extend this line to the real axis in each direction. This extension passes through a (perhaps doubly infinite) ordered sequence of fundamental regions $\ldots R_{-2}, R_{-1}, R_0, R_1, R_2, \ldots$.*[14] See figure 9.
Now for each R_n in the sequence, there is a $T_n \in \Gamma$ such that $T_n : R_n \to R_{n-1}$. We define the continued fraction of ℓ mod $\Gamma = CF_\Gamma(\ell) = \{\ldots, T_{-1}, T_0, T_1, \ldots\}$.

Let us take a look at the situation for $\Gamma(1) = SL(2,Z)$. We agree to denote by U, an element of $\Gamma(1)$ that pairs two "long" sides of a petal (and fixes p/q) and by V an element of $\Gamma(1)$ that pairs a broken side of a petal (and fixes the break point). For instance, $\begin{pmatrix} 1 & 1 \\ 0 & 1 \end{pmatrix}$ is a U and $\begin{pmatrix} 0 & 1 \\ -1 & 0 \end{pmatrix}$ is a V. See figure 10.

A moment's thought reveals that the line ℓ must enter the flower at p/q through a V paired side, pass across a number of U paired sides and exit through a V paired side, if it doesn't terminate at p/q. How many U paired sides does it go through? As we have seen in §2, $p/q = p_n/q_n$ for some n (here we think of ℓ as terminating at ζ). The mapping V_n takes p_n/q_n to ∞, and the flower at p_n/q_n to the flower at ∞ (see figures 3 and 4). Moreover the euclidean radius of $V_n(\ell)$ is about a_{n+1}, meaning $V_n(\ell)$ passes through about a_{n+1} U-paired sides at ∞. We now see that $CF_{\Gamma(1)}(\ell)$

$$= (\ldots, V, \underbrace{U, \ldots, U}_{a_1\text{-times}}, V, \underbrace{U, \ldots, U}_{a_2\text{-times}}, V, \underbrace{U, \ldots, U}_{a_3\text{-times}}, \ldots).$$

That is, the nth

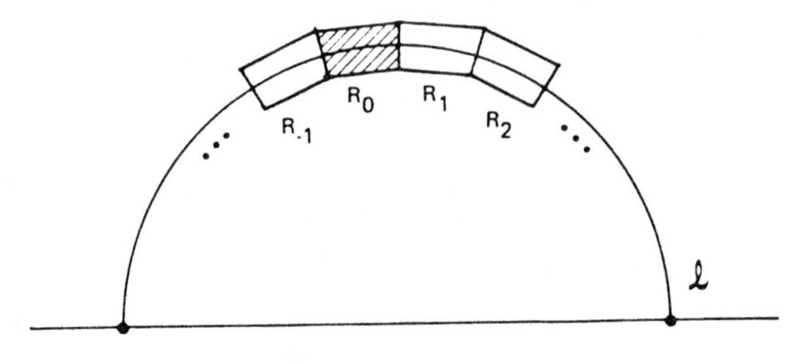

FIGURE 9

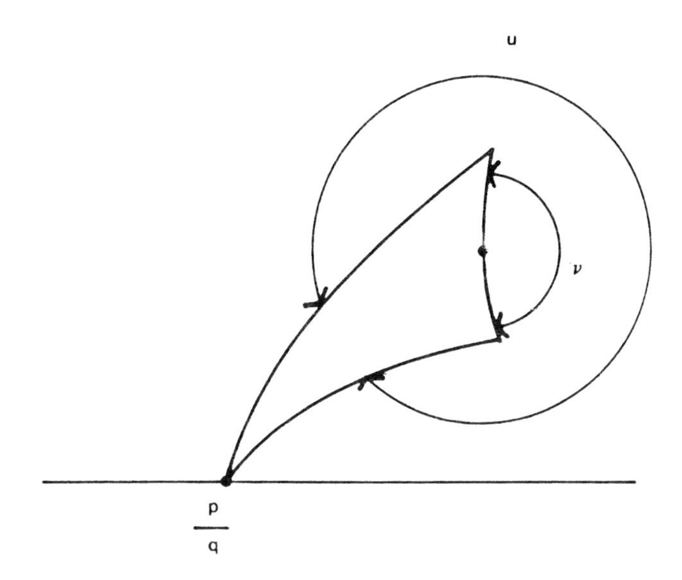

FIGURE 10

partial denominator of ζ reads off the number of U transformations necessary to get out of the n-th flower. To make the correspondence between the standard CF's and $CF_{\Gamma(1)}$ more complete, one should note that $\zeta = [a_0, a_1, a_2, \ldots]$ should be thought of as corresponding to the vertical line ending at ζ and is thus left-finite (as it ends at ∞). Also, a_0 tells the number of U's necessary to get to R_0.

We shall now describe two results for $\Gamma(1)$ that can be obtained using this mechanism. For complete details see [Sh1] and [Sh2], respectively.*[15]

1. There are uncountably many trajectories whose paths are dense in $S = H/\Gamma(1)$ but not dense in phase space of S.

2. There are uncountably many real numbers ζ such that $|\Delta(\zeta+iy)| \to \infty$ as $y \to 0$. (Here $\Delta(z) = e^{2\pi iz} \prod_{m=1}^{\infty} (1-e^{2\pi imz})^{24}$ is the unique modular form of weight -12. Notice that if ζ were rational, the above limit would be zero.)

We now turn to a discussion of arbitrary Fuchsian groups. Since some of the arguments below do not appear elsewhere, we shall give a fuller account. Accordingly, we must introduce more precise definitions. A Fuchsian group is of the _first_ (resp. _second_) kind if its fundamental region described in §1 does not (resp. does) have any free sides. An element V of $SL(2,\mathbb{R})$ may have one or two fixed points. If it has one fixed point, that fixed point is on the real-line (or ∞) and V is called parabolic. If it has two unequal fixed points, they may both be real (or ∞) and V is called _hyperbolic_, or they may be complex conjugates and V is called _elliptic_. The _limit set_ L of a Fuchsian group Γ is $\{z_0 | \exists z_1 \in H, A_i \in \Gamma, A_i \neq A_j (i \neq j), \text{ s.t. } A_i(z_1) \to z_0\}$. Such z_0 are necessarily real (or ∞). $H^+ - L$ is called $\mathcal{O}$, the _ordinary set_ of Γ.

Our objective is to explicitly construct dense trajectories on surfaces represented by arbitrary Fuchsian groups of the first kind and their analogues for groups of the second kind. We begin with some lemmas.

LEMMA 1. (See Hedland [He], p. 248.) Let Γ be a Fuchsian group and ζ_1, ζ_2 be two limit points. Let θ_1 and θ_2 be neighborhoods in $\mathbb{R} \cup \{\infty\}$ of ζ_1, ζ_2 resp. Then there exists a hyperbolic transformation in Γ with one fixed point in θ_1 and one fixed point in θ_2.

Proof. A conjugation argument shows that we may assume $0 \in \theta_1$ and Γ contains a hyperbolic element B fixing 0 and ∞. Write $B = \begin{pmatrix} \rho & 0 \\ 0 & \rho^{-1} \end{pmatrix}$ where we may assume $\rho > 1$. Since $\infty \in L$ we may find an element $A = \begin{pmatrix} a & b \\ c & d \end{pmatrix} \in \Gamma$ with $A(\infty) = a/c (\neq 0,\infty) \in \theta_2$. (See Beardon [Be], p. 43 for a check list including this and other properties of the ordinary and limit sets.) A direct calculation shows that the fixed points of $A \circ B^m$ $(m > 0)$ are

$$\frac{a - \rho^{-2m}\bar{d}+\bar{a}\sqrt{1+o(1)}}{2c} \to \begin{cases} \frac{a}{c}, & + \text{ sign} \\ 0, & - \text{ sign} \end{cases}$$

as $m \to \infty$. Since $0 \in \theta_1$ and $a/c \in \theta_2$ we are done. (Note $ac \neq 0$ as $a/c \neq 0,\infty$. $o(1)$ means a term which $\to 0$ as $m \to \infty$.)

Consider a finite section of a trajectory ℓ through R_0. This determines a finite piece of $CF_\Gamma(\ell)$: $(T_{-m},\dots,T_{-1},T_0, T_1,\dots,T_n)$. There are many other trajectories through R_0 which contain this same finite piece. The set of all of them may be thought of as the trajectories emanating from one interval 0_1, on the real axis and terminating in another 0_2. See figure 11. In this way, each finite section determines two open sets.

LEMMA 2. Let ℓ_1 and ℓ_1^* be two trajectories ending, on either end, in limit points. Take finite sections $[T_{-m},\dots,T_n]$ and $[T^*_{-m^*},\dots,T^*_{n^*}]$. Let the intervals corresponding to these be $0_1,0_2$ and $0_1^*,0_2^*$. Then there is an element $V \in \Gamma$ with $V(0_i^*) \cap 0_i \neq \emptyset$, $i = 1,2$.

Proof. As in Lemma 1, we may choose $A \in \Gamma$ with $A(0_2^*) \cap 0_2 \neq \emptyset$. (Recall 0_2 has a limit point.) By Lemma 1,

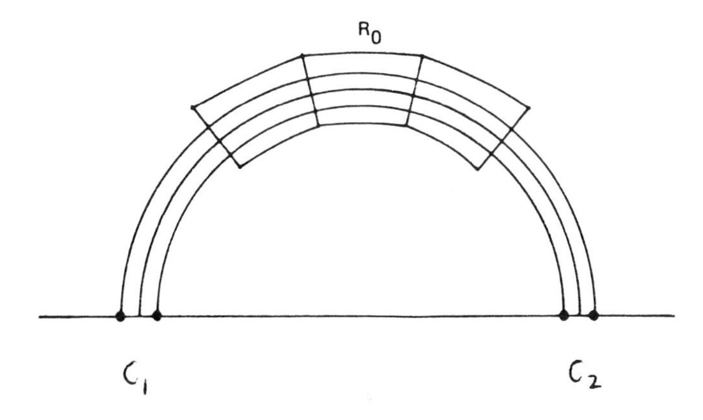

FIGURE 11

there is a $B \in \Gamma$ with fixed points $h_1 \in A(O_1^*)$ and $h_2 \in A(O_2^*) \cap O_2$. Since $h_1 \in A(O_1^*)$, there is an m such that (a): $B^m \circ (O_1^*) \cap O_1 \neq \emptyset$. Also, the fixed points of B^m are those of B, so $B^m \circ \{A(O_2^*) \cap O_2\} \cap \{A(O_2^*) \cap O_2\} \neq \emptyset$. This last relation implies (b): $B^m \circ A(O_2^*) \cap O_2 \neq \emptyset$. Now (a) and (b) establish the lemma with $V = B^m \circ A$.

Now starting with two trajectories ending in limit points and two "neighborhoods" about them we have shown how a subneighborhood of the first may be mapped by Γ into the second neighborhood. This is an effective procedure, all the calculations have been explicit. Iterating this procedure on a countable number of second neighborhoods, we come to

THEOREM 1. Given any trajectory connecting two limit points we can construct a trajectory arbitrarily close to it that can be mapped by Γ arbitrarily close to any trajectory connecting two limit points.

REMARKS. 1. "Arbitrarily close" means in the sense of these O intervals on the real axis, not the hyperbolic metric.

2. For groups of the first kind this just says that the "dense trajectories" are themselves dense in the set of trajectories (phase space).

3. The proof as it is does not show that almost all trajectories are dense for groups of the first kind. This is because if we examine the set of trajectories that do not contain a given sequence $[T_{-m},\ldots,T_n]$ or any Γ equivalent in its CF_Γ, we have an invariant Cantor set. The problem is to show that it has measure zero.

4. This approach is modeled on Artin's in [A]. In his case, it is amusing to note, Lemma 2 reduces to the triviality: given two sequences of integers $(a_1,\ldots,a_n)$ and $(b_1,\ldots,b_m)$, there is a CF containing the sequence $(a_1,\ldots,a_n,b_1,\ldots,b_m)$.

Also, he can rely on a number theoretic theorem of Burstin [Bu] to give the result of remark 3, above.

Before leaving groups of the first kind, we should note that if one could show (as I hope to) that the Cantor set in remark 3 was not empty one would have achieved a dramatic generalization of Myrberg's result asserting the uncountability of the non-dense trajectories for Γ finitely generated by parabolics. Again, in Artin's case this is trivial. It is easy to produce uncountably many CFs such that $a_n \neq 5$, any n.

While the above remark indicates that this method is of interest for groups of the first kind for reasons other than computational, it is for groups of the second kind that some new vistas seem to unfold.

As we have said in §3, for groups of the second kind, the classical theory simply states that almost all trajectories eventually leave any compact set on S. But are there any interesting ones that don't? Theorem 1 asserts that there are. Indeed it says that there exists trajectories that approximate any h-line connection two limit points. I.e., there exists a trajectory that traces out the boundary of the _Nielsen convex region_*[16] N _of_ Γ. In this regard these are a few of the several directions that may be pursued: (i) determine the distribution in N of the path of these "non-escaping" trajectories; (ii) find the nature of the set in N which is (perhaps eventually) free of them*[17]

NOTES

*[1] Actually since $\begin{pmatrix} -a & -b \\ -c & -d \end{pmatrix}$ and $\begin{pmatrix} a & b \\ c & d \end{pmatrix}$ give the same mapping, we should say $\begin{pmatrix} a & b \\ c & d \end{pmatrix} \in SL(2,\mathbb{R})/\{\pm I\}$.

*[2] Since the projection from H to S is not unique (it can be proceeded by any element of Γ, sometimes called in this context

a deck-transformation), we actually have (an equivalence) class of hyperbolic lines.

*[3] $\begin{pmatrix} p & \cdot \\ q & \cdot \end{pmatrix}$ means choose any integers for the dots which renders the matrix in $\Gamma(1)$. This can be done since $(p,q) = 1$.

*[4] In fact it is easy to see that this set of segments is independent of which ℓ in the set of hyperbolic lines representing the original trajectory on $H/\Gamma(1)$ we use.

*[5] I am indebted to G.V. Chudnovsky for this reference.

*[6] Actually, if ζ is rational, the sequence is finite. If it terminates in a_n we may always write ζ as the same sequence terminating in $a_n - 1,1$.

*[7] There are a finite number of segments rising vertically because ℓ emanates from figure 3.

*[8] Actually we must show there are uncountably many such ζ not equivalent under $SL(2,\mathbb{Z})$. This is easy to do using Hardy and Wright [HW], p. 142.

*[9] So beautiful in fact, that until quite recently the theory for Riemann surfaces was though to be complete.

*[10] Γ has no free sides does not imply H/Γ has finite area. This is the case, however, if we add the additional hypothesis that Γ is finitely generated.

*[11] Seidel [Se] used this technique to arrive at a weaker result in short order.

*[12] In this connection, see Hopf's comment in [Ho], top of p. 300.

*[13] Usually called a symbolic dynamics.

*[14] If ℓ happens to coincide with a side of an R_n, this sequence is not unique. We will not discuss the slight modifications necessary to handle this.

*[15] Result 2 below has recently been extended to more general Fuchsian groups (having one equivalence class of parabolic fixed points) by P. J. Nicholls in The Boundary Behavior of Automorphic Forms.

*[16] The Nielsen Convex Region is the interior of the set of lines connecting the limit points of a group of the second kind. It has recently enjoyed a resurgence due to Bers, Earle and Thurston.

*[17] Maskit has provided the author with a group which has such a free set in its Nielsen convex region.

REFERENCES

[A] E. Artin, Ein mechanisches System mit quasiergodischen Bahnen, Abh. Math. Sem. Hamburg 3(1923), 170-175.

[Be] A. Beardon, The Geometry of Discrete Groups, in Discrete Groups and Automorphic Functions, W.J. Harvey, ed., Academic Press, London, 1977.

[Bu] C. Burstin, Uber eine spezielle Klasse seeller periodischer Funktionen, Monat, für Math. 26 (1915), 229-262.

[F] L. Ford, Rational Approximations to Irrational Complex Numbers, Trans. Amer. Math. Soc. 19 (1918), 1-42.

[HW] G.H. Hardy and E.M. Wright, An Introduction to the Theory of Numbers, ed. 2, Oxford U. Press, Oxford, 1945.

[He] G.A. Hedlund, The Dynamics of Geodesic Flows, Bull. Amer. Math. Soc. 45 (1939), 241-260.

[Ho] E. Hopf, Fuchsian Groups and Ergodic Theory, Trans. Amer. Math. Soc. 39 (1936), 299-314.

[K] A. Ya. Khinchin, Continued Fractions, U. of Chicago Press, Chicago, 1964.

[L] J. Lehner, A Short Course in Automorphic Functions, Holt. Rinehart and Winston, New York, 1966.

[Mo] M. Morse, Recurrent Geodesics on a Surface of Negative Curvature, Trans. Amer. Math. Soc. 22 (1921), 84-100.

[My1] P.S. Myrberg, Einige Andwendungen der Kettenbrüche in der Theorie der binäsen quadratischen Formen und der elliptischen Modulfunktionen, Ann. Acad. Sci. Fenn. Ser. AI 23 (1924).

[My2] ____________, Ein Approximationssatz fur die Fuchsschen Gruppen, Acta Math. 57 (1931), 389-409.

[Se] W. Seidel, On a Metric Property of Fuchsian Groups, Proc. Nat. Acad. Sci. 21 (1935), 475-478.

[Sh1] M. Sheingorn, Boundary Behavior of Automorphic Forms and Transitivity for the Modular Group, Ill. J. of Math. 24 (1980), 440-451.

[Sh2] ____________, Transitivity for the Modular Group. Proc. Cambs. Phil. Soc. 88 (1980), 409-423.

ON THE ANALYTIC STRUCTURE OF DYNAMICAL SYSTEMS: PAINLEVÉ REVISITED

Michael Tabor*

Center for Studies of Nonlinear Dynamics
La Jolla Institute
La Jolla, California

1. INTRODUCTION

One of the main ideas underlying the work described in this seminar paper is due to the great Russian mathematician Sofya Kovalevskaya. Before going on to describe her work I would like to give a thumbnail sketch of the life and times of this very remarkable woman [1]. She was born in 1850, the middle child of a Russian general (retired) and landowner. (Her older sister was a beautiful and brilliant young woman who got herself involved in all sorts of radical causes and was at one time engaged to Dovstoevsky.) Sofya's great aptitude for learning was considerably thwarted by, as was fashionable at the time, a repressive English governess. Some of her early reading of mathematical texts had to be carried out under the bed clothes at night! When it was finally realized that she was something of a prodigy her father, very reluctantly, allowed her to undertake more formal studies. However, the

*Current Affiliation: Department of Applied Physics and Nuclear Engineering, Columbia University, New York, New York

only way a young lady could get away from home in those days (and hence continue her studies at the university) was to get married. Such a marriage was arranged, with this end in mind, and she and her young husband set out for Germany where they both hoped to study. Unfortunately it was virtually impossible for a woman in those days to attend a university in Germany, let alone anywhere else in Europe. Fortunately Weierstrass took her under his wing and gave her private lessons. In order to obtain a doctorate (which was then essentially impossible for a woman to do) she wrote three dissertations instead of the usual one. These were finally accepted by the University of Gottingen without her having to make a public thesis "defense" (women were barred from such activities). This was, in fact, just the beginning of a remarkable career. Among other things she was awarded the Bordin prize of the Paris Academy of Sciences in 1888 for her classic work on the rigid-body problem— her work was considered so outstanding that the prize money was trebled. She finally became a professor at the University of Stockholm and was, as such, one of the first women in Europe to hold a senior university faculty position. She died of pneumonia at the tragically young age of 41.

Her classic work on the rigid-body problem concerned the solution of the Euler-Poisson equations [2], [3]

$$
\begin{aligned}
A\frac{dp}{dt} &= (B-C)qr - \beta z_0 + \gamma y_0 \\
B\frac{dq}{dt} &= (C-A)pr - \gamma x_0 + \alpha z_0 \\
C\frac{dr}{dt} &= (A-B)pq - \alpha y_0 + \beta x_0 \\
\frac{d\alpha}{dt} &= \beta r - \gamma q \\
\frac{d\beta}{dt} &= \gamma p - \alpha r \\
\frac{d\gamma}{dt} &= \alpha q - \beta p
\end{aligned}
\tag{1.1}
$$

where (p,q,r) and (α,β,γ) are the components of the angular velocity and direction cosines respectively, (A,B,C) the moments of inertia and (x_0,y_0,t_0) the position of the center-of-gravity. The system has three "obvious" first integrals of the motion (i) energy, (ii) angular momentum and (iii) a constant that follows from trigonometric considerations $(\alpha^2 + \beta^2 + \gamma^2 = 1)$. With further standard simplifications the complete solution of the problem in terms of quadratures boils down to finding a fourth first integral of the motion. Apart from a trivial case and two special case solutions, due to Euler and to Lagrange, a general solution at that time seemed to be unobtainable.

Kovalevskaya's study of this problem, in the 1880s, was in a certain sense revolutionary in that it was the first time a real dynamical problem was solved by the use of the theory of functions of a complex variable. Motivated by the work of other mathematicians at that time, she asked the question: "under what conditions will the only movable singularities exhibited by the solution, in the complex time plane, be ordinary poles?" That is, given a singularity at $t = t_0$, when can the solutions to (1.1) be written in the form of a Laurent series, i.e.,

$$p(t) = \frac{1}{(t-t_0)^n} \Sigma_{m=0}^{\infty} a_m (t-t_0)^m \tag{1.2}$$

$$q(t) = \ldots, \text{ etc.}$$

She found that this only occurred for four special cases: the trivial case, the Euler case, the Lagrange case and one other—now known as the Kovalevskaya case. This suggested that for this new case there might also be a fourth integral of the motion. This she was indeed able to find; although it involved some pioneering and virtuoso work with hyperelliptic functions.

Kovalevskaya's remarkable results suggested that there might be some deep connection between the analytic structure of the solutions of differential equations and their integrability; although even to date there seem to be no general theorems about this.

For a while, there was a lot of activity investigating the analytic properties of differential equations. Painlevé studied the class of second-order differential equations

$$\frac{d^2y}{dx^2} = F\left(\frac{dy}{dx}, y, x\right) \tag{1.3}$$

where F is analytic in x and rational in y and dy/dx, and found that there were 50 types whose only movable singularities were poles (the "Painlevé property"). Forty-four of these equations have solutions in terms of known functions (elliptic functions) and the remaining six have become known as the <u>Painlevé trancendents</u> [4]. (Painlevé himself had a remarkable life. One of the less important events was his flight as history's first aviation passenger [5]!)

In recent years there has been some renewed interest in these matters. In certain studies of the integrability of partial differential equations and the inverse scattering transform method, as well as some other areas (see, for example, the paper by Flaschka and Newell [6] and the remarkable series by Jimbo et al. [7]), the Painlevé transcendents keep on popping up. Here we describe some recent studies of the analytic structure of dynamical systems of current interest. These include the Henon-Heiles Hamiltonian [8]

$$H = \frac{1}{2}(P_x^2 + P_y^2 + x^2 + y^2) + x^2y - \frac{1}{3}y^3 \tag{1.4}$$

which has become a standard model for nonintegrable Hamiltonians and the Lorentz system [9]

$$\frac{dX}{dr} = \sigma(Y-X)$$

$$\frac{dY}{dt} = -XZ + RX - Y \tag{1.5}$$

$$\frac{dZ}{dt} = XY - BZ$$

which is a popular model for studying convective instability.

This seminar paper is a synthesis (essentially a "cut-and-paste" job) of a series of papers [10,11,12,13] written in collaboration with my colleagues John Weiss, Y. F. Chang and John M. Greene. To these coworkers I would like to express my most sincere thanks for such a stimulating and enjoyable collaboration. Support from the Office of Naval Research (ONR Contract N-00014-79-C-0537) and the Department of Energy (DOE Contract 10923) is gratefully acknowledged.

2. PAINLEVÉ ANALYSIS FOR THE HENON-HEILES SYSTEM

We write the Henon-Heiles Hamiltonian in the general form

$$H = \frac{1}{2}(P_x^2 + P_y^2 + x^2 + y^2) + Dx^2y - \frac{C}{3}y^3 \tag{2.1}$$

In the case D = C = 1 (2.1) reduces to its standard form [8]. The second order (Newtonian) equations of motion are

$$\ddot{x} = -x - 2Dxy \tag{2.2a}$$

$$\ddot{y} = -y - Dx^2 + Cy^2 \tag{2.2b}$$

We determine the leading order behavior of the solution at a singularity at time $t = t_*$ by making the substitution

$$x = a(t-t_*)^{\alpha}, \quad y = b(t-t_*)^{\beta}$$

and equating most singular terms. This leads to the pair of equations

$$\alpha(\alpha-1)a(t-t_*)^{\alpha-2} = -2Dab(t-t_*)^{\alpha+\beta} \tag{2.3a}$$

$$\beta(\beta-1)b(t-t_*)^{\beta-2} = Cb^2(t-t_*)^{2\beta} - Da^2(t-t_*)^{2\alpha} \tag{2.3b}$$

with the two sets of solutions:

CASE 1. $\alpha = -2 \qquad a = \pm\frac{3}{D}\sqrt{2 + 1/\lambda}$

$\beta = -2 \qquad b = \frac{-3}{D}$

where for notational convenience we set $\lambda = D/C$, and

CASE 2. $\alpha = \frac{1}{2} \pm \frac{1}{2}\sqrt{1 - 48\lambda} \qquad a = \text{arbitrary},$

$\beta = -2 \qquad b = \frac{6}{C}$

Since the most singular behavior supported by the equations of motion is t^{-2}, both branches of the case 2 singularities can only exist for $\lambda > -\frac{1}{2}$. For the Painleve property to be satisfied all leading-order behaviors must be integers, and this places restrictions on the values of λ in case 2. The first few values of λ leading to integer α in this case are $\lambda = -1/6, -1/2, -1, -5/3$ ... etc. Typically, case 2 introduces irrational values of λ and for $\lambda > 1/48$ the order becomes complex. In the standard case $C = D = 1 (\lambda = 1)$ we have

$$\alpha = \frac{1}{2} \pm \frac{i}{2}\sqrt{47} \qquad a = \text{arbitrary}, \tag{2.4a}$$

$$\beta = -2 \qquad b = 6 \tag{2.4b}$$

In order to proceed with Painlevé analysis we have to look for the so called resonances [14], i.e. the conditions under which arbitrary parameters may enter into a general power series expansion about $t = t_*$. Since we have two second order equations the solution must be characterized by four constants of integration. One of these is provided by the singularity (hopefully pole) position $t = t_*$. Starting with the case 1 leading orders and following the procedure of Ablowitz et al., we now set

$$x = \pm\frac{3}{D}\sqrt{2 + 1/\lambda}\; t^{-2} + pt^{-2+r} \tag{2.5a}$$

$$y = \frac{3}{D}\, t^{-2} + qt^{-2+r} \tag{2.5b}$$

where p and q are the arbitrary parameters (whose values are fixed by the constants of integration) and for notational convenience we have set $t_* = 0$. These expansions are substituted into the equations of motion (2.2) with only the most singular (dominant) terms included, i.e.

$$\ddot{x} = -2Dxy \tag{2.6a}$$

$$\ddot{y} = -Dx^2 + Cy^2 \tag{2.6b}$$

Setting up the ensuing linear equations for p and q one finds, after a little analysis, that these will be arbitrary if

$$\begin{vmatrix} (3-r)(2-r) - 6 & \pm 6\sqrt{2+1/\lambda} \\ \pm 6\sqrt{2 + 1/\lambda} & (3-r)(2-r) + 6/\lambda \end{vmatrix} = 0 \qquad (2.7)$$

Setting

$$\Theta = (3-r)(2-r) \qquad (2.8)$$

one finds two possible solutions

$$\Theta = 12 \qquad (2.9a)$$

and

$$\Theta = -6(1+1/\lambda) \qquad (2.9b)$$

These values of Θ determine the values of r (and hence the powers of t) at which the resonances occur.

For $\Theta = 12$ we find

$$r = -1 \text{ or } 6 \qquad (2.10)$$

The root $r = -1$ is always present in such analyses and represents the arbitrariness of t_* [14]. This, together with the root $r = 6$, provides us with two of the arbitrary parameters. For $\Theta = -6(1+1\lambda)$ we find

$$r = \frac{5}{2} \pm \frac{1}{2}\sqrt{1 - 24(1+1/\lambda)} \qquad (2.11)$$

From this results we see that four-parameter solutions can exist for $\lambda > 0$ or $\lambda < -\frac{1}{2}$. Furthermore the resonances are complex when $\lambda > 0$ or $\lambda < -24/23$ and the imaginary part becomes infinite when $\lambda \to 0^+$.

The resonance analysis may be repeated using the case 2 leading orders. Now the dominant terms in the equations of motion are

$$x = -2Dxy \tag{2.12a}$$

$$y = Cy^2 \tag{2.12b}$$

The analysis proceeds exactly as before and yields the roots

$$r = -1 \text{ and } 6 \tag{2.13a}$$

and

$$r = 0 \text{ and } r = \mp\sqrt{1 - 48\lambda} \tag{2.13b}$$

The upper and lower signs of the last members of (2.13b) are associated with those of the leading order behavior α. In either case the two values of $r + \alpha$ calculated from (2.13b) are equal to the two values of α. The root $r = 0$ corresponds to the arbitrariness of the associated leading order coefficient. Four-parameter solutions can only exist when $\lambda > -1/2$. For $\lambda > 1/48$ the leading orders and resonances are complex (the imaginary parts become infinite when $\lambda \to +\infty$). When $-1/2 < \lambda < 1/48$ the negative branch $\alpha_- = 1/2 - 1/2\sqrt{1 - 48\lambda}$ can define a four-parameter solution, in this range α_- and $r_+ = +\sqrt{1 - 48\lambda}$ are real. Finally, when $\lambda = 0$ the singularity in the x-variable disappears and the equations of motion are integrable.

In order to determine those λ-values for which the Painlevé property is satisfied, we require that all leading orders and resonances, for both case 1 and case 2, are integers. The only values of λ for which this can occur are

$$\lambda = -1/6, -1/2, -1$$

However, at this stage, we have only determined that the movable signularities that occur for these values of λ are not algebraic branch points. There is still the possibility of logarithmic terms entering the expansion, and each case must be checked (by examining the associated recursion relations) for this eventuality. A detailed discussion of the role of logarithmic corrections is discussed, in the context of the Lorenz system [10], in Sec. 4.

The value $\lambda = -1$ gives the roots $r = -1,2,3,6$ for the resonances of the case 1 singularities. A detailed analysis of the expansion about the singularity demonstrated that the solution is Painlevé (single-valued) and depends on four arbitrary parameters. This implies that the system is integrable and in this case the integrals of motion have been known for some time [15].

The value $\lambda = -1/2$ is rather peculiar in that the coefficient in the first term of (2.5a) (case 1) vanishes. The resonances for case 1 are $r = -1,0.5,6$. The root $r = 0$ corresponds to the vanishing of the coefficient. What happens is that at $\lambda = -1/2$ the case 1 singularity merges with the positive ($\alpha = 3$) branch of the case 2 singularity. The negative branch ($\alpha = -2$) is undefined at this point. Thus the "leading orders" are: $x = at^3$, $y = -3t^{-2}$, where a is arbitrary. There is one resonance at $r = 6$ that introduces one further parameter. Detailed analysis of the expansion about a singularity shows that this is a three-parameter, Painlevé solution and, hence, not the general four-parameter form of the solution.

Finally, we consider the value $\lambda = -1/6$. The resonances of the case 1 singularity are $r = -3,-1,6,8$. This implies, and detailed calculation confirms, that the case 1 singularities are associated with a three-parameter, Painlevé form of the solution. On the other hand, for the case 2 singularities ($\alpha = -1,2$), we find a four-parameter, Painlevé form of solution associated with negative ($\alpha = -1$) branch. By numerical investigation of this case we have found only this four-parameter form of the solution to be present. Motivated by this numerical coincidence, John Greene was able to identify the additional integral of motion for this case, thereby confirming its integrability. The two integrals of motion are the energy:

$$E = H = \frac{1}{2}(P_x^2 + P_y^2 + Ax^2 + By^2) + x^2y + 2y^3 \qquad (2.14)$$

(where for generality we include the variable linear frequencies A and B) and the quantity

$$G = x^4 + 4x^2y^2 - 4\dot{x}(\dot{x}y - \dot{y}x) + 4Ax^2y + (4A - B)(\dot{x}^2 + Ax^2) \qquad (2.15)$$

The Painlevé properties of this case, i.e., $\lambda = -1/6$, and the case $\lambda = -1$ have been derived in [16]. This reference also uses, successfully, the Painlevé analysis to determine the integrability of the Toda lattice.

We now introduce the concept of a canonical resonance. In the normal search for resonances one starts with equations (2.5) which utilize the most singular leading order behaviors; in this case $\alpha = \beta = -2$ (case 1). One then proceeds to find the powers of t (that is $r - 2$) at which the parameters p, q enter the expansion. From (2.11) we see that this power is

$$\frac{1}{2} \pm \frac{1}{2}\sqrt{1 - 24(1+1/\lambda)} \qquad (2.16)$$

By canonical resonance we mean those cases when the power of t at which the resonance occurs is identical to the second possible leading order behavior (case 2). Comparing the square roots in (2.17) and case 2 the only values of λ for which this can occur are

$$\lambda = 1 \text{ and } \lambda = -1/2$$

The case $\lambda = -1/2$ results in a leading order/resonance at $\alpha = 3$ (the root $\alpha = -2$ is discarded). The case $\lambda = 1$ corresponds to the imaginary leading-order given in (2.4a). The significance of this canonical resonance is that the associated analytic structure has a particularly symmetric form. This idea will be illustrated below.

We conclude this section by briefly describing some of the results of our numerical investigations of the analytic structure in the complex t-plane. Many numerical integration techniques proceed with little knowledge of the precise positions and/or orders of the singularities encountered in the complex solution plane. Here we use a Taylor series method that yields detailed information concerning the singularity nearest to the point of integration [17]. The method is automatic in that one only needs to enter a statement of the o.d.es and such control parameters as initial conditions and path of integration. All of the results discussed below were obtained with this method, hereafter referred to as the ATSMCC method [17]. Applied to the Henon-Heiles system ATSMCC was able to locate the positions of the singularities to a high degree of accuracy and evaluate their orders in agreement with our leading-order analysis to four-figure accuracy or better.

Here we will only describe the structure of the singularities that occurs when $\lambda = 1$ (canonical resonance). When the solution is expanded at various points along the real-time axis there is found a nonuniform row of seemingly isolated singularities [see fig. 1]. (We specify the initial data

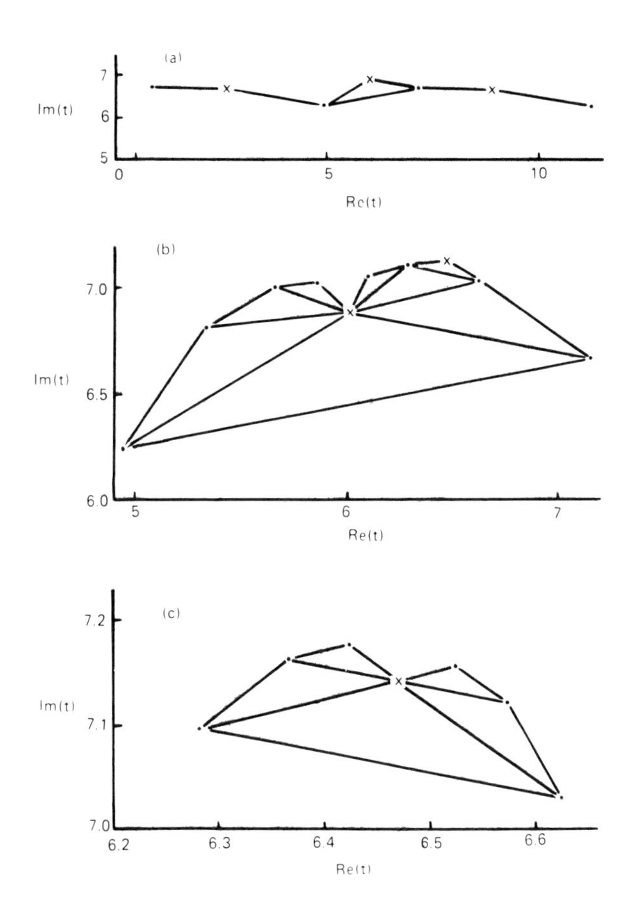

Figure 1. Analytic continuation of x(t) for $\lambda = 1$. (a) Sequence of singularities found from the real axis and one singularity found at the first stage of analytic continuation. = singularity of (leading order -2 and x = singularity of order $\pm$ 47. (b) Boxed region of (a) in more detail showing double spiral of singularities about apex of "triangle." (c) Boxed region of (b) in more detail showing self-similar nature of the double spiral of singularities. Analytic continuation of y(t) is identical but all singularities now have order -2.

so that the motion is bounded for real time and the singularities are a finite distance from the real-time axis.) However, when the path of integration is deformed into the complex-time plane and passes between two of the singularities observed from the real-time axis, there is found a third singularity located at the apex of an (approximately) isosceles triangle whose base is the line joining the two singularities that are on either side of the path of analytic continuation (see fig. 1]. If this base consists of two order -2 singularities, the singularity at the apex is of order 1/2. On the other hand, if the base consists of an order -2 and an order 1/2 (order refers to the real part of the leading order) singularity, there is found an order -2 singularity at the apex. The base angle is found to be approximately 25°.

Furthermore, when one integrates between _any_ pair of singularities that are observed to be "neighboring" during the process of analytic continuation, the above construction is repeated. Several levels of structure are implies by this "self-similar" process. One is that the set of singularities consists of a closed, perfect set with no isolated points on the multi-sheeted Riemann surface. Another is that about any singularity there emanates a double spiral (one clockwise, one anticlockwise) [see fig. 1]. Finally, since the base between the neighboring singularities is contracting geometrically at the successive stages of the analytic continuation process, it is impossible to continue the solution beyond more than a given finite distance in any direction beyond a pair of "base" singularities. That is, assuming one does not retrace the original path, any path of analytic continuation between a pair of singularities (on the same side of the real axis) will be trapped in a geometrically converging web of singularities that creates a _natural boundary_ of the solution. Using the self-similar nature of the above construction, we have estimated the "fractal dimension" of the singular set to be 1.1419 (this calculation is described in ref. [11]).

What is particularly striking about the singular structure for $\lambda = 1$ is its highly symmetric form when compared to those found for other values of λ, and we will show that this is a consequence of the value $\lambda = 1$ corresponding to a "canonical resonance." As the parameter λ is varied the natural boundary described above undergoes some remarkable changes. These are described in detail in [12]. We only mention here that for those λ values for which the system is integrable ($\lambda = -1, -1/6, 0$) the singularity structure becomes a single sheeted, regular lattice of poles, i.e., just like that found for the elliptic functions.

3. EXPANSIONS ABOUT A SINGULARITY, ASYMPTOTIC BEHAVIOR AND NATURAL BOUNDARIES

In this section we examine the expansions of the solutions about a singularity for Henon-Heiles system. Here we work with the Hamiltonian

$$H = \frac{1}{2}(P_x^2 + P_y^2 + x^2 + y^2) + \lambda x^2 y - \frac{1}{3}y^3 \qquad (3.1)$$

and equations of motion

$$x = -x - 2\lambda xy \qquad (3.2a)$$

$$y = -y - \lambda x^2 + y^2 \qquad (3.2b)$$

Restating the results of the previous section we have two possible leading orders:

CASE 1. $\alpha = -2,\qquad a = \frac{\pm 3}{\lambda}\sqrt{2 + 1/\lambda}$

$\beta = -2,\qquad b = \frac{-3}{\lambda}$

CASE 2. $\alpha = \frac{1}{2} \pm \frac{1}{2}\sqrt{1 - 48\lambda}\qquad a = \text{arbitrary}$

$\beta = -2,\qquad b = 6$

For case 1, the resonances occur at

$$r = -1, 6, \frac{5}{2} \pm \frac{1}{2}\sqrt{1 - 24(\frac{1}{\lambda} + 1)} \tag{3.3}$$

and for case 2 at

$$r = 0, -1, \pm\sqrt{1 - 48\lambda}, 6 \tag{3.4}$$

In subsequent discussions, it will sometimes be convenient to refer to the case 1 singularities of integer leading order as 'regular poles' and the other type of singularity (α, case 2) as the 'irregular poles.'

The general expansion of the solution about a singularity set at $t_* = 0$) takes the form of a double series. For case 1 leading orders,

$$x(t) = t^{-2}\Sigma_{k=0}^{\infty}\Sigma_{j=0}^{\infty} a_{kj}\tau^k t^j + t^{-2}\Sigma_{k=1}^{\infty}\Sigma_{j=0}^{\infty}\bar{a}_{kj}\tau^{-k}t^j, \tag{3.5a}$$

$$y(t) = t^{-2}\Sigma_{k=0}^{\infty}\Sigma_{j=0}^{\infty} b_{kj}\tau^k t^j + t^{-2}\Sigma_{k=1}^{\infty}\Sigma_{j=0}^{\infty}\bar{b}_{kj}\tau^{-k}t^j, \tag{3.5b}$$

where

$$\tau = t^{\alpha},\qquad \alpha = \frac{1}{2} + \frac{1}{2}\sqrt{1 - 24(\frac{1}{\lambda} + 1)} \tag{3.5c}$$

$$\bar{\tau} = t^{\bar{\alpha}}, \qquad \bar{\alpha} = \frac{1}{2} - \frac{1}{2}\sqrt{1 - 24(\frac{1}{\lambda} + 1)} \tag{3.5d}$$

and

$$a_{00} = \frac{\pm 3}{\lambda}\sqrt{2 + 1/\lambda}, \qquad b_{00} = \frac{-3}{\lambda}$$

For case 2 leading orders

$$x(t) = \tau \, \Sigma_{k=1}^{\infty}\Sigma_{j=0}^{\infty} a_{kj}\tau^k t^j + \bar{\tau}\, \Sigma_{k=0}^{\infty}\Sigma_{j=0}^{\infty} \bar{a}_{kj}\bar{\tau}^k t^j \tag{3.6a}$$

$$y(t) = t^{-2}\, \Sigma_{k=0}^{\infty}\Sigma_{j=0}^{\infty}\, b_{kj}\tau^k t^j \tag{3.6b}$$

$$+ \; t^{-2}\, \Sigma_{k=1}^{\infty}\Sigma_{j=0}^{\infty}\, \bar{b}_{kj}\bar{\tau}^k t^j$$

where

$$\tau = t^{\alpha}, \qquad \alpha = \frac{1}{2} + \frac{1}{2}\sqrt{1 - 48\lambda} \tag{3.6c}$$

$$\bar{\tau} = t^{\bar{\alpha}}, \qquad \bar{\alpha} = \frac{1}{2} - \frac{1}{2}\sqrt{1 - 48\lambda} \tag{3.6d}$$

and

$$a_{00},\ \bar{a}_{00} \text{ are arbitrary}, \qquad b_{00} = 6$$

These double series are valid, for both case 1 and case 2, so long as the α and $\bar{\alpha}$ are not rationally related. The exceptions to this occur for the λ values, $\lambda = -1, -1/2, -1/6$ for case 1 and $\lambda = -1, -1/2, -1/6, -6$ for case 2. We also note that the series (3.5) and (3.6) are only formal expansions (albeit self-consistent) and at this stage we know little about their convergence properties.

Substitution of the series expansions for the case 1 singularities into the equations of motion (3.2) leads (after much tedious manipulation) to the following set of recursion relations:

$$(\alpha k + j - 2)(\alpha k + j - 3)a_{kj} + a_{kj-2} \tag{3.7a}$$
$$+ 2\lambda \Sigma_{\ell=0}^{k} \Sigma_{m=0}^{j} \bar{a}_{k-\ell\ j-m} \bar{b}_{\ell m}$$
$$+ 2\lambda \Sigma_{n=1}^{j} \Sigma_{m=0}^{j-n} \{a_{k+n,m} \bar{b}_{n,j-n-m} + b_{k+n,m} \bar{a}_{n,j-n-m}\}$$
$$= 0$$

$$(\bar{\alpha} k + j - 2)(\bar{\alpha} k + j - 3)\bar{a}_{kj} + a_{kj-2} \tag{3.7b}$$
$$+ 2\lambda \Sigma_{\ell=0}^{k} \Sigma_{m=0}^{j} \bar{a}_{k-\ell,j-m} \bar{b}_{\ell m}$$
$$+ 2\lambda \Sigma_{n=0}^{j} \Sigma_{m=0}^{j-n} \{\bar{a}_{k+n,m} b_{n,j-n-m} + \bar{b}_{k+n,m} a_{n,j-n-m}\}$$
$$= 0$$

$$(\alpha k + j - 2)(\alpha k + j - 3)b_{kj} + b_{kj-2} \tag{3.7c}$$
$$+ \Sigma_{\ell=0}^{k} \Sigma_{m=0}^{j} \{\lambda a_{k-\ell,j-m} a_{\ell m} - b_{k-\ell,j-m} b_{\ell m}\}$$
$$+ \Sigma_{n=1}^{j} \Sigma_{m=0}^{j-n} \{\lambda a_{k+n,m} \bar{a}_{n,j-n-m} - b_{k+n,m} \bar{v}_{n,j-n-m}\}$$
$$= 0$$

$$(\alpha k + j - 2)(\bar{\alpha} k + j - 3)\bar{b}_{kj} + \bar{b}_{kj-2} \tag{3.7d}$$
$$+ \Sigma_{\ell=0}^{k} \Sigma_{m=0}^{j} \lambda \bar{a}_{k-\ell,j-m} \bar{a}_{\ell m} - \bar{b}_{k-\ell,j-m} \bar{b}_{\ell m}$$
$$+ \Sigma_{n=0}^{j} \Sigma_{m=0}^{j-n} \{\lambda \bar{a}_{k+n,m} a_{n,j-n-m} - \bar{b}_{k+n,m} b_{n,j-n-m}\}$$
$$= 0$$

It can be verified that the expansions defined by equations (3.7a-d) are consistent and well defined. The parameters of the expansion are found, in accordance with the resonance condition (3.3), to be:

(i) $b_{12} = \pm\sqrt{2 + 1/\lambda\ a_{12}}$ $\qquad \bar{b}_{12} = \pm\sqrt{2 + 1/\lambda\ a_{12}}$

where a_{12} and $\bar{a}_{12}$ are arbitrary and

$$\text{(ii)}\quad b_{06} \text{ is determined by } \frac{42}{\lambda}\left(3 + \frac{1}{\lambda}\right)b_{06} = H \tag{3.8}$$

where H is the Hamiltonian (3.1), i.e., the total energy. By detailed consideration of the recursion relations, one finds that the nonzero coefficients form a certain pattern. Furthermore, one may show that the following set of coefficients

$$a_{j,2j} \text{ and } b_{j,2j} \qquad \text{for } j = 0,1,2,\ldots$$

or

$$\bar{a}_{j,2j} \text{ and } \bar{b}_{j,2j} \qquad \text{for } j = 0,1,2$$

define a <u>closed</u> set of recursion relations. For example, setting

$$\theta_j = a_{j,2j} \qquad \psi_j = b_{j,2j}$$

it is easy to show that

$$(\alpha j + 2j - 2)(\alpha j + 2j - 3)\theta_j + 2\lambda\ \Sigma_{m=0}^{j}\ \theta_{j-m}\psi_m = 0$$

$$(\alpha j + 2j - 2)(\alpha j + 2j - 3)\psi_j + \Sigma_{m=0}^{j}\{\lambda\theta_{j-m}\theta_m - \psi_{j-m}\psi_m\} = 0$$

As we shall see, these closed sets of relations may be used to study the asymptotic properties ($|t| \ll 1$) of the series expansions near a singularity.

One may also derive the recursion relations for the expansions associated with the case 2 singularities (3.6). These are:

$$[\alpha(k+1) + j][\alpha(k+1) + j - 1]a_{kj} + a_{kj-2} + 2\lambda \Sigma_{\ell=0}^{k} \Sigma_{m=0}^{j} a_{k-\ell,j-m} b_{\ell m} + 2\lambda \Sigma_{n=1}^{j} \Sigma_{m=0}^{j-n} \{a_{k+n,m} \bar{b}_{n,j-n-m} + b_{k+n,m} \bar{a}_{n,j-n-m}\} = 0 \tag{3.9a}$$

$$[\bar{\alpha}(k+1) + j][\bar{\alpha}(k+1) - j]\bar{a}_{kj} + a_{k,j-2} + 2\lambda \Sigma_{\ell=0}^{k} \Sigma_{m=0}^{j} a_{k-\ell,j-m} b_{\ell m} + 2\lambda \Sigma_{n=0}^{j} \Sigma_{m=0}^{j-n} \{\bar{a}_{k+n,m} b_{n,j-n-m} + \bar{b}_{k+n,m} a_{n,j-n-m}\} = 0 \tag{3.9b}$$

$$(\alpha k + j - 2)(\alpha k + j - 3)b_{kj} + b_{kj-2} + \lambda \Sigma_{\ell=0}^{k-2} \Sigma_{m=0}^{j-4} a_{k-\ell-2,j-m-4} a_{\ell m} - \Sigma_{\ell=0}^{k} \Sigma_{m=0}^{j} b_{k-\ell,j-m} b_{\ell m} + \lambda \Sigma_{n=0}^{j-5} \Sigma_{m=0}^{j-n-5} a_{k+n,m} \bar{a}_{n,j-n-m-5} - \Sigma_{n=1}^{j} \Sigma_{m=0}^{j-n} b_{k+n,m} \bar{b}_{n,j-n-m} = 0 \tag{3.9c}$$

$$(\bar{\alpha}k + j - 2)(\bar{\alpha}k + j - 3)\bar{b}_{kj} + \bar{b}_{k,j-2} \tag{3.9d}$$

$$+ \lambda \Sigma_{\ell=0}^{k-2}\Sigma_{m=0}^{j-4} \bar{a}_{k-\ell-2,j-m-4}\bar{a}_{\ell m}$$

$$- \Sigma_{\ell=0}^{k}\Sigma_{m=0}^{j} \bar{b}_{k-\ell,j-m}\bar{b}_{\ell m}$$

$$+ \lambda \Sigma_{n=0}^{j-5}\Sigma_{m=0}^{m-n-5} \bar{a}_{k+n,m}a_{n,j-n-m-5}$$

$$- \Sigma_{n=0}^{j}\Sigma_{m=0}^{j-n} \bar{b}_{k+n,m}b_{n,j-n-m} = 0$$

As before one can determine the parameters of this expansion. These are:

(i) a_{00} and $\bar{a}_{00}$ are arbitrary and

(ii) b_{06} is determined by $-84b_{06} = H$ (3.10)

The closed sets of recursion relations is now associated with the coefficient

$a_{2j,4j}$ and $b_{2j,4j}$ for $j = 0,1,2,...$

or

$\bar{a}_{2j,4j}$ and $\bar{b}_{2j,4j}$ for $j = 0,2,1,...$

For example, setting

$\theta_j = a_{2j,4j}$ $\psi_j = b_{2j,4j}$

one finds that

$$(\alpha(2j+1) + 4j)(\alpha(2j+1) + 4j - 1)\theta_j$$

$$+ 2\lambda \Sigma_{m=0}^{j} \theta_{j-m}\psi_m = 0$$

$$(\alpha(2j) + 4j - 2)(\alpha(2j) + 4j - 3)\psi_j$$

$$+ \Sigma_{m=0}^{j}\{\lambda\theta_{j-m}\theta_m - \psi_{j-m}\psi_m\} = 0$$

In order to investigate the asymptotic properties of the series expansions we introduce certain generating functions. (Such an analysis was first carried out in our earlier paper on the Lorenz system [10].) For case 1 these are

$$\Theta(X) = \Sigma_{j=0}^{\infty} a_{j,2j}X^j \tag{3.11a}$$

$$\Psi(X) = \Sigma_{j=0}^{\infty} b_{j,2j}X^j \tag{3.11b}$$

where

$$X = t^{\alpha+2}, \quad \alpha = \frac{1}{2} + \frac{1}{2}\sqrt{1 - 24(1/\lambda+1)} \tag{3.12}$$

and for case 2

$$\Theta(X) = \Sigma_{j=0}^{\infty} a_{2j,4j}X^j \tag{3.13a}$$

$$\Psi(X) = \Sigma_{j=0}^{\infty} b_{2j,4j}X^j \tag{3.13b}$$

where

$$X = t^{2(\alpha+2)} \qquad \alpha = \frac{1}{2} + \frac{1}{2}\sqrt{1 - 48\lambda} \tag{3.14}$$

Using the (closed recursion relations for the coefficient sets

$(a_{j,2j}, b_{j,2j})$ and $(a_{2j,4j}, b_{2j,4j})$, the generating functions may be shown to satisfy the equations, for case 1,

$$(\alpha + 2)^2 X(X\Theta')' - 5(\alpha + 2)X\Theta' + 6\Theta + 2\lambda\Theta\Psi = 0 \quad (3.15a)$$

$$(\alpha + 2)^2 X(X\Psi')' - 5(a + 2)X\Psi' + 6\Psi + \lambda\Theta^2 - \Psi^2 = 0 \quad (3.15b)$$

and for case 2

$$4(\alpha + 2)^2 X(X\Theta')' + 4(\alpha + 2)(\alpha - \frac{1}{2})X\Theta' + \alpha(\alpha - 1)\Theta \quad (3.16a)$$

$$+ 2\lambda\Theta\Psi = 0$$

$$4(\alpha + 2)^2 X(X\Psi')' - 10(\alpha + 2)X\Psi' + 6\Psi + \lambda X\Theta^2 \quad (3.16b)$$

$$- \Psi^2 = 0$$

where primes denote differentiation with respect to X. These differential equations may now be used to analytically continue the functions $\Theta(X)$ and $\Psi(X)$.

The equations (3.15) and (3.16) may also be obtained in a more direct way. For case 1, substitution of

$$x(t) = \frac{1}{t^2}\Theta(X) \quad \text{and} \quad y(5) = \frac{1}{t^2}\Psi(X) \quad (3.17)$$

where the variable X is defined in (3.12), into the equations of motion (3.2), yields exactly equations (3.15) in the limit $|t| \to 0$; the contribution from the linear terms vanishing in this limit. Similarly, for case 2, the substitution into (3.2) of

$$x(t) = t^{\alpha}\Theta(X) \quad \text{and} \quad y(t) = \frac{1}{t^2}\Psi(X) \quad (3.18)$$

where now the variable X is defined by (3.14), yields equations (3.16). Again the linear terms in (3.2) vanish in the limit $|t| \to 0$. (It is amusing to note that the term λx^2 in (3.2b) contributes to the equation for $\Psi(X)$ although it does not contribute to the resonance calculation (2.12,13).)

From the above it appears that the part of the solution associated with the closed sets of recursion relations represents the asymptotic behavior of the solution near a singularity ($|t| \to 0$). This connection is explored at greater length in [13].

The types of singularities that $\Theta(X)$ and $\Psi(X)$ can display are easily determined by applying a standard leading order analysis. Considering first the case 1 equations (3.15), we set

$$\Theta(X) \simeq A(X - X_0)^{\gamma} \qquad \Psi(X) \simeq B(X - X_0)^{\delta}$$

where X_0 is the singularity position, and find that there are two possible cases:

$$\text{CASE a.} \quad \gamma = -2 \qquad A = \pm\frac{3}{\lambda}\sqrt{2 + 1/\lambda}\; X_0^2\,(2 + \alpha)^2$$
$$\delta = -2 \qquad B = -\frac{3}{\lambda}X_0^2(2 + \alpha)^2 \tag{3.19}$$

$$\text{CASE b.} \quad \gamma = \frac{1}{2} \pm \frac{1}{2}\sqrt{1 - 48\lambda} \qquad A = \text{arbitrary}$$
$$\delta = -2 \qquad B = 6X_0^2(2 + \alpha)^2 \tag{3.20}$$

The identical two cases are also found for the case 2 equations (3.16), except for slight changes in the leading order coefficients A and B. Thus, $\Theta(X)$ and $\Psi(X)$ display exactly the same sort of singularities in the X-plane as do $x(t)$ and $y(t)$

in the t-plane. Therefore, close to a given singularity, the singularity structure of $x(t)$ and $y(t)$ may be determined by studying the singularities of $\Theta(X)$ and $\Psi(X)$. The key is to correctly map the singularities from the X-plane to the t-plane.

To see how this is done, we first consider the case 2 equations (3.16) which correspond to the asymptotics about an irregular pole (which is placed at the origin of the X-plane and will, of course, map to the "origin" of the t-plane set at $t = t_*$ (= 0). Since we are unable to solve equations (3.16) analytically, we have to resort to a numerical solution; this always indicates that the singularity nearest $X = 0$ is of the regular variety. This singularity position, X_0, can be multiplied by the unit phase factor, i.e., we can say that the singularity is at

$$X = X_0 \equiv X_0 e^{2\pi in} \qquad n = 0,1,2,...$$

We recall that for case 2 the variable X is $X = t^{2(2+\alpha)}$ and $\alpha = +i/2\sqrt{48\lambda - 1}$ ($\lambda > 1/48$). Therefore, the corresponding singularity position in the t-plane is given by

$$t_0 = X_0^{1/2(2+\alpha)} \exp[\frac{\pi \min}{2(2+\alpha)}] \tag{3.21}$$

$$= X_0^{1/2(2+\alpha)} \exp\{n\pi[\frac{5i - \sqrt{48\lambda-1}}{2(12\lambda+6)}]\}$$

Thus, each pole in the X-plane yields an equiangular spiral [20] of poles in the t-plane; one pole for each value of $n = 0,1,2,...$. These poles have an angular displacement about the central (irregular) pole given by

$$\Delta\theta = \frac{5\pi}{2(12\lambda+6)} \tag{3.22a}$$

and with radial decrement

$$\Delta|t| = \exp[\frac{-n\pi\sqrt{48\lambda-1}}{2(12\lambda+6)}], \qquad n = 0,1,2,... \tag{3.22b}$$

In the canonical resonance case $\lambda = 1$, these quantitites are

$$\Delta\theta = \frac{5\pi}{36} = 25^{\circ} \tag{3.23a}$$

$$\Delta|t| = \exp[-\frac{n\pi\sqrt{47}}{36}] \tag{3.23b}$$

This mapping is shown in fig. 2 and can be compared with some numerical results in fig. 1. (The reason why we observe double spirals will be explained shortly.)

We notice in these figures that the singularities all seem to lie on the corners of exactly isosceles triangles. That this is indeed almost exactly so comes about as the result of an amusing coincidence. Returning to fig.2 in order to demonstrate that the triangle OAB is isosceles, we require that $OB \cos\theta = \frac{1}{2} OA$. From (3.21) we deduce that this can only be so if $\cos(\frac{5\pi}{36}) = \frac{1}{2}\exp[\frac{\pi\sqrt{47}}{36}]$. The actual numerical values are $\cos(\frac{5\pi}{36}) = 0.90631...$ and $\frac{1}{2}\exp[\frac{\pi\sqrt{47}}{36}] = 0.90948...$! Thus, for all practicaly purpose the triangles may indeed be taken to be isosceles.

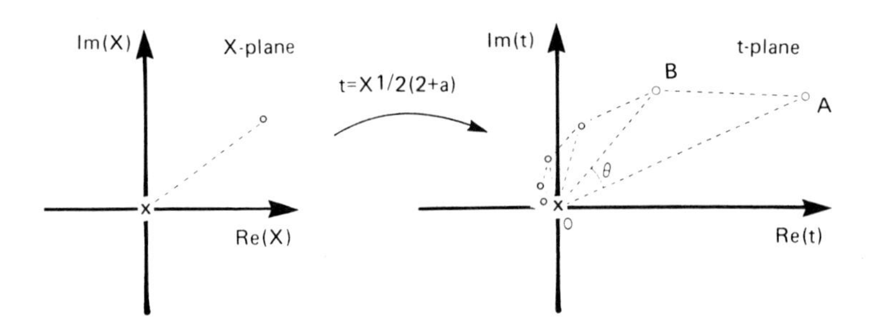

Figure 2. Mapping of a regular pole (o) and an irregular pole (x), from the X-plane to the t-plane. $\theta = 25^{\circ}$.

The same arguments may be applied to singularities of $\Theta(X)$ and $\Psi(X)$ associated with the case 1 equations (3.15). However, here, recall that the variable is now $X = t^{2+\alpha}$ and $\alpha = \frac{1}{2} + \frac{i}{2}\sqrt{24(1+1/\lambda)-1}$. Thus, about the central (regular) pole the singularity positions map onto the t-plan as

$$t_0 = X^{1/(2+\alpha)} \exp\{2n\pi[\frac{5i+\sqrt{24(1+1/\lambda)-1}}{2(12\lambda+6)}]\} \quad (3.24)$$

In the canonical resonance case $\lambda = 1$ we therefore obtain a spiral with angle

$$\Delta\theta = \frac{5\pi}{18} \quad (3.25a)$$

and radial decrement

$$\Delta|t| = \exp\{\frac{-n\pi\sqrt{47}}{18}\} \qquad n = 0,1,2,... \quad (3.25b)$$

It would appear, then, that around a regular pole the spiral angle is twice that about an irregular pole. However, when we investigate the situation numerically, in the X-plane, we find that near $X = 0$ regular and irregular singularities always appear in diametrically opposite pairs. Thus when one of these <u>pairs</u> is mapped onto the t-plan we again obtain the highly symmetric 25° spiral; but now with alternating regular and irregular poles. This is shown in fig. 3 and can be compared with some numerical results shown in fig. 4.

It sould now be abundantly clear what the significance of the canonical resonance is. Given the emperical findings just described above, the canonical resonance yielding complex powers results in the asymptotic singularity structure, about any given regular <u>or</u> irregular pole, having the identical geometry. Hence the observed, highly symmetric structure.

We also note that the whole of the above analysis can be repeated using the variables $\bar{X} = t^{2+\bar{\alpha}}$, with

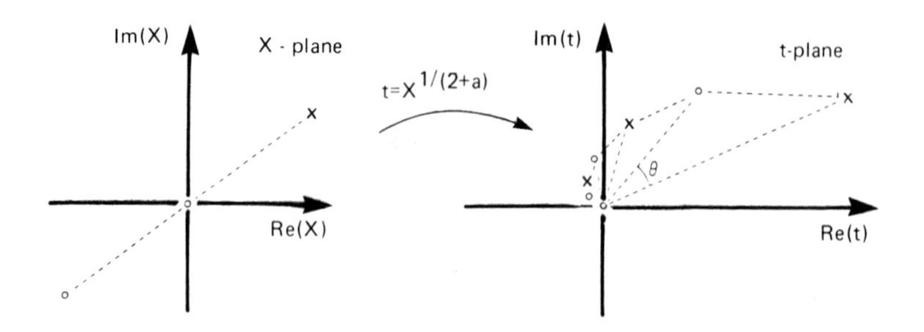

Figure 3. Mapping of a regular pole (o) and an irregular pole (x), about a regular pole, from the X-plane to the t-plane. θ = 25°.

$\bar{\alpha} = \frac{1}{2} - \frac{i}{2}\sqrt{24(1+1/\lambda)-1}$, for the asymptotics about the regular poles and $\bar{X} = t^{2(2+\bar{\alpha})}$, with $\bar{\alpha} = \frac{1}{2} - \frac{i}{2}\sqrt{48\lambda-1}$ for the asymptotics about the irregular poles. Everything is the same as before except that the spirals are now in the opposite direction. Thus around any given pole there is a double spiral of singularities. This is exactly what we observe.

Exactly the same type of asymptotic analysis carried out for the singularities in the t-plane can be carried out for the

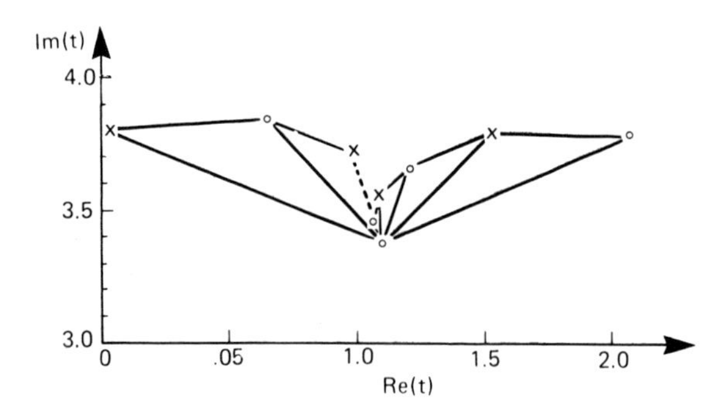

Figure 4. Adaptation of the fundamental domain F' in the z-plane and the square-symmetric period lattice in the u-plane to suit A. Schmidt's matrix group. Note that the symbols have special definitions for Section 7 which apply only here.

Actual numerical example of singularity structure in the t-plane showing double spirals of alternating regular (o) and irregular (x) poles about a regular pole.

singularities in the x-plane. The same mathematical structure is produced with each singularity in the x-plane being found to have its own double spiral structure of singularities. This self-similar structure makes it easy to understand how the singularities in the original t-plane become dense and form a natural boundary. These results are also strongly suggestive of there being some form of renormalization group present.

In [13] we have made a detailed study of the analytic structure of other Hamiltonian systems and noted the frequent occurrence of spirals of singularities. It is amusing (if not fitting) to note that we find these spirals in the original rigid-body problem studied by Kovalevskaya.

4. ANALYTIC STRUCTURE OF THE LORENZ SYSTEM: LOGARITHMIC SINGULARITIES

We first of all introduce the scaling [18]

$$X \to \frac{X}{\epsilon} \qquad Y \to \frac{Y}{\sigma\epsilon^2} \qquad Z \to \frac{Z}{\sigma\epsilon^2} \qquad t \to \epsilon t \qquad \epsilon = \frac{1}{\sqrt{\sigma R}}$$

thereby transforming the Lorenz equations (1.5) to the form

$$\frac{dX}{dt} = Y - \sigma\epsilon X$$

$$\frac{dY}{dt} = -XZ + X - \epsilon Y$$

$$\frac{dZ}{dt} = XY - \epsilon BZ \tag{4.1}$$

In the limit $\epsilon \to 0$ ($R \to \infty$) these equations reduce to a conservative integrable system and the solutions can be expressed in terms of the Jacobi elliptic functions [18]. These functions are doubly periodic (i.e. periodic in both real and imaginary directions) and have singularities, which are simple poles, arranged on an (infinite) periodic lattice in the complex t-plane.

We consider the leading order behavior of a singularity at $t = t_*$ by setting

$$X = \frac{a}{(t-t_*)^{\alpha}} \qquad Y = \frac{b}{(t-t_*)^{\beta}} \qquad Z = \frac{c}{(t-t_*)^{\gamma}} \tag{4.2}$$

from which it is easily deduced that

$$\alpha = 1 \qquad \beta = 2 \qquad \gamma = 2 \tag{4.3}$$

and

$$a = \pm 2i \qquad b = \pm 2i \qquad c = -2 \tag{4.4}$$

To examine the behavior of the solution in theneighborhood of the singularity at t_* we make the ansatz

$$X = \frac{2i}{(t-t_*)} \Sigma_{j=0}^{\infty} a_j (t-t_*)^j$$

$$Y = \frac{-2i}{(t-t_*)^2} \Sigma_{j=0}^{\infty} b_j (t-t_*)^j \tag{4.5}$$

$$Z = \frac{-2}{(t-t_*)^2} \Sigma_{j=0}^{\infty} c_j (t-t_*)^j$$

On substitution of these expansions into eqns. (2.1) we obtain the following sets of relationships between the coefficients:

$$a_0 = b_0 = c_0 = 1 \tag{4.6}$$

which follows trivially from eqns. (2.3);

$$a_1 = \frac{(3\sigma-2B-1)\varepsilon}{6}, \quad b_1 = -\sigma\varepsilon, \quad c_1 = \frac{(B-1-3\sigma)\varepsilon}{3} \tag{4.7}$$

and for $j = 2,3,4,\ldots,$

$$\begin{pmatrix} j-1 & 1 & 0 \\ 2 & j-2 & 2 \\ 2 & 2 & j-2 \end{pmatrix} \begin{pmatrix} a_j \\ b_j \\ c_j \end{pmatrix} = \begin{pmatrix} -\sigma\varepsilon a_{j-1} \\ -2\Sigma_{k=1}^{j-1} a_{j-k} c_k - a_{j-2} - \varepsilon b_{j-1} \\ -2\Sigma_{k=1}^{j-1} a_{j-k} b_k - \varepsilon B c_{j-1} \end{pmatrix} \tag{4.8}$$

Owing to the form of the coefficient matrix in the recursion relations (2.8) consistency conditions must be imposed when $j = 2$, and $j = 4$ (for these values it has no unique inverse). If these conditions, which impose restrictions on the parameters (σ,ε,B), are satisfied we can solve for the coefficient sets (a_2,b_2,c_2) and (a_4,b_4,c_4) and hence for all (a_j,b_j,c_j). However, the solutions for $j = 0,2$ will depend on an arbitrary parameter; that is (a_2,b_2,c_2) and (a_4,b_4,c_4) will be determined up to a vector that belongs to the null space of their respective coefficient matrices. Thus the general solution at a singularity will depend on three arbitrary parameters, one each for $j = 2$ and $j = 4$ and the actual value of t_*; providing, of course, the conditions on the parameters (σ,ε,B) are met. These conditions are for $j = 2$

$$\varepsilon^2\{6\sigma^2 - \sigma B - 2\sigma\} = B(B-1)\varepsilon^2 \tag{4.9}$$

and for $j = 4$

$$\frac{\varepsilon^2}{9}(B-1)\{57(\sigma-1) - 15(B-1) + 24\} = \varepsilon^2\sigma(2\sigma-1) \tag{4.10a}$$

and

$$2(1+B-\sigma)a_1^2c_1\varepsilon + \varepsilon^2\{\frac{2\sigma-B-5}{3}\}\{\frac{B(1-\sigma\varepsilon^2)}{2} + 2\sigma a_1^2 + Ba_1c_1\} = 0 \qquad (4.10b)$$

There are two conditions when $j = 4$ (apart from the "trivial" one $\varepsilon = 0$) since the inclusion of an arbitrary parameter in the solution for (a_2,b_2,c_2) "splits" the condition for (a_4,b_4,c_4). We could have used the consistency condition at $j = 4$ to fix the "arbitrary" parameter and relax condition (2.9). In this way a two parameter branch of the general solution at a singularity could be found.

In general, the conditions (2.9) and (2.10a) specify σ and B while (2.10b) determines ε. For these values of (σ,ε,B) the ansatz (2.5) is valid, i.e. the solution satisfies the Painlevé condition. These values have been given by Segur [19] who notes that the Lorenz system satisfies the Painlevé condition when:

(i) $\sigma = 0$. The equations are linear and this case is not included in the ansatz (2.5).

(ii) $\sigma = 1/2$, $B = 1$, $R = 0$ $(\varepsilon = \infty)$. The equations have two exact integrals and the solution reduces to elliptic functions.

(iii) $\sigma = 1$, $B = 2$, $R = 1/9$ $(\varepsilon = 3)$. There is one first integral and the equations reduce to the second Painlevé transcendent.

(iv) $\sigma = 1/3$, $B = 0$, $R(\varepsilon)$ arbitrary. There is one first integral and the equations reduce to the third Painlevé transcendent.

The other cases mentioned by Segur are

(v) $B = 1$, $R = 0$ $(\varepsilon = \infty)$, σ arbitrary. There exists a first integral but the Painlevé condition is not satisfied.

(vi) $B = 2\sigma$, $R(\varepsilon)$ arbitrary. There exists a first integral but again the Painlevé condition is not satisfied.

For case (vi) we note that when $B = 2\sigma$ condition (2.9) is satisfied but conditions (2.10) are not. We interpret this as implying that the satisfaction of one consistency condition indicates the existence of one first integral. In view of this we suggest, that since (2.9) is also satisfied for $B = 1-3\sigma$, the additional case

(vii) $B = 1 - 3\sigma$, $R(\varepsilon)$ arbitrary. The consistency condition (2.9) is satisfied and a first integral may exist (which we have not yet been able to identify) but the Painlevé condition is not satisfied.

We also note that condition (v) is not related to the consistency conditions and is presumably a property of a more general form of solution which we now discuss.

The expansion about a movable singularity is, in general, not of the Painlevé type and the resulting psi-series contains logarithmic terms. The form of this expansion for the Lorenz system is (where, for notational simplicity we set the pole position $t_* = 0$)

$$X = \frac{2i}{t} \Sigma_{k=0}^{\infty}\Sigma_{j=0}^{\infty} a_{kj} t^j (t^2 \ln t)^k$$

$$Y = \frac{-2i}{t} \Sigma_{k=0}^{\infty}\Sigma_{j=0}^{\infty} b_{kj} t^j (t^2 \ln t)^k$$

$$Z = \frac{-2}{t^2} \Sigma_{k=0}^{\infty}\Sigma_{j=0}^{\infty} c_{kj} t^j (t^2 \ln t)^k \tag{4.11}$$

A straightforward but tedious calculation gives the recursion relations

$$\begin{pmatrix} 2k+j-1 & 1 & 0 \\ 2 & 2k+j-2 & 2 \\ 2 & 2 & 2k+j-2 \end{pmatrix} \begin{pmatrix} a_{kj} \\ b_{kj} \\ c_{kj} \end{pmatrix} =$$

$$\begin{pmatrix} -(k+1)a_{k+1j+2} - \sigma\varepsilon a_{kj-1} \\ -(k+1)b_{k+1j-2} - a_{kj-2} - \varepsilon b_{kj-1} - 2\Sigma_{m=1}^{j} a_{kj-m}c_{om} \\ \quad - 2\Sigma_{m=0}^{j-1} a_{oj-m}c_{km} - 2\Sigma_{\ell=1}^{k-1}\Sigma_{m=0}^{j} a_{k-\ell j-m}c_{km} \\ -(k+1)c_{k+1j-2} - \varepsilon Bc_{kj-2} - 2\Sigma_{m=1}^{j} a_{kj-m}b_{om} \\ \quad - 2\Sigma_{m=0}^{j-1} a_{oj-m}b_{km} - 2\Sigma_{\ell=1}^{k-1}\Sigma_{m=0}^{j} a_{k-\ell j-m}b_{\ell m} \end{pmatrix} \qquad (4.12)$$

To solve for a given (a_{kj}, b_{kj}, c_{kj}) one must know, in general, the coefficients $(a_{\alpha\beta}, b_{\alpha\beta}, c_{\alpha\beta})$ for all (α,β) in the range

$$0 \leq \beta \leq j$$

$$0 \leq 2\alpha+\beta \leq 2k+j \qquad (4.13)$$

Thus for any coefficient set (a_{kj}, b_{kj}, c_{kj}), $k,j \neq 0$, the recursion relations are not closed.

The coefficient martix in (4.12) is singular when

(i) $k = j = 0$

(ii) $2k + j = 2$ i.e. $k = 1,\ j = 0$
$k = 0,\ j = 2$

(iii) $2k + j = 4$ i.e. $k = 2,\ j = 0$
$k = 1,\ j = 2$
$k = 0,\ j = 4$

At every point where the coefficient matrix is singular, consistency conditions are imposed on the solution and vectors belonging to the null space of the solution are introduced. The cases $k = 0$, $j = 2$ and $k = 0$, $j = 4$ have already been described in the previous section.

The recursion relations for all coefficient sets (a_{k0}, b_{k0}, c_{k0}) are closed and take the form

$$\begin{pmatrix} 2k-1 & 1 & 0 \\ 2 & 2k-2 & 2 \\ 2 & 2 & 2k-2 \end{pmatrix} \begin{pmatrix} a_{k0} \\ b_{k0} \\ c_{k0} \end{pmatrix} = \begin{pmatrix} 0 \\ -2\Sigma_{\ell=1}^{k-1} a_{k-\ell 0} c_{\ell 0} \\ -2\Sigma_{\ell=1}^{k-1} a_{k-\ell 0} b_{\ell 0} \end{pmatrix} \quad (4.14)$$

The consistency conditions required for $k = 0,1,2$ are trivial, i.e. they are identically satisfied without restrictions on (σ, ε, B) or previously introduced eigenvectors. It is easy to show that

$$(a_{00}, b_{00}, c_{00}) = (1,1,1) \quad (4.15a)$$

$$(a_{10}, b_{10}, c_{10}) = (\lambda, -\lambda, -\lambda) \quad (4.15b)$$

$$(a_{20}, b_{20}, c_{20}) = (\gamma, -3\gamma, 2\gamma) + (0,0,\lambda^2) \quad (4.15c)$$

where λ and γ are parameters whose values are fixed in the following manner.

The value of λ is determined by the consistency conditions introduced at $k = 0$, $j = 2$. Explicitly, we have [cf. Eq. (4.8)]

$$\begin{pmatrix} 1 & 1 & 1 \\ 2 & 0 & 2 \\ 2 & 2 & 0 \end{pmatrix} \begin{pmatrix} a_{02} \\ b_{02} \\ c_{02} \end{pmatrix} + \begin{pmatrix} \lambda \\ -\lambda \\ -\lambda \end{pmatrix} = \begin{pmatrix} -\sigma\varepsilon a_{01} \\ -a_{00} - \varepsilon b_{01} - 2a_{01}c_{01} \\ -\varepsilon B c_{01} - 2a_{01}b_{01} \end{pmatrix} \quad (4.16)$$

which yields

$$3\lambda = 2a_{01}(b_{01}-\sigma\varepsilon) + \varepsilon Bc_{01} \tag{4.17}$$

and using the values of (a_{01}, b_{01}, c_{01}) as given in Eq. (4.7) we obtain

$$\lambda = \frac{\varepsilon^2}{9}\{B(B-1) - 6\sigma^2 + \sigma B + 2\sigma\} \tag{4.18}$$

If the consistency condition (4.9) is satisfied then $\lambda = 0$. In this case the logarithmic terms in the psi-series enter as powers of $t^4 \ln t$. We will return to this point later.

In a similar manner the parameter γ is determined by the consistency conditions at $k = 1$, $j = 2$ which yield

$$\gamma = \frac{\lambda\varepsilon^2}{5}\{\frac{\sigma}{3}(2\sigma-1) + (\frac{5-22\sigma}{3})(\frac{B-1}{3}) + 6(\frac{B-1}{3})^2\} \tag{4.19}$$

We note that γ is of order λ and that it does not depend on the eigenvector introduced at $k = 0$, $j = 2$. The eigenvector at $k = 1$, $j = 2$ is specified by the consistency conditions at $k = 0$, $j = 4$. Finally we recall that the eigenvectors introduced at $k = 0$, $j = 2$ and $k = 0$, $j = 4$ are, in general, complex constants of integration. The above considerations show that the recursion relations (4.2) are well defined and consequently (4.1) represents the general form of the formal expansion about a singularity.

In order to investigate the leading order logarithmic terms in the general expansion (4.1) we introduce the generating functions

$$\theta(x) = \Sigma_{k=0}^{\infty} a_{k0}x^k \tag{4.20a}$$

$$\Phi(x) = \Sigma_{k=0}^{\infty} b_{k0}x^k \tag{4.20b}$$

$$\Psi(x) = \Sigma_{k=0}^{\infty} c_{k0} x^k \tag{4.20c}$$

where

$$x = t^2 \ln t \tag{4.21}$$

Using the recursion relations (4.4) we can deduce

$$2x \frac{d\Theta}{dx} - \Theta + \Phi = 0 \tag{4.22a}$$

$$2x \frac{d\Phi}{dx} - 2\Phi + 2\Theta\Psi = 0 \tag{4.22b}$$

$$2x \frac{d\Psi}{dx} - 2\Psi + 2\Theta\Phi = 0 \tag{4.22c}$$

from which we obtain the following (closed) differential equation for θ

$$2x^2 \frac{d^2\Theta}{dx^2} = x \frac{d\Theta}{dx} - 3\lambda x\Theta + \Theta(\Theta^2 - 1) \tag{4.23}$$

where

$$\Phi = \Theta - 2x \frac{d\Theta}{dx}$$

and

$$\Psi = \Theta^2 - 3\lambda x$$

We note the special values at $x = 0$

$$\Theta = 1, \quad \frac{d\Theta}{dx} = \lambda, \quad \frac{d^2\Theta}{dx^2} = 2\gamma \tag{4.24}$$

Before attempting to solve Eq. (4.23) it is amusing to determine whether it is actually of the Painlevé type. We make the ansatz

$$\Theta(x) = \frac{2x_0}{(x-x_0)} \Sigma_{j=0}^{\infty} \theta_j (x-x_0)^j \tag{4.25}$$

and find that

$$\theta_0 = 1, \ \theta_1 = \frac{3}{4x_0}, \ \theta_2 = -\frac{1}{16x_0^2} + \frac{\lambda}{4x_0} \tag{4.26}$$

$$\theta_3 = \frac{1}{32x_0^3} + \frac{21}{32}\frac{\lambda}{x_0^2}$$

The coefficient θ_4 is arbitrary and the corresponding consistency condition requires that $\lambda = 0$. Thus, in general, Eq. (4.23) is not of the Painlevé type. In the special case of $\lambda = 0$, as mentioned earlier, the Eqs. (4.1) involve powers of $t^4 \ln t$. Some of the properties of the leading order logarithmic terms in this representation are described in Appendix 1 of [10].

In order to solve Eq. (4.23) we make the substitution

$$\Theta = x^{1/2} f(x^{1/2}) \tag{4.27}$$

which yields

$$f'' = 2f^3 - 6\lambda f \tag{4.28}$$

where primes denote differentiation with respect to the argument of f. Using the special values given in (4.24) we find

$$(f')^2 = f^4 - 6\lambda f^2 + 7\lambda^2 - 10\gamma \tag{4.29}$$

which can be solved in terms of the Jacobi elliptic functions. Let

$$\alpha = 3\lambda + \sqrt{2\lambda^2 + 10\gamma}$$

$$\beta = 3\lambda - \sqrt{2\lambda^2 + 10\gamma}$$

and define the (squared) modulus

$$k^2 = 1 - (\frac{\beta}{\alpha})^2 \tag{4.30}$$

where we assume α, $\beta > 0$. The solution of (4.29) is

$$f(x) = \frac{\alpha}{\text{sn}(\alpha x, k)}$$

where sn is the Jacobi elliptic function and hence

$$\Theta(x) = \frac{\alpha x^{1/2}}{\text{sn}(\alpha x^{1/2}, k)} \tag{4.31}$$

Using $x = t^2 \ln t$, we find:

$$x(t) = \frac{2i}{t}\{\frac{\alpha t(\ln t)^{1/2}}{\text{sn}(\alpha t(\ln t)^{1/2}, k)}\} + O(t) \tag{4.32}$$

When α and β are not both positive, or not real, the solution of (4.29) will take different forms that are all expressible in terms of combinations of Jacobi elliptic functions. Rather than write out each case, we note that (α, β) are real when:

$$2\lambda^2 + 10\gamma \geq 0 \tag{4.33}$$

Furthermore, the modulus, k, of the elliptic functions used to express the solution is either zero or one when:

$$2\lambda^2 + 10\gamma = 0$$

or (4.34)

$$7\lambda^2 - 10\gamma = 0$$

Since

$$sn(z,0) = \sin(z)$$

and

$$sn(z,1) = \sinh(z)$$

we might expect special types of solutions when $k^2 = 0$ and $k^2 = 1$. By algebraic simplification:

$$7\lambda^2 - 10\gamma = \frac{7\lambda\varepsilon^2}{9}\{(B-1)^2 + \sigma+1)(B-1) + 3\sigma(1-2\sigma)\} \tag{4.35}$$

and for $\lambda\varepsilon^2 \neq 0$, $7\lambda^2 - 10\gamma = 0$ if:

$$B = 1 + 9\sigma \tag{4.36}$$

or

$$B = \frac{2}{5}(3\sigma + 1) \tag{4.37}$$

In addition, we find that:

$$2\lambda^2 + 10\gamma = \lambda\varepsilon^2\{\frac{14}{3}(B-1) + 4 - 14\sigma\}(\frac{B-1}{3}) = 0 \tag{4.38}$$

which, for $\lambda\varepsilon^2 \neq 0$, is satisfied when

$$B = 1 \tag{4.39}$$

or

$$B = 3\sigma + 1/7 \tag{4.40}$$

We recall that there exists an integral of motion when $B = 1$, σ is arbitrary, and $R = 0$ ($\varepsilon = \infty$). Of all the known integrals this alone appeared to be unrelated to the structure of the singularities. Our analysis suggests that the existence of this integral may, in fact, be related to the conditions $k^2 = 0$ and $k^2 = 1$. However, it is important to emphasize that the above analysis, although interesting in its own right, only indicates that the special values of B given in (4.36) through (4.40) may result in the corresponding solutions having some preferred property <u>near</u> a singularity. At this stage the most we can say concerning the existence of integrals of the motion is that the conditions (4.36), (4.37), (4.39), (4.40) probably provide restrictions on the parameter space in which these integrals may be found.

5. CONCLUSIONS

Clearly the analytic structure of the systems discussed here is very rich but I want to conclude with a few of the many questions that the above results raise.

(i) Most important of all, what, in fact, is the connection between the Painlevé property and integrability?

(ii) For what classes of system does this connection apply? Can it be extended to rational branch points? (See [12], [21].)

(iii) What is the role of the solutions with less than the requisite number of parameters, e.g., the three-parameters solution of the Henon-Heiles system? Do these solutions also have to satisfy the Painlevé property to ensure integrability?

(iv) To what extent do the consistency conditions determine the "degree" of integrability? In the Lorenz system we noted that the satisfaction of the first consistency condition (which causes the logarithmic corrections to enter later in the series) seemed to be related to the existence of one "integral" of the motion. Also, for the Henon-Heiles system the onset of widespread chaos, as a function of $\lambda (\lambda < 0)$ seemed to coincide with the breakdown of the last consistency condition for the associated recursion relations [21].

(v) To what extent do the different analytic structures influence the real-time solutions? For example, for the Henon-Heiles system the singularities, as a function of λ, can be variously rational branch points, logarithmic branch points, self-similar natural boundaries or (for special values) ordinary poles. Is there some quantifiable difference (in the power spectrum, maybe) that reflects these differences? Or, for example, will these different types of structure result in the formation of multi-sheeted, "local" integrals of the motion; the jumping between these different sheets thereby producing a mechanism for generating "chaos".

REFERENCES

[1] S. Kovalevskaya, A Russian Childhood, edited by Beatrice Stillman, Springer-Verlag, New York (1978).

[2] V. V. Golubev, Lectures on Integration of the Equations of Motion of a Rigid Body about a Fixed Point. State Publishing House, Moscow, 1953.

[3] Eugene Leimanis, The General Problem of the Motion of Coupled Bodies about a Fixed Point, Springer-Verlag, New York, 1965.

[4] Einar Hille, Ordinary Differential Equations in the Complex Plane, John Wiley and Sons, New York, 1976.

[5] David Chudnovsky, private communication 109th and Broadway (1981).

[6] H. Flaschka and A. C. Newell, Comm. Math. Phys. 76, 65 (1980).

[7] M. Jimbo et al., series of papers in Physica D (1980-81).

[8] M. Henon and C. Heiles, Astron. J. 69, 73 (1964).

[9] Lorenz, E. N., J. Atmos, Sci. 20, 130 (1963).

[10] M. Tabor and J. Weiss, Analytic Structure of the Lorenz System, Phys. Rev. A., 24, 2157 (1981).

[11] Y. F. Chang, M. Tabor, J. Weiss and G. Corliss, On the Analytic Structure of the Henon-Heiles System, Phys. Lett. A., 85A, no. 4, 211 (1981).

[12] Y. F. Chang, M. Tabor and J. Weiss, Analytic Structure of the Henon-Heiles Hamiltonian in Integrable and Non-Integrable Regimes (1981) accepted for publication in J. Math. Phys. (Nov. 1981).

[13] Y.F. Chang, J. M. Greene, M. Tabor and J. Weiss, The Analytical Structure of Dynamical Systems and Self-similar Natural Boundaries, Submitted to J. Math. Phys. (1981).

[14] M. J. Ablowitz, A. Ramani and H. Segur, J. Math. Phys. 21, 715 (1980).

[15] Y. Aizawa and N. Saito, J. Phys. Soc. Japan 32, 1636 (1972).

[16] T. Bountis, H. Segur and F. Vivaldi, Integrable Hamiltonian Systems and the Painleve Property (1981) submitted to Phys. Rev. A.

[17] Y. F. Chang and G. Corliss, J. Inst. Maths. Applics. 25, 349 (1980).

[18] K. A. Robbins, SIAM J. Appl. Math. 36, 457 (1979).

[19] H. Segur, Solitons and the Inverse Scattering Transform, Lectures given at the International School of Physics, Enrico Fermi, Varenna, Italy, July 7-19, 1980.

[20] Equiangular spirals of the form described here have been observed in many other contexts. They were first discussed by Descartes in 1638, and their frequent occurrence in nature is described by D'Arcy Thompson in his book On Growth and Form.

[21] J. Weiss, Analytic Structure of the Henon-Heiles System. To be published in Mathematical Methods in Hydrodynamics and Integrability in Related Dynamical Systems, Eds. M. Tabor and Y. Treve (AIP converence proceedings, 1982).

APPENDIX I: TRAVAUX DE J. DRACH (1919)

David V. Chudnovsky and Gregory V. Chudnovsky

Department of Mathematics
Columbia University
New York, New York

In this note we reproduce Drach's [1], [2] original contribution establishing the relationship between complete integrability and spectral theory. Drach's results can be compared with the modern treatment of the same class of equations [8]-[10]. In [11] we reported Drach's contribution immediately after we cam across his works.

J. Drach in his papers [1]-[2] in C.R. Acad. Sci. 1918-1919 had described the class of completely integrable equations, now known as "stationary Korteweg-de Vries equations."

J. Drach starts from the linear differential equation of the second order

$$\frac{d^2y}{dx^2} = [\varphi(x) + h]y \qquad (1)$$

In his studies J. Drach was following his teacher Darboux, who first found (in 1879-1882 see [3]) the Darboux-Bäcklund transformation of the equation (1). This Darboux transformation is reproduced in the Ince book [4], Ch. V, and has the following form.

Let y_0 satisfie (1) with a given h_0: $y_0'' = (\varphi + h)y$, and let y satisfie (1) with h. Then the function

$$z = y' - yy_0'/y_0$$

satisfies an equation similar to (1) (with a new potential):

$$z'' = (\psi(x) + h)z \quad \text{for} \quad \psi = 2(y_0'^2/y_0^2 - h) - \varphi$$

Cf. [7] for the relation of Darboux-Bäcklund transformation to the inverse scattering method.

As in the modern studies, see Kruskal et al. [5] or [6], Drach transforms (1) into a Riccati equation

$$\rho' + \rho^2 = \varphi(x) + h \tag{2}$$

J. Drach writes the following linear differential equation of the third order for the resolvent $R = R(x;h)$:

$$R''' - 4R'(\varphi + h) - 2R\varphi' = 0 \tag{3}$$

Any function satisfying the equation (3) is called in modern terminology a resolvent of (1), cf. [8]. Drach remarkes that (3) always implies the following nonlinear equation for $R(x,h)$:

$$(R')^2 - 2RR'' + 4R^2(\varphi + h) = \Omega(h) \tag{4}$$

for some (independent of x) function $\Omega(h)$.

The relation of resolvent in (4) with the Riccati equation (2) is based on the following identity: the solutions of ρ of (2) can be represented as

$$\rho_{\pm} = \frac{R' \pm \sqrt{\Omega(h)}}{2R}$$

Already in his first paper on this subject J. Drach [1] (1918) proposes the following:

Problem: Let $R(x,h)$ be a resolvent satisfying (3) where $R(x;h)$ is a polynomial in h of degree n:

$$R = R_0 h^n + R_1 h^{n-1} + \ldots \tag{5}$$

What equations on $\varphi(x)$ are equivalent to (3) after the substitution of (5) into (3)?

In his papers of 1919 J. Drach [1], [2], had proved that the corresponding system of ordinary differential equations on $\varphi(x)$ is completely integrable. He found first integrals of this system of equations, denoted by Drach as E_{2n+1}, and wrote down explicit solutions in terms of hyperelliptic integrals.

This systems of differential equations is equivalent to the class of stationary n-th order Korteweg-de Vries (KdV) equations in the sense of [8], [10].

Indeed, all resolvents are linearly related to the standard resolvent

$$\bar{R}(x,h) = h^{-1/2} \Sigma_{n=0}^{\infty} \bar{R}_n[\varphi] h^{-n}$$

satisfying the normalized equation

$$(\bar{R}')^2 - 2\bar{R}\bar{R}'' + 4(\varphi + h)\bar{R}^2 = 1 \tag{6}$$

The relation of a resolvent $R(x,h)$ from (4) with the normalized resolvent $\bar{R}(x,h)$ from (6) is simple:

$$R(x,h) = \sqrt{\Omega(h)}\ \bar{R}(x,h)$$

Here $\bar{R}_n[\varphi]$ are all polynomials in $\varphi, \varphi', \varphi'', \ldots$ determined by inductions from (3), (4): the first of them are:

$$\bar{R}_0 = 1/2, \quad \bar{R}_1 = -1/4\varphi, \quad \bar{R}_2 = 1/16(3\varphi^2 - \varphi''), \ldots$$

Now we will be following Drach's paper [2], (presented to the Academy on 17 Fevrier 1919).

1.

We consider the case when the Riccati equation (3) admit solutions $\frac{R' \pm \sqrt{\Omega(h)}}{2R}$ for $R(x,h)$ being a polynomial in h of degree n:

$$R = R_0 h^n + R_1 h^{n-1} + \ldots \tag{5}$$

and $\Omega(h)$ a polynomial in h of the degree $2n + 1$ with constant coefficients.

THEOREM 1. (J. Drach). Let us consider the polynomial $R = h^n + \ldots$ in (5) satisfying the resolvent equation (3)

$$R''' - 4R'(\varphi + h) - 2R\varphi' = 0$$

This equation defines a nonlinear ordinary differential equation E_{2n+1} of the order $2n + 1$ on $\varphi(x)$:

$$(E_{2n+1})\, Q(\varphi, \varphi', \varphi'', \ldots, \varphi^{(2n+1)}) = 0$$

depending on n+1 arbitrary constants $c, \ldots, c_n$. This equation (E_{2n+1}) is completely integrable for an $c_0, \ldots, c_n$.

This equations (E_{2n+1}) can be also rewritten in terms of canonical resolvent $\bar{R}$. Let $\Omega(h)$ be the polynomial of the degree $2n + 1$ in h, corresponding to $R(x;h)$ in (3) and (5):

$$R'^2 - 2RR'' + 4R^2(\varphi + h) = \Omega(h) \quad (6)$$

Let us denote:

$$\sqrt{\Omega(h)} = h^{n+1/2} \Sigma_{k=0}^{\infty} c_k h^{-k}$$

Then we have $R(x;h) = \sqrt{\Omega(h)}\ \bar{R}(x;h)$. The system (E_{2n+1}) is then equivalent to the following equation of the order $2n + 1$ depending on constants $c_0, \ldots, c_n$ only:

$$(E_{2n+1})\ \Sigma_{i=0}^{n+1} c_{n+1-i} \bar{R}_i[\varphi] = 0 \quad (7)$$

The corresponding proof is trivial: if $R(x,h) = \Sigma_{j=0}^{n} R_j h^j$, then

$$R_k = \Sigma_{\ell=0}^{n-k} c_{n-(\ell+k)} \bar{R}_\ell$$

Now one only substitute $k = -1$, (i.e. R is a polynomial).Q.E.D.

Following [5], [6], [8]-[11] an equation

$$\varphi_t = \frac{\partial}{\partial x} \bar{R}_n[\varphi]$$

or any equation

$$\varphi_t = \frac{\partial}{\partial x} R(x,h)$$

where $R(x,h)$ is of a resolvent degree $n - 1$ in h, is called now an $(n-1)$-th Korteweg-de-Vries equation. E.g. for $n = 2$, we have an equation

$$\varphi_t = 1/16\ (6\varphi\varphi_x - \varphi_{xxx})$$

equivalent to KdV. In the stationary case $\varphi_t = 0$ these equations are called "stationary KdV's".

The simplest equation is the elliptic function equation (n = 1):

$$3\varphi^2 - \varphi'' + d_1\varphi + d_0 = 0$$

How to find the first integrals of (E_{2n+1}) and its solutions? The answer is given by the second and third Drach's statements.

Let $\Omega(h)$ be the constant function in x satisfying (4) for a given solution of (3).

THEOREM 2. (J. Drach). Let R(x,h) be the solution of (3), which is polynomial in h of the degree n: see (5) and let $\Omega(h)$ be the corresponding polynomial of the degree 2n + 1 in h satisfying (4). Let

$$\Omega(h) = \Sigma_{j=0}^{2n+1} J_j h^j$$

Then $J_n, J_{n+1}, \ldots, J_{2n+1}$ are functions in n + 1 constants $c_0, \ldots, c_n$. The other quantities $d_0, d_1, \ldots, d_{n-1}$ are first integrals of the system E_{2n+1}, being differential polynomials in $\varphi, \varphi', \ldots$.

Proof. Indeed, as simple computations show,

$$\Omega(h) = h(\Sigma_{k=0}^{n} h^k c_{n+1-k})^2 + \Omega_{n-1}(h)$$

for $\Omega_{n-1}(h)$ being a polynomial of the degree n - 1.

The fact that $d_0, d_1, \ldots, d_{2n+1}$ are then differential polynomials in φ follows from (4), because $\Omega(h)$ is the differential polynomial in R, i.e. the polynomial in $\varphi, \varphi', \ldots$ and $c_0, \ldots, c_n$. Q.E.D.

Finally, J. Drach finds explicitly the exact solutions of (E_{2n+1}) and, in the "modern language", action-angle variables.

Let us denote

$$R(x;h) = R_0(h - \omega_1)\ldots(h - \omega_n)$$

where ω_i are algebraic functions of $\varphi, \varphi', \ldots, \varphi^{(2n-2)}$. Then

$$-R'/R = \Sigma_{i=1}^{n} \frac{\omega_i'}{h-\omega_i}$$

and according to (6), R divides $R'^2 - \Omega(h)$. In this way we obtain n equations

$$\frac{\varepsilon_i R_0 \omega_i'}{\sqrt{\Omega(\omega_i)}} = \frac{1}{(\omega_i - \omega_1)\ldots * \ldots(\omega_i - \omega_n)}: \; i = 1,\ldots,n$$

Here $\varepsilon_i = \pm 1$ $(i = 1,\ldots,n)$ and it corresponds to one of the two sheets of the Riemann surface of the curve $y^2 - \Omega(x) = 0$ on which ω_i is lying.

Now J. Drach finds using the Liouville method the canonical variables $u_0, u_1, \ldots, u_{n-1}$, cannonically conjugate to (action variables) $d_0, d_1, \ldots, d_{n-1}$. Here they are in Drach's notations

$$u_\ell = \Sigma_{i=1}^{n} \int_\infty^{\omega_i} \frac{\omega_i^\ell d\omega_i}{\sqrt{\Omega(\omega_i)}}: \; \ell = 0,1,\ldots,n-1$$

Then by the original system of differential equations and classical arguments,

$$\frac{du_\ell}{dx} = 0: \; \ell = 0,1,\ldots,n-2$$

$$\frac{du_\ell}{dx} = \frac{1}{R_0}: \ell = n - 1$$

or $u_\ell = \text{const}: \ell = 0,1,\ldots,n-2,\ u_{n-1} = \frac{1}{R_0}x + u'_{n-1},\ u'_{n-1} = \text{const}.$

In order to get the expression for $\varphi(x)$, J. Drach only notices that

$$\omega_1 + \ldots + \omega_n = -\frac{R_1}{R_0} = \frac{c_0\varphi}{4} - \frac{c_1}{2}$$

As a consequence (on p. 339 of [2]) J. Drach gives

COROLLARY (J. Drach). Any solution $\varphi(x)$ of (E_{2n+1}) is a uniform, meromorphic function of x, quasi-periodic with $2n$ periods and expressed in terms of inversions of hyperelliptic integrals having, in general, genus equal to n.

At the same time J. Drach [2] considered those cases when the genus is smaller than n. He found that Darboux-Backlund transformation does not change the genus.

E.g. if one replaces φ using the Darboux transformation into

$$\psi = 2(\rho_0^2 - h_0) - \varphi$$

for $\rho_0' + \rho_0^2 = \varphi + h_0$, then the resolvent R of degree n in h, changes into a resolvent of degree $n + 1$. But $\Omega(h)$ gets only additional factor $(h - h_0)^2$, which does not change the genus of the curve $y^2 - \Omega(x) = 0$.

In the modern literature this corresponds exactly to the addition of solitons to "finite-band" solutions. In order to get the multisoliton solutions one starts from $\varphi = 0$ and arrives to the rational curve with

$$\Omega(h) = (h - h_0)^2 \ldots (h - h_{n-1})^2(h - h_n)$$

REFERENCES

[1] J. Drach, C.R. Acad. Sci. Paris, t. 167 (1918), pp. 743-746; C.R. Acad. Sci. Paris, t. 168 (1919), pp. 47-50.

[2] J. Drach, C.R. Acad. Sci. Paris, t. 168 (1919), pp. 337-340.

[3] G. Darboux, C.R. Acad. Sci. Paris, t. 94 (1882), pp. 1465-1467.

[4] E.L. Ince, Ordinary differential equations, 1926, Dover repring, N.Y., 1956, Ch. 5.

[5] C.S. Gardner, J.M. Greene, M.D. Kruskal, R.M. Miura, Phys. Rev. Lett. 19 (1967), 1095-1097.

[6] V.E. Zakharov, L.D. Faddeev, Funct. Anal. Appl. 5 (1971), 280-287.

[7] H. Flaschka, D.V. McLaughlin, Lecture Notes Math., v. 515, N.Y., 1976, pp. 253-295.

[8] I.M. Gel'fand,L.A. Dikij, Russian Math. Survey 30 (1975), 37-100.

[9] I.M. Gel'fand, L. A. Dikij, Funct. Anal. Appl. 11 (1977), 93-104.

[10] S.P. Novikov, B.A. Dubrovin, V.B. Matveev, Russian Math. Survey 31 (1976), 59-100.

[11] D.V. Chudnovsky, in Bifurcation phenomena in mathematical physics and related topics, D. Reidel Publishing Company, Boston, 1980, 385-449.

APPENDIX II: SUR L'INTEGRATION PAR QUADRATURES DE L'ÉQUATION $\frac{d^2y}{dx^2} = [\varphi(x) + h]y$

M. Jules Drach

ANALYSE MATHÉMATIQUE. — *Sur l'intégration par quadratures de l'équation* $\frac{d^2y}{dx^2} = [\varphi(x) + h]y$. Note de M. JULES DRACH.

Nous avons indiqué récemment [1] quels sont les cas généraux de réduction du *groupe de rationalité* de l'équation

$$\text{(H)} \qquad \frac{d^2y}{dx^2} = [\varphi(x) + h]y,$$

où h est *un paramètre arbitraire*. Les plus intéressants d'entre eux sont ceux où l'équation de Riccati

$$\rho' + \rho^2 = \varphi + h$$

[et par suite aussi l'équation (H)] s'intègre par *quadratures*; nous allons montrer comment on détermine la fonction φ dans tous ces cas.

1. L'équation de Riccati admet ici deux solutions $\frac{R' \pm \sqrt{\Omega}}{2R}$, où R est un polynome en h, de degré n

---

[1] *Comptes rendus*, t. 168, 1919, p. 47.

C. R., 1919, 1er *Semestre*. (T. 168, N° 7.)

Reprinted with the permission of Gauthier-Villars from Les Comte Rendus de l'Académie des Sciences de 1919, no. 7, pages 337 to 340.

$$R = h^n + R_1 h^{n-1} + \dots$$

et Ω un polynome en h de degré $(2n+1)$ *à coefficients constants.*

La fonction R satisfait à l'équation du troisième ordre

$$(A) \qquad R''' - 4R'(\varphi + h) - 2R\varphi' = 0$$

ce qui donne, pour déterminer φ, une équation d'ordre $(2n+1)$, E_{2n+1}, dépendant de n constantes arbitraires $c_2, \dots, c_n$. C'est cette équation que nous avons réussi à intégrer.

En observant que (A) est équivalente à son adjointe, on en conclut l'intégrale première quadratique

$$(B) \qquad R'^2 - 2RR'' + 4R^2(\varphi + h) = \Omega$$

Les n premiers coefficients de Ω sont des fonctions de $c_1, \dots, c_n$; les $(n+1)$ derniers $d_1, \dots, d_{n+1}$ sont arbitraires et représentent autant d'intégrales de E_{2n+1}, *entières* en $\varphi, \varphi', \dots, \varphi^{(2n+1)}$. Il reste donc à intégrer une équation d'ordre n, avec $(2n+1)$ constantes.

2. Posons

$$R = h^n + R_1 h^{n-1} + \dots = (h - \omega_1)(h - \omega_2)\dots(h - \omega_n)$$

les ω_i sont des fonctions algébriques de $\varphi, \varphi', \dots, \varphi^{2n-2}$ et l'on aura

$$-\frac{R'}{R} = \frac{\omega'_1}{h - \omega_1} + \dots + \frac{\omega'_n}{h - \omega_n}$$

En observant que les racines $h = \omega_i$ annulent $R'^2 - \Omega$ d'après (B), c'est-à-dire l'un des facteurs $R' + \sqrt{\Omega}$, $R' - \sqrt{\Omega}$, on pourra écrire les n équations

$$\frac{\varepsilon_i \omega'_i}{\sqrt{\Omega_i}} = \frac{1}{(\omega_i - \omega_1)\dots(\omega_i - \omega_n)} \qquad (i = 1, \dots, n)$$

où $\Omega_i = \Omega(\omega_i)$, $\varepsilon_i = \pm 1$.

On déduit de là les n intégrales cherchées u_i sous la forme

$$\int_{\alpha_1}^{\omega_1} \frac{\omega_1^\lambda \, d\omega_1}{\sqrt{\Omega_1}} + \dots + \int_{\alpha_n}^{\omega_n} \frac{\omega_n^\lambda \, d\omega_n}{\sqrt{\Omega_n}} = u_\lambda \qquad (\lambda = 0, 1, \dots, n-2)$$

$$\int_{\alpha_1}^{\omega_1} \frac{\omega_1^{n-1} \, d\omega_1}{\sqrt{\Omega_1}} + \dots + \int_{\alpha_n}^{\omega_n} \frac{\omega_n^{n-1} \, d\omega_n}{\sqrt{\Omega_n}} = x + u_{n-1}$$

à condition de fixer les chemins suivis dans l'intégration lorsque $\varepsilon_i = -1$ de manière à remplacer l'intégrale I rectiligne par $2A - I$, A désignant une constante convenable; ce qui donne $\left(\frac{n}{2}\right)$ systèmes, différents de forme.

3. La fonction φ, qui est donnée par

$$\omega_1 + \omega_2 + \ldots + \omega_n = -R_1 = -C_1 + \frac{1}{2}\varphi$$

et les fonctions symétriques élémentaires des ω [de même que celles des $\sqrt{\Omega_i}$] sont donc des *fonctions abéliennes* (hyperelliptiques) des arguments $u_0, \ldots, u_{n-1} + x$, c'est-à-dire des fonctions *uniformes*, *méromorphes*, de ces éléments possédant $2n$ systèmes de périodes ([1]). La fonction φ de la variable x n'est pas en général périodique, mais elle reprend sa valeur quand on ajoute simultanément à x, $u_0, \ldots, u_{n-2}$ des périodes correspondantes : ceci définit le *groupe de monodromie* des intégrales u_i. Le cas $n = 1$ donne pour φ la fonction inverse de l'intégrale elliptique de première espèce.

Un raisonnement analogue, mais portant sur les autres racines $\xi_1, \ldots, \xi_{n+1}$ de l'équation $R'^2 - \Omega = 0$ donnerait $(n+1)$ intégrales transcendantes de E_{2n+1} ; le théorème d'Abel montre aisément que ces intégrales s'expriment avec les précédentes et les intégrales algébriques $d_1, \ldots, d_{n+1}$.

Les intégrales fondamentales de l'équation (H) sont $\sqrt{R}\, e^{\frac{1}{2}\sqrt{\Omega}\int\frac{dx}{R}}$; on observera que l'on a

$$\sqrt{\Omega}\int\frac{dx}{R} = \sum \int_{x_i}^{\omega_i} \frac{\sqrt{\Omega(h)}}{(h-\omega_i)}\frac{d\omega_i}{\sqrt{\Omega(\omega_i)}}$$

Cette somme d'intégrales de troisième espèce est la somme d'un logarithme de fonction abélienne et d'une fonction abélienne. Les transformations subies par les intégrales de (H) se déduisent de là.

Ajoutons que l'équation (H) admet *en elle-même* la transformation

$$Y = [\theta(h) - R']y + 2Ry'$$

où $\theta(h)$ est un polynome en h à coefficients constants.

4. L'étude des équations (H) qui s'intègrent par quadratures n'est donc rien moins que celles de *fonctions abéliennes*, et de leurs *dégénérescences* lorsque Ω a des facteurs multiples (étude faite en détail par MM. Émile Picard et Painlevé, pour $n = 2$).

La transformation de Darboux qui remplace φ par $2(\rho_0^2 - h_0) - \varphi$ où $\rho_0' + \rho_0^2 = \varphi + h_0$, et R par un polynome de degré $(n+1)$ en h, ne conduit pas à des transcendantes *propres* à l'indice $(n+1)$; elle multiplie Ω par $(h - h_0)^2$, ce qui ne modifie pas le *genre*. Inversement si Ω contient le facteur $(h - h_0)^2$, la transformation de Darboux relative à h_0 et à $\frac{R'(h_0)}{2R(h_0)}$

([1]) Weierstrass, *Journal de Crelle*, t. 47 et 52.

abaisse d'une unité l'indice n. Les fonctions φ, *impropres* à l'indice n, s'obtiennent par quadratures superposées à partir de fonctions *propres* à un indice inférieur; elles sont uniformes mais *non méromorphes*.

Enfin pour les valeurs $h = h_i$ qui annulent Ω, on a des équations $y'' = (\varphi + h_i)y$ qui admettent une solution $\sqrt{R(h_i)}$ *abélienne*, donc uniforme en $u_0, \ldots, u_{n-1} + x$.

On reconnaît ici l'extension naturelle des recherches mémorables d'Hermite et de M. Émile Picard sur l'équation de Lamé (1), pour laquelle $\varphi = n(n+1)k^2 sn^2 x$; et des travaux ultérieurs de Brioschi, Elliot, Fuchs, Darboux sur des équations analogues. Remarquons que, lorsqu'on choisit φ pour variable indépendante, $\frac{d\varphi}{dx}$ est bien *uniforme* en x, mais *transcendante* en φ, sauf dans le cas de fonctions φ dérivées des fonctions elliptiques, considéré seul jusqu'à présent.

APPENDIX III: ON THE THEORY OF PERIODIC AND LIMIT-PERIODIC JACOBIAN MATRICES

P.B. Naĭman

Periodic and limit-periodic Jacobian matrices and the continued fractions associated with them have for some time been the subject of a series of investigations (cf. [1] and the bibliography presented there). The present note develops an algebraic approach to these matrices, which, when combined with the operator theoretic methods [2,3], leads to some new results.

1. What we call a generalized Jacobian matric of order $2m$ is an infinite matrix $A = \| a_{jk} \|$ $(j, k = \pm 0, 1, 2, \cdots)$ with complex entries satisfying the conditions $a_{jk} = 0$ for $|j-k| > m$ and $a_{jk} \neq 0$ for $|j-k| = m$. For $m = 1$, the generalized Jacobian matrices are all ordinary Jacobian matrices. The following theorem is the discrete analogue of a theorem of Burchnall and Chaundy [4] on linear differential expressions.

Theorem 1. *If two generalized Jacobian matrices A and B, of orders $2m$ and $2n$ respectively, commute, then they satisfy an algebraic relation $D(A, B) = 0$ of degree $2n$ in A and degree $2m$ in B.*

The proof is similar to that presented in [4] for the continuous case.

Let E_n be the matrix all of whose entries are zero except $e_{k,k-n} = e_{k,k+n} = 1$ $(k = \pm 0, 1, 2, \cdots)$. It is easily seen that the spectrum $S(E_n)$ is the interval $-2 \leq \mu \leq 2$ and that each point $\lambda \in S(E_n)$ has multiplicity $2n$. If a generalized Jacobian matrix A has period n, then it obviously commutes with E_n, and so $D(A, E_n) = 0$. In the simplest situation of an ordinary n-periodic complex-symmetric Jacobian matrix

$$T = \| t_{jk} \| \quad (t_{jk} = t_{kj};\ j, k, = \pm 0, 1, 2, \cdots)$$

the polynomial $D(\lambda, \mu)$ is a perfect square, whence

Theorem 2. *Any complex-symmetric n-periodic Jacobian matrix T satisfies an nth degree algebraic equation*

Reprinted from Soviet Math. Doklady, v. 3, No. 2, pp. 383-385, 1962, by the permission of The American Mathematical Society.

$$P(T) = E_n \tag{1}$$

If the matrix T is real, then the coefficients of the polynomial $P(\lambda)$ are likewise real. Since we know the spectrum $S(E_n)$, relation (1) permits a complete characterization tobe given of the spectrum $S(T)$. In this connection, for the special case of a real matrix, a known result on the structure of the spectrum obtains [1].

Theorem 3. *The spectrum $S(T)$ of a complex-symmetric n-periodic Jacobian matrix T coincides with the complete inverse image Γ of the interval $-2 \leq \mu \leq 2$ under the transformation*

$$P(\lambda) = \mu \tag{2}$$

and consequently consists of n algebraic arcs which may occasionally have common endpoints.

If the matrix T is real, then whenever $-2 \leq \mu \leq 2$, every solution λ of equation (2) is real, so that $S(T)$ is a system of n intervals which may occasionally have common endpoints. Moreover, every point $\lambda \in S(T)$ has multiplicity two.

Proof. The relation $S(T) \subset \Gamma$ follows from (1). If the matrix T is real, the multiplicity of $S(T)$ does not exceed two, so that the multiplicity of the left hand side of equation (1) is not larger than $2n$. Moreover, since the multiplicity of each point of the spectrum $S(E_n)$ is $2n$, all the contentions of the theorem for the real case follow from equation (1).

In the general case of a complex matrix T, it is not hard, following the well-known procedure of Floquet, to establish that a necessary and sufficient condition for the boundedness of at least one of the solutions $\mathbf{y} = \{y_k\}_{k=-\infty}^{\infty}$ of the finite difference equation

$$t_{k,k-1}y_{k-1} + t_{k,k}y_k + t_{k,k+1}y_{k+1} = \lambda y_k \tag{3}$$

for a given complex number λ is that $-2 \leq P(\lambda) \leq 2$, so that the set Γ is a system of n curvilinear stability zones of equation (3). Now let $\lambda \in \Gamma$. If in this connection $\{y_k\}_{k=-\infty}^{\infty} \in l_2$, the number λ must obviously be an eigenvalue of T. If, however, $\{y_k\}_{k=-\infty}^{\infty} \overline{\in} l_2$, then $\sum_{k=-\infty}^{\infty} |y_k|^2 = \infty$, and the sequence of vectors $\mathbf{z}_N$ with coordinates

$$z_{N,k} = \begin{cases} y_k & (|k| \leq N) \\ 0 & (|k| > N) \end{cases}$$

satisfies

$$\lim_{N\to\infty} \frac{\| T\mathbf{z}_N - \lambda \mathbf{z}_N \|}{\| \mathbf{z}_N \|} = 0$$

so that $\lambda \in S(T)$. Thus $\Gamma \subset S(T)$, and the theorem is proved.

2. In the real case, we number the lacunae in $S(T)$ with indices $1, 2, \cdots, n+1$, beginning with the lacuna running from $-\infty$, and we perturb T by the addition of a completely continuous real diagonal matrix K. According to a well-known theorem of H. Weyl on completely continuous perturbations, the spectrum $S(T+K)$ will consist of the system of intervals described in Theorem 3 and of a bounded set of eigenvalues which may be concentrated only at the ends of the lacunae. Obviously, the corresponding perturbation

$$Q = P(T+K) - P(T)$$

of the matrix $P(T)$ is a completely continuous generalized Jacobian matrix of order $2n-2$. Thus the polynomial $P(\lambda)$ transforms the limit-periodic matrix $T+K$ into the limit-constant matrix

$$P(T+K) = E_n + Q \tag{4}$$

By virtue of relation (4), the conditions for the finiteness and infiniteness of the sets of points in the spectrum $S(T+K)$ lying in the union of all odd or even lacunae of the spectrum $S(T)$ are the same as the conditions for the finiteness and infiniteness of the sets of points in the spectrum of the limit-constant matrix $E_n + Q$ lying to the left of the point $\mu = -2$ or to the right of the point $\mu = 2$. Determination of the latter conditions is carried out as in [3], and leads to the following results.

Let σ_r denote the sum of the absolute values of all the entries in the rth row of the matrix Q except for the diagonal element q_r, and put

$$\omega_r' = \sigma_r - q_r, \quad \omega_r'' = \sigma_r + q_r$$

In addition, let S' and S'' denote the sets of points of the spectrum introduced by the perturbation K into the union of all the odd, resp. even, lacunae of the spectrum $S(T)$. With no loss of generality, we may suppose that

$$(-1)^n t_{12} \cdot t_{23} \cdot \ldots \cdot t_{n,n+1} > 0$$

Theorem 4. *If*

$$\limsup_{|r| \to \infty} r^2 \omega_r' < \frac{n^2}{4}$$

then S' is finite. If

$$\limsup_{|r| \to \infty} r^2 \omega_r'' < \frac{n^2}{4}$$

then S'' is finite. If one of the $2n$ series

$$\sum_{r=0}^{\pm\infty} q_{j+nr} \quad (j = 1, 2, \cdots, n) \tag{5}$$

diverges to $+\infty$, then S' is infinite. If one of the $2n$ series (5) diverges to $-\infty$, then S'' is infinite.

Using Theorem 4, we can obtain conditions in terms only of the entries k_r in the perturbation K. For example, if $k_r = o(\frac{1}{r^2})$, the set $S' + S''$ is finite. In the special case $n = 2$, supposing, to be definite, that $t_{12} - t_{10} > 0$, we arrive, in this way, at the following result:

Theorem 5. *If one of the two series $\sum^{\pm\infty} k_{2r-1}$ consists entirely of positive terms and diverges, then the perturbation K introduces an infinite set of eigenvalues into the internal lacuna of a 2-periodic matrix T. And if one of the two series $\sum^{\pm\infty} k_{2r}$ consists entirely of positive terms and diverges, then the perturbation K introduces an infinite set of eigenvalues into the union of the external lacunae.*

A similar result holds for any n. In the cases when almost all $k_r \geq 0$ (or ≤ 0) the spectrum introduced by the perturbation K cannot be concentrated at the left (right) ends of the lacunae.

Received 21/OCT/61

BIBLIOGRAPHY

[1] Ja. L. Geronimus, Izv. Akad. Nauk SSSR Ser. Mat. 5 (1941), 203.

[2] I. M. Glazman, Dokl. Akad. Nauk SSSR 118 (1958), 423.

[3] P. B. Naĭman, Izv. Vysš. Učebn. Zaved. Matematika 1959, no. 1 (8), 129.

[4] E. L. Ince, *Ordinary differential equations*, Dover, New York, 1944.

Translated by:
F. E. J. Linton